CRC SERIES IN CHROMATOGRAPHY

Gunter Zweig and Joseph Sherma
Editors-in-Chief

GENERAL DATA AND PRINCIPLES
Editors
Gunter Zweig, Ph.D.
U.S. Environmental Protection Agency
Washington, D.C.
Joseph Sherma, Ph.D.
Lafayette College
Easton, Pennsylvania

CARBOHYDRATES
Editor
Shirley C. Churms, Ph.D.
Research Associate
C.S.I.R. Carbohydate Chemistry Research Unit
Department of Organic Chemistry
University of Cape Town, South Africa

DRUGS
Editor
Ram Gupta, Ph.D.
Department of Laboratory Medicine
St. Joseph's Hospital
Hamilton, Ontario, Canada

CATECHOLAMINES TERPENOIDS
Editor
Carmine J. Cosica, Ph.D.
Professor of Biochemistry
St. Louis University School of Medicine
St. Louis, Missouri

STEROIDS
Editor
Joseph C. Touchstone, B.S., M.S., Ph.D.
Professor
School of Medicine
University of Pennsylvania
Philadelphia, Pennsylvania

PESTICIDES
Editors
Joseph Sherma, Ph.D.
Lafayette College
Easton, Pennsylvania
Joanne M. Follweiler, Ph.D.
Lafayette College
Easton, Pennsylvania

LIPIDS AND TECHNICAL LIPID DERIVATIVES
Editor
H. K. Mangold
Executive Director and Professor
Federal Center for Lipid Research
4400 Munster, Germany

HYDROCARBONS
Editors
Willie E. May, Ph.D.
Center for Analytical Chemistry
U.S. Department of Commerce
Washington, D.C.
Walter L. Zielinski, Jr., Ph.D.
Air Program Manager
National Bureau of Standards
Washington, D.C.

INORGANICS
Editor
M. Qureshi, Ph. D.
Professor, Chemistry Section
Zakir Husain College of Engineering and Technology
Aligarh Muslim University
India

PHENOLS AND ORGANIC ACIDS
Editor
Toshihiko Hanai, Ph.D.
University of Montreal
Quebec, Canada

AMINO ACIDS AND AMINES
Editor
Dr. S. Blackburn
Leeds, England

POLYMERS
Editor
Charles G. Smith
Dow Chemical, USA
Midland, Michigan

PLANT PIGMENTS
Editor
Dr. Hand-Peter Kost
Botanisches Institut der Universitat Munchen
Munchen, West Germany

CRC Handbook of Chromatography: Amino Acids and Amines

Volume I

Author

Stanley Blackburn, Ph.D., C. Chem., F.R.S.C.
Leeds, England

CRC Series in Chromatography
Editors-in-Chief

Gunter Zweig, Ph.D.
School of Public Health
University of California
Berkeley, California

Joseph Sherma, Ph.D.
Professor of Chemistry
Lafayette College
Easton, Pennsylvania

CRC Press, Inc.
Boca Raton, Florida

Library of Congress Cataloging in Publication Data
Main entry under title:

  (CRC series in chromatography)
  Includes index.
  Contents: v. 1. Amino acids and amines.
  1. Chromatographic analysis—Handbooks, manuals,
etc. I. Blackburn, S. (Stanley) II. Title:
C.R.C. handbook of chromatography. III. Series.
QD79.C4C7   1983   543'.089   82-9561
ISBN 0-8493-3064-5 (v. 1)

This book represents information obtained from authentic and highly regarded sources. Reprinted material is quoted with permission, and sources are indicated. A wide variety of references are listed. Every reasonable effort has been made to give reliable data and information, but the author and the publisher cannot assume responsibility for the validity of all materials or for the consequences of their use.

All rights reserved. This book, or any parts thereof, may not be reproduced in any form without written consent from the publisher.

Direct all inquiries to CRC Press, Inc., 2000 Corporate Blvd., N.W., Boca Raton, Florida, 33431.

© 1983 by CRC Press, Inc.

International Standard Book Number 0-8493-3064-5

Library of Congress Card Number 82-9561
Printed in the United States

CRC HANDBOOK OF CHROMATOGRAPHY

SERIES PREFACE

The present volume by Dr. S. Blackburn on the subject of the chromatography of amino acids and amines represents an immense effort by this editor and his editorial board to bring together in one volume all the latest chromatographic data for amino acids and amines. We believe that Dr. Blackburn has succeeded very well in organizing this rather complex subject by first tabulating chromatographic data (liquid, gas, paper, and thin-layer chromatography) followed by detailed instructions on the preparation of derivatives, detection reagents for PC and TLC, and a valuable listing of commercial sources for chromatographic equipment and supplies. Dr. Blackburn has confined his treatment of the subject to amino acids and amines, but in a subsequent volume will cover the subject of chromatography of peptides.

Other volumes in the series of Handbook of Chromatography already published include Drugs, Carbohydrates, Polymers, and Phenols and Organic Acids; Future volumes will cover the subjects of Lipids and Fatty Acids; Terpenes and Catecholamines; Pesticides; Inorganics; Steroids; Plant Pigments; and, as already mentioned, Peptides. We invite our readers to communicate with us or volume editors concerning errors, omissions, additions, or suggestions for future volumes. Again, we want to thank Dr. S. Blackburn, the volume editor, for his outstanding contribution, and the staff of CRC Press for their excellent cooperation and production.

Gunter Zweig
Joseph Sherma
Series Editors

PREFACE

During the last decade or so the use of chromatographic techniques in chemical and biochemical investigations has increased at a phenomenal rate. This growth has been reflected by an increase in the size and number of scientific journals devoted almost entirely to chromatographic methods. It would now be impractical to include chromatographic data covering the wide range of chemical compounds in one or two volumes, as was done earlier in Volumes I and II of the *Handbook of Chromatography*.

The present volume is therefore restricted to chromatographic data on amino acids and amines. It supplements the data given in Volume I of the earlier handbook, and covers the literature since 1970. A section on detection reagents, principally for paper and thin-layer chromatography is included. Another section describes methods of sample preparation for chromatography and methods of protein hydrolysis. Commercial sources of chromatographic materials are listed in a further section. A book directory lists books which should serve as a good source for more detailed reading. A number of articles give information on methods of chromatographic separation applicable to amino acids. These supplement the information on chromatographic methods and techniques given in Volume II of the earlier handbook.

My thanks are due to my collaborators Dr. Z. Deyl and Dr. A. J. Smith for their invaluable help.

S. Blackburn

THE EDITORS-IN-CHIEF

Gunter Zweig, Ph.D., received his undergraduate and graduate training at the University of Maryland, College Park, where he was awarded the Ph.D. in biochemistry in 1952. Two years following his graduation, Dr. Zweig was affiliated with the late R. J. Block, pioneer in paper chromatography of amino acids. Zweig, Block, and Le Strange wrote one of the first books on paper chromatography which was published in 1952 by Academic Press and went into three editions, the last one authored by Gunter Zweig and Dr. Joe Sherma, the co-Editor-in-Chief of this series. *Paper Chromatography* (1952) was also translated into Russian.

From 1953 till 1957, Dr. Zweig was research biochemist at the C. F. Kettering Foundation, Antioch College, Yellow Springs, Ohio, where he pursued research on the path of carbon and sulfur in plants, using the then newly developed techniques of autoradiography and paper chromatography. From 1957 till 1965, Dr. Zweig served as lecturer and chemist, University of California, Davis and worked on analytical methods for pesticide residues, mainly by chromatographic techniques. In 1965, Dr. Zweig became Director of Life Sciences, Syracuse University Research Corporation, New York (research on environmental pollution), and in 1973 he became Chief, Environmental Fate Branch, Environmental Protection Agency (EPA) in Washington, D.C.

During his government career, Dr. Zweig continued his scientific writing and editing. Among his works are (many in collaboration with Dr. Sherma) the now 11-volume series on *Analytical Methods for Pesticides and Plant Growth Regulators* (published by Academic Press); the pesticide book series for CRC Press; co-editor of *Journal of Toxicology and Environmental Health;* co-author of basic review on paper and thin-layer chromatography for *Analytical Chemistry* from 1968 to 1980; co-author of applied chromatography review on pesticide analysis for *Analytical Chemistry,* beginning in 1981.

Among the scientific honors awarded to Dr. Zweig during his distinguished career are the Wiley Award in 1977, Rothschild Fellowship to the Weizmann Institute in 1963/64; the Bronze Medal by the EPA in 1980.

Dr. Zweig has authored or co-authored over 75 scientific papers on diverse subjects in chromatography and biochemistry, besides being the holder of three U.S. patents.

At the present time (1980/81), Dr. Zweig is Visiting Scholar in the School of Public Health, University of California, Berkeley, where he is doing research on farmworker safety as related to pesticide exposure.

Joseph Sherma, Ph.D., received a B.S. in chemistry from Upsala College, East Orange, N.J. in 1955 and a Ph.D. in analytical chemistry from Rutgers University in 1958. His thesis research in ion exchange chromatography was under the direction of the late William Rieman III. Dr. Sherma joined the faculty of Lafayette College in September 1958, and is presently full professor there in charge of two courses in analytical chemistry. At Lafayette he has continued research in chromatography and has additionally worked a total of 12 summers in the field with Harold Strain at the Argonne National Laboratory, Illinois, James Fritz at Iowa State University, Ames, Gunter Zweig at Syracuse University Research Corporation, New York, Joseph Touchstone at the Hospital of the University of Pennsylvania, Philadelphia, Brian Bidlingmeyer at Waters Associates, Framingham, Mass., and Thomas Beesley at Whatman Inc., Clifton, N.J.

Dr. Sherma and Dr. Zweig [who is now with U.S. Environmental Protection Agency (EPA)] co-authored Volumes I and II of the *CRC Handbook of Chromatography,* a book on paper chromatography, and 6 volumes of the series *Analytical Methods for Pesticides and Plant Growth Regulators.* Other books in the pesticide series and further volumes of the *CRC Handbook of Chromatography* are being edited with Dr. Zweig, and Dr. Sherma will co-author the handbook on pesticide chromatography. A book on quantitative TLC (published by Wiley-Interscience, New York) was edited jointly with Dr. Touchstone. Dr. Sherma has been co-author of 7 biennial reviews of liquid chromatography (1968 to 1980) and the 1981 review of pesticide analysis for the journal *Analytical Chemistry*. Dr. Sherma has authored major invited chapters and review papers on chromatography and pesticides in *Chromatographic Reviews* (analysis of fungicides), *Advances in Chromatography* (analysis of fungicides), *Advances in Chromatography* (analysis of nonpesticide pollutants), Heftmann's *Chromatography* (chromatography of pesticides), Race's *Laboratory Medicine* (chromatography in clinical analysis), *Food Analysis: Principles and Techniques* (TLC for food analysis), *Treatise on Analytical Chemistry* (paper and thin layer chromatography), and *CRC Critical Reviews in Analytical Chemistry* (pesticide residue analysis). A general book on thin layer chromatography co-authored by Dr. Sherma is now in press at Marcel Dekker.

Dr. Sherma spent 6 months in 1972 on sabbatical leave at the EPA Perrine Primate Laboratory, Perrine, Fla., with Dr. T. M. Shafik, and two additional summers (1975, 1976) at the U.S. Department of Agriculture (USDA) in Beltsville, Md., with Melvin Getz doing research on pesticide residue analysis methods development. He spent 3 months in 1979 on sabbatical leave with Dr. Touchstone developing clinical analytical methods. A total of more than 200 papers, books, book chapters, and oral presentations concerned with column, paper, and thin layer chromatography of metal ions, plant pigments, and other organic and biological compounds; the chromatographic analysis of pesticides; and the history of chromatography have been authored by Dr. Sherma, many in collaboration with various co-workers and students. His major research area at Lafayette is currently quantitative TLC (densitometry), applied mainly to clinical analysis and pesticide residue determinations.

Dr. Sherma has written an analytical quality control manual for pesticide analysis under contract with the U.S. EPA and has revised this and the EPA Pesticide Analytical Methods Manual under a 4-year contract (EPA) jointly with Dr. M. Beroza of the Association of Official Analytical Chemists (AOAC). Dr. Sherma has also written an instrumental analysis quality assurance manual and other analytical reports for the U.S. Consumer Product Safety Commission, and is currently preparing a manual on the analysis of food additives for the U.S. Food and Drug Administration, both of these projects also in collaboration with Dr. Beroza of the AOAC.

Dr. Sherma taught the first prototype short course on pesticide analysis with Henry Enos of the EPA for the Center for Professional Advancement. He is editor of the Kontes TLC quarterly newsletter and also teaches short courses on TLC for Kontes and the Center for Professional Advancement. He is a consultant for several industrial companies and federal agencies on chemical analysis and chromatography and regularly referees papers for analytical journals and research proposals for government agencies.

Dr. Sherma has received two awards for superior teaching at Lafayette College and the 1979 Distinguished Alumnus Award from Upsala College for outstanding achievements as an educator, researcher, author, and editor. He is a member of the American Chemical Society, Sigma Xi, Phi Lambda Upsilon, Society for Applied Spectroscopy, and the American Institute of Chemists.

THE AUTHOR

Dr. Stanley Blackburn gained his Honours B.Sc. degree in Chemistry at the University of Leeds, England and his Ph.D. degree in Organic Chemistry from the same university. He is a former research scientist at the Wool Industries Research Association, Leeds, where his research interests included the development of chromatographic techniques, the structure and amino acid sequence of the proteins of wool keratin, and the end group determination of peptides and proteins.

Dr. Blackburn is a Chartered Chemist, a Fellow of the Royal Society of Chemistry, and a member of the Biochemical Society and the American Chemical Society. His current work is centered on scientific writing and documentation. He has written more than 30 scientific papers dealing with protein analysis and structure and is the author of several texts, including *Amino Acid Determination; Methods and Techniques, Protein Sequence Determination; Methods and Techniques and Enzyme Structure and Function,* all published by Marcel Dekker Inc.

ADVISORY BOARD

A. J. Smith, Ph.D.
Lecturer of Biochemistry
Birmingham Dental School
Birmingham, England

Zdenek Deyl, Dr. Sc., Ph.D.
Institute of Physiology
Czechoslovakian Academy of Science
Praha, Czechoslovakia

TABLE OF CONTENTS

Section I: Tables

I.I. Gas Chromatography Tables... 3
I.II. Liquid Chromatography Tables.. 57
I.III. Paper Chromatography Tables.. 103
I.IV. Thin-Layer Chromatography Tables 109

Section II: Techniques

II.I. Procedures for the Determination of Amino Acids by Liquid
      Chromatography... 141
II.II. Ion Exchange Resins for Amino Acid Determinants.......................... 157
II.III. The Gas Chromatographic Determination of Amino Acids 159
II.IV. The Determination of Amino Acid Enantiomers by Gas Chromatography..... 165
II.V. Detection Systems for Amino Acids in Automatic Analyzers 169

Section III: Detection Reagents

III.I. Detection Reagents for Paper Chromatography and Thin-Layer
      Chromatography... 177
III.II. Summary Tables for Detection Reagents 185

Section IV: Guide to Methods of Sample Preparation Including Derivatization......... 197

Section V: Products and Sources of Chromatographic Materials....................... 223

Section VI: Chromatography Book Directory... 259

Section VII: Reviews of Chromatographic Methods and Equipment..................... 267

Section VIII: Indexes

Author Index 271
Compound Index 279
Subject Index... 293

Section I
Tables

I.I. Gas Chromatography Tables
I.II. Liquid Chromatography Tables
I.III. Paper Chromatography Tables
I.IV. Thin-Layer Chromatography Tables

Section I.1.

GAS CHROMATOGRAPHY TABLES

Wherever possible, tables are arranged according to classes of chemical compounds. This was not always possible when different types of compounds were chromatographed under the same experimental conditions. The reader is referred to the compound index for specific compounds which may appear in different tables.

Table GC 1
ENANTIOMERS OF N-TRIFLUOROACETYL-AMINO ACID ISOPROPYL ESTERS AND AMIDES

Column packing	P1	P2
Temperature (°C)	180	180
Gas: Flow rate (mℓ/min)	He, 0.6	He, 0.6
Column		
Length (m)	40	40
Diameter (mm I.D.)	0.25	0.25
Form	Capillary	Capillary
Material	G	G
Detector	FID	FID

N-TFA isopropyl ester	t_r(mins)[a]	
D-Alanine	1.00	
L-Alanine	1.07	
D-Valine	1.12	
L-Valine	1.19	
D-Leucine	1.93	
L-Leucine	2.05	
D-Proline	3.40	
L-Proline	3.40	
D-Proline[b]		6.85
L-Proline[b]		6.95
D-Proline[c]		25.56
L-Proline[c]		26.32
D-Methionine	9.88	
L-Methionine	10.50	
D-Phenylalanine	12.48	
L-Phenylalanine	13.20	

N-TFA isopropylamide		
D-Alanine	13.80	
L-Alanine	14.30	
D-Valine	18.02	
L-Valine	18.67	
D-Leucine	24.40	
L-Leucine	24.71	
D-Proline	9.60	
L-Proline	10.77	
D-Proline[b]		27.53
L-Proline[b]		35.10
D-Proline[c]		134.90
L-Proline[c]		191.30
D-Methionine	92.50	
L-Methionine	95.10	
D-Phenylalanine	118.0	
L-Phenylalanine	120.0	

[a] Time from solvent peak.
[b] 130°.
[c] 100°.

Table GC 1 (continued)
ENANTIOMERS OF N-TRIFLUOROACETYL-AMINO ACID ISOPROPYL ESTERS AND AMIDES

Column packing P1 = OA-300 (N,N'-(2,4-(6-ethoxy-1,3,5-triazinediyl))bis (L-valyl-L-valyl-L-valine isopropyl ester)).

P2 = OA-400(N,N',N''-(2,4,6-(1,3,5-triazinetriyl))tris (Nα-lauroyl-L-lysine-tert-butylamide)).

REFERENCE

1. Oi, N., Horiba, M., and Kitahara, H., *J. Chromatogr.*, 202, 299, 1980.

Table GC 2
ENANTIOMERS OF N-TRIFLUOROACETYL-AMINO ACID ISOPROPYL ESTERS

Column packing	P1
Gas, Flow rate (mℓ/min)	He, 3
Column	
Length (m)	50
Diameter (mm I.D.)	0.5
Form	Capillary
Material	SS
Detector	FID

Amino acid	t_r(min)	Temp(°C)
D-Alanine	4.76	120
L-Alanine	5.50	
D-Threonine	6.74	120
L-Threonine	7.40	
D-Valine	7.70	120
L-Valine	8.50	
Glycine	9.12	120
D-Alloisoleucine	11.78	120
L-Alloisoleucine	13.06	
D-Isoleucine	13.06	120
L-Isoleucine	14.44	
D-Serine	12.30	120
L-Serine	13.56	
D-Leucine	16.08	120
L-Leucine	19.80	
D-Proline	19.60	120
L-Proline	20.08	
D-Aspartic Acid	37.12	120
L-Aspartic Acid	38.68	
D-Cystine	18.98	140
L-Cystine	22.36	
D-Methionine	10.54	180
L-Methionine	11.24	
D-Glutamic acid	12.90	180
L-Glutamic acid	13.68	
D-Phenylalanine	16.07	180
L-Phenylalanine	16.85	
D-Ornithine	20.08	200
L-Ornithine	21.54	

Table GC 2 (continued)
ENANTIOMERS OF N-TRIFLUOROACETYL-AMINO ACID ISOPROPYL ESTERS

D-Lysine	28.60	200
L-Lysine	30.40	
D-Tryptophan	62.52	200
L-Tryptophan	65.08	
D-α-Aminobutyric acid	6.34	120
L-α-Aminobutyric acid	7.28	
D-α-Aminopentanoic acid	11.70	120
L-α-Aminopentanoic acid	13.74	
D-α-Aminohexanoic acid	6.48	150
L-α-Aminohexanoic acid	7.30	
D-α-Aminoheptanoic acid	10.75	150
L-α-Aminoheptanoic acid	11.85	
D-α-Aminooctanoic acid	12.30	160
L-α-Aminooctanoic acid	13.30	
D-*tert*-Leucine	8.06	120
L-*tert*-Leucine	8.48	
D-Phenylglycine	19.90	150
L-Phenylglycine	20.76	
L-γ-Aminovaleric acid	31.80	120
D-γ-Aminovaleric acid	33.26	
L-γ-Amino-δ-methylhexanoic acid	61.78	120
D-γ-Amino-δ-methylhexanoic acid	66.50	
L-γ-Amino-ε-methylheptanoic acid	110.50	120
D-γ-Amino-ε-methylheptanoic acid	118.20	
α-Methylvaline (I)	10.88	110
α-Methylvaline (II)	11.40	
α-Methylnorvaline (I)	9.26	110
α-Methylnorvaline (II)	9.26	
α-Methylleucine (I)	12.36	110
α-Methylleucine (II)	12.36	
α-Methylnorleucine (I)	16.04	110
α-Methylnorleucine (II)	16.04	

Column packing P1 = N-docosanoyl-L-valine-2-(2-methyl)-n-heptadecyl-amide.

REFERENCE

1. Charles, R. and Gil-Av, E., *J. Chromatogr.*, 195, 317, 1980.

Reproduced by permission of Elsevier Scientific Publishing Company.

Table GC 3
ENANTIOMERS OF N(O)-TRIFLUOROACETYL-AMINO ACID ISOPROPYL ESTERS

	P1		P2	
Column packing	P1		P2	
Gas	He		He	
Column				
Length(m)	17		17	
Diameter(mm I.D.)	0.25		0.25	
Form	Coiled capillary		Coiled capillary	
Material	Glass[a]		Glass[a]	
Detector	FID		FID	

Amino acid	Temp(°C)	t_r(min)	Temp(°C)	t_r(min)
D-Alanine	110	4.6	110	3.2
L-Alanine		5.3		3.6
D-Valine	110	6.6	110	4.7
L-Valine		7.6		5.2
D-Threonine	110	7.3	110	4.9
L-Threonine		8.1		5.5
Glycine	110	9.4	110	6.2
D-Alloisoleucine	110	9.8	110	6.8
L-Alloisoleucine		11.3		7.8
D-Isoleucine	110	10.8	110	7.4
L-Isoleucine		12.3		8.4
D-Leucine	110	14.5	130	4.3
L-Leucine		17.6		4.9
D-Serine	110	15.2	130	4.3
L-Serine		16.6		4.6
D-Proline	110	16.0	130	5.1
L-Proline		16.3		5.1
D-Aspartic Acid	150	8.5	150	5.7
L-Aspartic Acid		8.8		5.9
D-Methionine	150	14.8	150	9.8
L-Methionine		16.3		10.7
D-Glutamic Acid	170	8.9	170	6.1
L-Glutamic Acid		9.4		6.5
D-Phenylalanine	170	9.7	170	6.7
L-Phenylalanine		10.4		7.1
D-Tyrosine			170	14.5
L-Tyrosine				15.7
D-Ornithine			170	32.9
L-Ornithine				35.4
D-Lysine			170	47.7
L-Lysine				51.5

[a] Pretreated with 0.5% HF for 1 hr and silanized with 3-aminopropyltriethoxysilane. A methylene chloride solution containing about 12% of the phase was passed through the capillary column, which was then conditioned at 170°C overnight with a flow of He as carrier gas.

Column packing P1 = N-lauroyl-L-valyl-L-valine cyclohexyl ester.
P2 = N-lauroyl-L-valyl-L-leucine cyclohexyl ester.

REFERENCE

1. Abe, I. and Musha, S., *J. Chromatogr.*, 200, 195, 1980.

Reproduced by permission of Elsevier Scientific Publishing Company.

Table GC 4
ENANTIOMERS OF N-TRIFLUOROACETYL AMINO ACID ISOPROPYL ESTERS

Column packing		P1	P1	P1	P1	P2	P2	P3
Temperature (°C)		100	110	120	130	110	120	100
Gas		He,12psig	He,12psig	He,12psig	He,12psig	He,12psig	He,12psig	He,12psig
Column								
Length (ft)		400	400	400	400	400	400	400
Diameter (inches)		0.02	0.02	0.02	0.02	0.02	0.02	0.02
Material		SS	SS	SS	SS	SS	SS	SS
Detector		FID	FID	FID	FID	FID	FID	FID
Amino acid	Isomer	*N-TFA-L-norleucine isopropyl ester						
Alanine	D	0.246	0.268	0.290	0.312	0.289	0.315	0.231
	L	0.268	0.289	0.310	0.330	0.305	0.329	0.254
Amino-n-butyric acid	D	0.341	0.363	0.380	0.409	0.476	0.390	0.330
	L	0.372	0.391	0.406	0.434	0.507	0.411	0.362
Valine	D	0.366	0.389	0.412	0.436	0.382	0.418	0.364
	L	0.399	0.419	0.440	0.462	0.409	0.443	0.397
Norvaline	D	0.563	0.583	0.603	0.623	0.597	0.628	0.553
	L	0.611	0.628	0.643	0.661	0.631	0.659	0.604
Leucine	D	0.764	0.771	0.787	0.788	0.844	0.759	0.749
	L	0.839	0.838	0.844	0.838	0.900	0.801	0.832
Isoleucine	D	0.535	0.556	0.572	0.600	0.580	0.546	0.535
	L	0.591	0.607	0.616	0.639	0.625	0.579	0.592
Norleucine	D	0.911	0.921	0.931	0.941	0.940	0.948	0.903
	L	1.000	1.000	1.000	1.000	1.000	1.000	1.000
Tert-leucine	D	0.362	0.384	0.400	0.428	0.387	0.381	0.364
	L	0.381	0.402	0.416	0.442	0.407	0.397	0.381

Column packing	P3	P3	P3	P4	P4	P4
Temperature (°C)	110	120	130	100	110	120	130
Gas	He,12 psig	He,12 psig	He,12 psig	He,12 psig	He,12 psig	He,12 psig
Column						
Length (ft)	400	400	400	400	400	400
Diameter (inches)	0.02	0.02	0.02	0.02	0.02	0.02
Material	SS	SS	SS	SS	SS	SS

Detector			FID	FID	FID	FID	FID	FID	FID
Amino acid		Isomer	ⁿ-TFA-L-norleucine isopropyl ester						
Alanine		D	0.253	0.273	0.299	0.228	0.249	0.272	0.293
		L	0.275	0.293	0.318	0.249	0.270	0.290	0.311
Amino-n-butyric acid		D	0.354	0.362	0.396	0.327	0.351	0.375	0.397
		L	0.384	0.388	0.420	0.356	0.378	0.400	0.420
Valine		D	0.386	0.409	0.435	0.366	0.389	0.413	0.435
		L	0.416	0.437	0.459	0.396	0.416	0.437	0.458
Norvaline		D	0.573	0.592	0.613	0.551	0.571	0.592	0.612
		L	0.619	0.633	0.650	0.600	0.616	0.633	0.647
Leucine		D	0.760	0.752	0.744	0.748	0.759	0.769	0.777
		L	0.833	0.813	0.829	0.828	0.831	0.830	0.829
Isoleucine		D	0.558	0.567	0.599	0.541	0.565	0.587	0.608
		L	0.610	0.612	0.640	0.595	0.614	0.631	0.646
Norleucine		D	0.920	0.924	0.935	0.904	0.914	0.927	0.937
		L	1.000	1.000	1.000	1.000	1.000	1.000	1.000
Tert-leucine		D	0.389	0.401	0.430	0.368	0.390	0.415	0.446
		L	0.404	0.414	0.442	0.380	0.402	0.424	0.446

Column packing P1 = N-trifluoroacetyl-L-α-amino-n-butyryl-L-α-amino-n-butyric acid cyclohexyl ester
P2 = N-trifluoroacetyl-L-alanyl-L-alanine cyclohexyl ester
P3 = N-trifluoroacetyl-L-norvalyl-L-norvaline cyclohexyl ester
P4 = N-trifluoroacetyl-L-norleucyl-L-norleucine cyclohexyl ester

Note: The columns were conditioned and maintained at 10°C above the m.p. of the respective phase.

REFERENCE

1. Parr, W. and Howard, P.Y., *Anal. Chem.*, 45, 711, 1973.

Reprinted with permission from *Anal. Chem.*, 45, 711, 1973. Copyright (1973) American Chemical Society.

Table GC 5
ENANTIOMERS OF N-TRIFLUOROACETYL AMINO ACID ISOPROPYL ESTERS

Column packing	P1	P2	P3	P4	P5	P6	P7	P8	P9	P10	P11	P12
Temperature (°C)	130	130	130	130	150	130	130	130	130	130	130	130
Gas	He,10 psi	He,10 psi	He,10 psi	He,10 psi	He,10 psi	He,10 psi	He,10 psi	He,10 psi	He,10 psi	He,10 psi	He,10 psi	He,10 psi
Column												
Length (ft)	150	150	150	150	150	150	150	150	150	150	150	150
Diameter (inches I.D.)	0.02	0.02	0.02	0.02	0.02	0.02	0.02	0.02	0.02	0.02	0.02	0.02
Material	SS	SS	SS	SS	SS	SS	SS	SS	SS	SS	SS	SS
Detector	FID	FID	FID	FID	FID	FID	FID	FID	FID	FID	FID	FID
Amino acid	$\alpha^a_{L/D}$											
Alanine	1.323	1.270	1.174	1.282	1.206	1.162	1.174	1.150	1.270	1.142	1.228	1.145
Aspartic acid	1.083	1.070	1.036	1.097	1.044	1.048	1.045	1.041	1.051	N.R.^{c}	1.063	1.043
α-Aminobutyric acid	1.350	1.259	1.105	1.268	1.179	1.143	1.144	1.126	1.212	1.068	1.181	1.119
Glutamic acid	1.285	1.234	1.153	1.231	1.168	1.137	1.136	1.122	1.169	1.106	1.181	1.109
α-Aminohexanoic acid	1.395	1.299	1.180	1.307	1.206	1.154	1.167	1.144	1.237	1.102	1.186	1.127
Leucine	1.499	1.375	1.246	1.376	1.251	1.216	1.223	1.190	1.329	1.172	1.280	1.166
tert-Leucine	1.136	1.109	N.R.^{b}	1.110	1.079	1.050	N.R.	1.064	1.110	N.R.	1.054	1.058
Methionine	1.382	1.301	1.189	1.282	1.199	1.148	1.178	1.146	1.218	1.137	1.187	1.181
α-Aminooctanoic acid	1.405	1.282	1.180	1.307	1.203	1.153	1.171	1.135	1.246	1.110	1.181	1.118
Phenylalanine	1.347	1.293	1.170	1.304	1.201	1.149	1.172	1.143	1.222	1.165	1.205	1.125
Phenylglycine	1.210	1.118	1.078	1.121	1.084	1.060	1.064	1.068	1.093	1.062	1.076	1.039
Proline	1.055	N.R.^{c}	N.R.	N.R.	N.R.	N.R.^{c}	N.R.	1.031	1.038	N.R.	N.R.^{c}	N.R.^{c}
Serine	1.198	1.183	1.107	1.183	1.122	1.098	1.102	1.112	1.117	1.122	1.140	1.110
Threonine	1.239	1.203	1.162	1.218	1.126	1.119	1.141	1.150	1.170	1.134	1.147	1.143
α-Aminovaleric acid	1.391	1.295	1.175	1.294	1.204	1.173	1.161	1.145	1.234	1.099	1.201	1.127
Valine	1.308	1.223	1.067	1.252	1.163	1.125	1.114	1.127	1.175	N.R.	1.152	1.113

^{a} Calculated for 100% optical purity of the stationary phases at 130°C.
^{b} N.R. = no resolution.
^{c} Shoulder.

Table GC 6
ENANTIOMERS OF N-TRIFLUOROACETYL AMINO ACID METHYL ESTERS

Column packing	P1	P2	P3	P4	P5	P6	P7	P8	P9	P10	P11	P12
Temperature (°C)	130	130	130	130	150	130	130	130	130	130	130	130
Gas	He,10 psi	He,10 psi	He,10 psi	He,10 psi	He,10 psi	He,10 psi	He,10 psi	He,10 psi	He,10 psi	He,10 psi	He,10 psi	He,10 psi
Column												
Length (ft)	150	150	150	150	150	150	150	150	150	150	150	150
Diameter (inches I.D.)	0.02	0.02	0.02	0.02	0.02	0.02	0.02	0.02	0.02	0.02	0.02	0.02
Material	SS	SS	SS	SS	SS	SS	SS	SS	SS	SS	SS	SS
Detector	FID	FID	FID	FID	FID	FID	FID	FID	FID	FID	FID	FID

Amino acid	$\alpha^a_{L/D}$											
Alanine	1.323	1.221	1.123	1.221	1.168	1.136	1.125	1.130	1.262	1.098	1.166	1.121
Aspartic acid	1.080	1.057	N.R.[b]	1.051	N.R.	1.036	N.R.	1.044	1.033	N.R.	1.057	N.R.
α-Aminobutyric acid	1.302	1.223	1.064	1.218	1.158	1.112	1.114	1.119	1.202	N.R.	1.152	1.084
Glutamic acid	1.264	1.141	1.109	1.182	1.142	1.123	1.090	1.108	1.160	1.094	1.149	1.093
α-Aminohexanoic acid	1.344	1.236	1.132	1.248	1.175	1.138	1.121	1.129	1.232	1.072	1.176	1.087
Leucine	1.419	1.287	1.167	1.320	1.223	1.167	1.163	1.155	1.270	1.098	1.200	1.121
tert-Leucine	1.116	1.086	N.R.	1.089	N.R.	N.R.	N.R.	1.049	1.081	N.R.	N.R.	N.R.
Methionine	1.319	1.227	1.145	1.233	1.176	1.132	1.123	1.135	1.206	1.076	1.173	1.096
α-Aminooctanoic acid	1.333	1.231	1.113	1.249	1.174	1.134	1.128	1.124	1.229	1.063	1.131	1.084
Phenylalanine	1.304	1.225	1.137	1.249	1.178	1.138	1.122	1.142	1.232	1.058	1.190	1.107
Proline	1.071	1.056	N.R.	N.R.[c]	N.R.	N.R.	N.R.	1.034	1.061	N.R.	1.043	N.R.
Serine	1.167	1.133	1.078	1.156	1.077	1.080	1.057	1.089	1.098	1.047	1.096	1.087
Threonine	1.175	1.146	1.135	1.156	N.R.	1.086	N.R.	1.107	1.121	1.058	1.120	1.093

Column packing   See Table 6.

REFERENCE

1. Beitler, U. and Feibush, B., *J. Chromatogr.*, 123, 149, 1976.

Table GC6 (continued)
ENANTIOMERS OF N-TRIFLUOROACETYL AMINO ACID METHYL ESTERS

	P1	P2	P3	P4	P5	P6	P7	P8	P9	P10	P11	P12
α-Aminovaleric acid	1.345	1.247	1.141	1.241	1.180	1.156	1.145	1.129	1.240	1.068	1.162	1.093
Valine	1.232	1.176	N.R.	1.171	1.122	1.086	N.R.	1.082	1.139	N.R.	1.119	N.R.

^a Calculated for 100% optical purity of the stationary phases at 130°C.
^b N.R. = no resolution.
^c Shoulder.

Column packing P1 = N-lauroyl-L-valine *tert*-butylamide
P2 = N-lauroyl-L-valine neopentylamide
P3 = N-lauroyl-L-valine 6-undecylamide
P4 = N-lauroyl-L-valine cyclooctylamide
P5 = N-lauroyl-L-valine 1-adamantylamide
P6 = N-lauroyl-L-valine *n*-dodecylamide
P7 = N-isobutyryl-L-valine *n*-dodecylamide
P8 = N-pivaloyl-L-valine *n*-dodecylamide
P9 = N-tert-butylacetyl-L-valine *n*-dodecylamide
P10 = N-dipentylacetyl-L-valine *n*-dodecylamide
P11 = N-cyclooctanecarbonyl-L-valine *n*-dodecylamide
P12 = N-adamantane-1-carbonyl-L-valine *n*-dodecylamide

REFERENCE

1. Beitler, U. and Feibush, B., *J. Chromatogr.*, 123, 149, 1976.

Reproduced by permission of Elsevier Scientific Publishing Company.

Table GC 7
ENANTIOMERS OF N-TRIFLUOROACETYL-ALANINE AND -LEUCINE ESTERS

Column packing	P1	P2	P3	P4	P5	P6	P7	P8	P9	P10	P11	P12
Temperature (°C)	130	130	130	130	150	130	130	130	130	130	130	130
Gas	He,10 psi	He,10 psi	He,10 psi	He,10 psi	He,10 psi	He,10 psi	He,10 psi	He,10 psi	He,10 psi	He,10 psi	He,10 psi	He,10 psi
Column												
Length (ft)	150	150	150	150	150	150	150	150	150	150	150	150
Diameter (inches I.D.)	0.02	0.02	0.02	0.02	0.02	0.02	0.02	0.02	0.02	0.02	0.02	0.02
Material	SS	SS	SS	SS	SS	SS	SS	SS	SS	SS	SS	SS
Detector	FID	FID	FID	FID	FID	FID	FID	FID	FID	FID	FID	FID

Amino acid Ester	$\alpha^a_{L/D}$											
Alanine Methyl	1.323	1.221	1.123	1.221	1.168	1.136	1.125	1.130	1.262	1.098	1.166	1.121
Ethyl	1.331	1.233	1.141	1.258	1.181	1.151	1.129	1.137	1.268	1.106	1.200	1.127
n-Propyl	1.314	1.254	1.150	1.261	1.186	1.134	1.154	1.137	1.210	1.131	1.201	1.116
n-Butyl	1.313	1.258	1.148	1.268	1.181	1.153	1.160	1.141	1.230	1.106	1.207	1.120
Isopropyl	1.351	1.270	1.174	1.282	1.206	1.162	1.174	1.150	1.278	1.142	1.228	1.145
3-Pentyl	1.402	1.313	1.192	1.327	1.216	1.197	1.184	1.177	1.293	1.155	1.252	1.166
Cyclopentyl	1.321	1.249	1.128	1.257	1.185	1.141	1.138	1.121	1.218	1.089	1.188	1.117
Leucine Methyl	1.419	1.287	1.167	1.320	1.223	1.167	1.163	1.155	1.270	1.098	1.200	1.121
Ethyl	1.431	1.323	1.198	1.337	1.235	1.181	1.175	1.174	1.293	1.104	1.231	1.140
n-Propyl	1.433	1.329	1.206	1.323	1.217	1.193	1.164	1.176	1.286	1.091	1.233	1.126
n-Butyl	1.433	1.308	1.206	1.339	1.231	1.183	1.174	1.174	1.296	1.090	1.220	1.121
Isopropyl	1.499	1.375	1.246	1.376	1.251	1.216	1.223	1.190	1.329	1.172	1.280	1.166
3-Pentyl	1.576	1.422	1.264	1.435	1.298	1.246	1.235	1.217	1.393	1.184	1.336	1.200
Cyclopentyl	1.467	1.341	1.200	1.337	1.223	1.195	1.189	1.141	1.283	1.115	1.233	1.189

^a Calculated for 100% optical purity of the stationary phase at 130°C.

Column packing   See Table 6.

REFERENCE

1. Beitler, U. and Feibush, B., *J. Chromatogr.*, 123, 149, 1976.

Reproduced by permission of Elsevier Scientific Publishing Company.

Table GC 8
ENANTIOMERS OF N-TRIFLUOROACETYL AMINO ACID ESTERS

Column packing		P1	P1	P1	P1	P1	P1
Gas		na	na	na	na	na	na
Column							
Length (ft)		400	400	400	400	400	400
Diameter (inches I.D.)		0.02	0.02	0.02	0.02	0.02	0.02
Material		SS	SS	SS	SS	SS	SS
Detector		FID	FID	FID	FID	FID	FID
Alcohol residue		Me	Et	n-Pr	i-Pr	n-Bu	3-n-pentyl
Amino acid	Temp°C			t_r (min)			
Alanine	100	59.7	82.3	157.0	85.6	303.5	86.8
		56.7	78.3	149.0	82.2	288.5	83.0
	120	28.8	37.5	78.4	39.2	126.4	37.5
		27.6	36.0	75.0	37.8	120.9	36.0
Valine	100	71.1			104.3		
		67.4			98.8		
	120	33.8			36.0		
		32.5			34.5		
Leucine	100	200.0	270.0	549.0	320.0	1015.0	
		200.0	270.0	538.0	320.0	990.0	
	120	75.0	120.0		95.0		80.0
		75.0	120.0		95.0		80.0
Tert-Leucine	100	59.1			58.2		
		56.5			56.1		
	130				20.3		
					19.7		
Proline	100	203.1			170.0		
		195.5			166.0		
	130	61.0					
		59.2					
Phenylglycine	130				62.7		
					61.0		
Phenylalanine	130	270.0					
		270.0					
α-Methyl tert-leucine	100	82.2					
		80.7					
α-Methyl valine	100	72.9					
		72.3					
α-Methyl leucine	100	60.0					
		60.0					
α-Methyl norleucine	100	105.5					
		102.5					
α-Methyl norvaline	100	62.8					
		61.7					

Column packing   P1 = N-lauroyl-S-α-(1-naphthyl) ethylamine

REFERENCE

1. Weinstein, S., Feibush, B., and Gil-Av, E., *J. Chromatogr.*, 126, 97, 1976.

Table GC 9
ENANTIOMERS OF N-TRIFLUOROACETYL-DL-ALANINE ISOPROPYL ESTER

Column packing	P1		P2	
Gas; Flow rate (ml/min)	N_2,2		N_2,2	
Column				
Length (m)	40		40	
Diameter (mm)	0.25		0.25	
Form	Open tubular		Open tubular	
Material	G[a]		G[a]	
Detector	FID		FID	
Temperature (°C)	K_1[b]	K_2	K_1	K_2
130°	10.125	10.545	3.610	3.892
125°	11.699	12.188	4.220	4.583
120°	14.050	14.642	4.711	5.163
100°	0.470	0.634	10.828	12.052
95°	0.504	0.707	13.930	15.700
90°	0.568	0.826	17.129	19.442
70°			48.287	56.390

[a] Borosilicate glass.
[b] K_1 and K_2 are the respective capacity ratios of the enantiomers. The emergence time of methane, which was assumed to be nonretained, was used in calculating K.

Column packing   P1 = ureide of L-valine isopropyl ester.
                              P2 = N-TFA-L-valyl-L-valine cyclohexyl ester.

REFERENCE

1. Corbin, J. A. and Rogers, L. B., *Anal. Chem.*, 42, 1786, 1970.

Reprinted with permission from *Anal. Chem.*, 42, 1786, 1970. Copyright (1970) American Chemical Society.

Table GC 10
ENANTIOMERIC AMINO ACID DERIVATIVES

Column packing	P1		P2		P3		P4		P5		P6	
Temperature	T1		T2		T1		T2		T1		T2	
Gas; Flow rate (ml/min)	N₂,2		N₂,2		N₂,2		N₂,2		N₂,2		N₂,2	
Column												
Length (m)	40		40		40		40		40		40	
Diameter (mm)	0.25		0.25		0.25		0.25		0.25		0.25	
Material	Gᵃ		G		G		G		G		G	
Detector	FID		FID		FID		FID		FID		FID	

Amino acid derivative	K_1	α	K_1	α	K_1	α	K_1	α	K_1	α	K_1	α	K_1
N-TFAᵇ-DL-alanine isopropyl ester	9.27	1.126	4.18	1.112	8.71	1.118	3.65	1.095	13.04	1.114	5.95	1.090	
N-TFA-DL-valine ethyl ester	14.57	1.080	6.18	1.064	13.85	1.061	5.33	1.043	19.59	1.087	8.55	1.066	
N-TFA-DL-alanine ethyl ester	8.30	1.107	3.70	1.079	7.83	1.095	3.24	1.075	11.37	1.095	5.35	1.075	
N-PFPᶜ-DL-alanine ethyl ester							2.73	1.068					

Amino acid derivative	K_1	α	K_1	α	K_1	α	K_1	α	K_1	α
N-TFA-DL-alanine isopropyl ester	9.88	1.103	4.21	1.076	6.67	1.096	2.59	1.072	4.90	1.088
N-TFA-DL-valine ethyl ester	14.77	1.062	6.00	1.049	10.07	1.061	3.81	1.045		
N-TFA-DL-alanine ethyl ester	8.70	1.081	3.77	1.064	5.73	1.076	2.32	1.058	4.31	1.069
N-PFP-DL-alanine ethyl ester							3.79			1.063

[a] Borosilicate glass.
[b] n-trifluoroacetyl.
[c] n-pentafluoropropionyl.

Column packing	P1	= N-TFA-L-valyl-L-valine cyclohexyl ester
	P2	= N-TFA-L-valyl-L-leucine cyclohexyl ester
	P3	= N-TFA-L-leucyl-L-valine cyclohexyl ester
	P4	= N-TFA-L-leucyl-L-leucine cyclohexyl ester
	P5	= N-TFA-L-leucyl-L-leucyl-L-leucine cyclohexyl ester
	P6	= N-PFP-L-valyl-L-leucine cyclohexyl ester
Temperature	T1	= 90°C
	T2	= 110°C

REFERENCE

1. Corbin, J. A., Rhoad, J. E., and Rogers, L. B., *Anal. Chem.*, 43, 327, 1971.

Reprinted with permission from *Anal. Chem.*, 43, 327, 1971. Copyright (1971) American Chemical Society.

Table GC 11
AMINO ACIDS, N-TRIFLUOROACETYL n-BUTYL ESTERS

Column packing	P1	P2	P3	P4	P5	P6	P7	P8
Temperature	T1	T1	T1	T1	T2	T3	T4	T5
Gas; Flow rate (mℓ/min)	He,60	He,60	He,60	He,60	N_2,17	N_2,30	N_2,30	
Column								
Length (m)	1.83	1.83	1.83	1.83	1.83	1.83	1.83	1.8
Diameter (mm I.D.)	0.4	0.4	0.4	0.4	0.2	0.2	0.2	2
Material	G	G	G	G	G	G	G	G
Detector	FID	FID	FID	FID	FID	FID	FID	FID
Literature	1	1	1	1	2	2	2	3
Amino acid			'Internal standards^a			'r(mins)		'Arg
Alanine	0.192	0.195	0.186	0.168	15.71	7.23	3.65	0.46
γ-Aminobutyric acid	0.633	0.636	0.648	0.640				
Arginine							30.02	1.00
Aspartic acid	0.734	0.711	0.710	0.695		22.92	20.88	0.74
Cysteine	0.562	0.560	0.572	0.561	21.91	19.91	12.59	
Cystine						50.87	43.67	
Glutamic acid	0.873	0.853	0.849	0.841	31.94	26.51	24.65	0.83
Glycine	0.343	0.355	0.369	0.351	16.36	10.41	4.73	0.49
Histidine							32.35	0.79
Hydroxyproline	0.662	0.621	0.631	0.621		21.09	17.43	
Isoleucine	0.305	0.264	0.247	0.208	21.59	11.49	8.86	0.55
Leucine	0.364	0.329	0.319	0.276	21.26	12.97	8.86	0.55
Lysine	1.191	1.165	1.180	1.202	30.17	32.86	27.41	0.92
Methionine	0.734	0.733	0.741	0.737		20.61	18.52	0.69
Phenylalanine	0.765	0.752	0.755	0.749	29.74	22.17	20.88	0.72
Proline	0.547	0.520	0.517	0.489		13.77	15.96	0.65
Serine	0.421	0.431	0.448	0.433	16.57	17.15	7.58	0.57
Threonine	0.278	0.264	0.265	0.231	15.90	15.20	6.02	0.53
							31.28, 39.18	
Tryptophan	1.191	1.116	1.147	1.121		34.17		
Tyrosine	0.873	0.873	0.878	0.878	30.57	28.85	24.35	0.84
Valine	0.219	0.195	0.186	0.161		9.10	6.02	0.51

^{a} Tranexamic acid.

Column packing   P1 = 0.75% Silar 5CP on Gas-Chrom® Q, 100-120 mesh
                       P2 = 0.75% Silar 7CP on Gas-Chrom® Q, 100-120 mesh
                       P3 = 0.75% Silar 9CP on Gas-Chrom® Q, 100-120 mesh
                       P4 = 0.75% Silar 10C on Gas-Chrom® Q, 100-120 mesh
                       P5 = 10% Apiezon M on Chromosorb® W, 80-100 mesh
                       P6 = Tabsorb
                       P7 = Tabsorb HAC
                       P8 = 3% OV-210 on Gas-Chrom® Q, 100-120 mesh
Temperature   T1 = Programed, 2°C/min up to 240°C, maintained at 240°C for 15 min
                       T2 = Programed, 90°C for 6 min, then 6°/min to 260°C
                       T3 = Programed, 100°C for 7 min, then 4°C/min to 200°C
                       T4 = Programed, 110°C for 7 min, then 4°C/min to 210°C
                       T5 = Programed, 50°C then 8°C/min

REFERENCES

1. Nagy, S. and Hall, N. T., *J. Chromatogr.*, 177, 141, 1979.
2. Sarkar, S. K. and Malhotra, S. S., *J. Chromatogr.*, 170, 371, 1979.
3. Frick, W., Chang, D., Folkers, K., and Daves, G. D., Jr., *Anal. Chem.*, 49, 1241, 1977.

Table GC 12
NONPROTEIN AMINO ACIDS, N-TRIFLUOROACETYL n-BUTYL ESTERS

Column packing	P1
Temperature	T1
Gas, Flow rate (ml/min)	N_2, 40
Column	
Length (m)	1.5
Diameter (mm I.D.)	3
Material	Pyrex glass
Detector	FID

Amino acid	t_r (min-sec)
α-Aminoisobutyric acid	9-05
Alanine	10-35
Sarcosine	11-55
α-Aminobutyric acid	12-15
Valine	13-05
Glycine	14-05
Norvaline	15-30
Isoleucine	15-50
β-Aminoisobutyric acid	16-30
β-Alanine	17-05
Leucine	17-45
1-Amino-1-cyclopropanecarboxylic acid	18-00
Norleucine	18-35
Baikiain	18-35
Proline	20-20
Threonine	20-45
Serine	23-40
γ-Aminobutyric acid	24-30
L-Azetidine-2-carboxylic acid	24-30
N-Methylmethionine	27-35
Cysteine	27-45
Methionine	28-15
Hydroxyproline	29-05
Phenylalanine	29-30
Homoserine	30-00
5-Hydroxypipecolic acid	30-00
ε-Aminocaproic acid	30-30
Aspartic acid	30-50
Allohydroxyproline	31-10
Glutamic acid	34-10
Pyrolidine-2,5-dicarboxylic acid	34-25
2,4-Diaminobutyric acid	35-00
α-Aminoadipic acid	36-15
Tyrosine	36-30
α-Aminopimelic acid	37-55
Kainic acid	39-30
α,α′-Diaminopimelic acid	43-55
Methionine sulfone	47-35

Column packing   P1 = 0.325% EGA (ethyleneglycol adipate) on Chromosorb® W, AW, 80-100 mesh.

Temperature   T1 = Programed; 80°C, then 2°C/min for 25 min and 4°C/min for 20 min up to 210°C, 210°C for 6 min.

Table GC 12 (continued)
NONPROTEIN AMINO ACIDS, N-TRIFLUOROACETYL n-BUTYL ESTERS

REFERENCE

1. Amico, V., Oriente, G., and Tringali, C., *J. Chromatogr.*, 116, 439, 1976.

Reproduced by permission of the Elsevier Scientific Publishing Company.

Table GC 13
AMINO ACIDS, N-HEPTAFLUOROBUTYRYL ISOBUTYL ESTERS

Column packing	P1
Temperature	T1
Gas; Flow Rate (mℓ/min)	N_2, 18
Column	
Length (m)	6
Diameter (mm I.D.)	2
Form	Coiled
Material	glass
Detector	FID

Amino acid	t_r
Cysteic acid	12.2
Alanine	12.8
2-Aminoisobutyric acid	13.0
Glycine	13.2
2-Aminobutyric acid	14.6
Sarcosine	14.7
β-Alanine	15.1
Valine	15.5
DL-3-Aminobutyric acid	15.6
DL-3-Aminoisobutyric acid	15.6
Threonine	16.1
Norvaline	16.2
Serine	16.4
Leucine	17.0
Isoleucine	17.3
Norleucine (internal standard)	18.0
4-Aminobutyric acid	18.0
Homoserine	18.8
	19.0
Proline	19.0
DL-4-Amino-3-hydroxybutyric acid	19.4
Cysteine	19.4
S-Methylcysteine	19.5
3,4-Dihydroxyproline	19.7
DL-Pipecolic acid	20.0
4-Hydroxyproline	20.8
2,4-Diaminobutyric acid	21.0
Methionine	21.6
Methionine sulfoxide	21.6
Asparagine/aspartic acid	22.7

Table GC 13 (continued)
AMINO ACIDS, N-HEPTAFLUOROBUTYRYL ISOBUTYL ESTERS

Phenylalanine	23.5
Ornithine	23.9
Citruline	23.9
Glutamine/glutamic acid	24.8
5-Hydroxylysine (DL and DL-allo)	25.5
	25.6
Lysine	25.8
Tyrosine	26.2
DL-2-Aminoadipic acid	26.5
3,4-Dihydroxyphenylalanine	26.9
Arginine	27.6
S-Carboxymethylcysteine	28.0
Histidine	28.3
Tryptophan	29.9
DL-2,6-Diaminopimelic acid	29.9
3-Iodotyrosine	30.0
Cystathionine	32.4
Cystine	33.2
3,5-Diiodotyrosine	34.0
DL-Homocystine	36.5

Column packing P1 = 3% OV-101 on Gas-Chrom® Q, 80-100 mesh.
Temperature T1 = Programed: 80°C for 5 min, then 6°C/min to 250°C, maintained at 250°C.

REFERENCE

1. Siezen, R. L. and Mague, T. H., *J. Chromatogr.*, 130, 151, 1977.

Reproduced by permission of Elsevier Scientific Publishing Company.

Table GC 14
DERIVATIVES OF DL-AMINO ACIDS

Column packing	P1
Temperature	T1
Gas: Flow rate (cm/sec)	H_2, 50
Column	
Length (m)	22
Diameter (mm I.D.)	0.25
Form	Whiskered capillary
Material	G[a]
Detector	FID

Amino acid	TFA-i-Pr[b]	TFA-n-Bu	PFP-i-Pr[c]	PFP-n-Bu	HFB-i-Pr[d]	HFB-n-Bu[d]
D-Valine	4.00	8.96	3.00	7.02	2.90	6.40
L-Valine	4.40	9.42	3.28	7.40	3.10	6.72
Glycine	6.24	13.00	5.40	11.36	5.26	10.80
D-Leucine	7.80	14.70	6.24	12.16	5.90	11.30
L-Leucine	8.70	15.52	6.90	12.90	6.60	12.04
D-Proline	10.90	18.70	9.14	16.10	8.30	14.60
L-Proline	11.00	18.80	9.22	16.22	8.40	14.68
D-Aspartic Acid	17.48	33.10	15.22	29.96	14.16	28.20
L-Aspartic acid	17.80	33.22	15.44	30.12	14.40	28.40
D-Phenylalanine	25.70	33.60	22.90	30.50	21.80	28.74
L-Phenylalanine	26.40	34.08	23.56	31.00	22.44	29.22
D-Glutamic acid	26.26	41.64	23.30	38.26	22.00	36.00
L-Glutamic acid	26.80	42.10	23.90	38.62	22.60	36.40

[a] Borosilicate glass capillary, washed with methanol and CH_2Cl_2 and heated overnight with dry N_2 at 300°C. The surface is modified by whisker formation and deactivated with a compound like triethanolamine.
[b] N-trifluoroacetyl.
[c] N-pentafluoropropionyl.
[d] N-heptafluorobutyryl.

Column packing	P1	= L-valine-tert-butylamide grafted on to a modified polycyanopropylmethyl phenylmethyl silicone (OV-225).
Temperature	T1	= Programed from 90°C to 190°C at 2°C/min.

REFERENCE

1. Saeed, T., Sandra, P. and Verzele, M., *J. Chromatogr.*, 186, 611, 1979.

Reproduced by permission of Elsevier Scientific Publishing Company.

Table GC 15
ENANTIOMERIC N-TRIFLUOROACETYL AND N-PENTAFLUOROPROPIONYL AMINO ACID ISOPROPYL ESTERS

Column packing		P1	P1	P2	P2
Temperature (°C)		110	110	110	110
Gas		He, 10 psi	He, 10 psi	He, 10 psi	He, 10 psi
Column					
Length (ft)		400	400	400	400
Diameter (inches I.D.)		0.02	0.02	0.02	0.02
Material		SS	SS	SS	SS
Detector		FID	FID	FID	FID
Amino acid	Enantiomer	$t_r^{a,c}$	$t_r^{b,d}$	$t_r^{a,c}$	$t_r^{b,d}$
Alanine	D	0.287	0.304	0.278	0.297
	L	0.321	0.338	0.308	0.330
Valine	D	0.462	0.466	0.452	0.461
	L	0.502	0.508	0.495	0.502
Glycine		0.528	0.591	0.532	0.623
Threonine	D	0.576	0.521	0.475	0.437
	L	0.634	0.566	0.522	0.474
Isoleucine	D	0.672	0.658	0.662	0.670
	L	0.793	0.775	0.793	0.800
Leucine	D	0.900	0.896	0.883	0.884
	L	1.000	1.000	1.000	1.000
Serine	D	0.977	1.062	0.823	0.898
	L	1.057	1.145	0.893	0.959
Proline	D	1.318	1.416	1.850	2.130
	L	1.371	1.487	1.850	2.130
Cysteine	D	2.666	2.866	2.582	2.753
	L	2.890	3.083	2.739	2.977

[a] N-trifluoroacetyl derivatives.
[b] N-pentafluoropropionyl derivatives.
[c] Relative to N-TFA-L-leucine isopropyl ester.
[d] Relative to N-PFP-L-leucine isopropyl ester.

Column packing   P1 = N-TFA-L-valyl-L-valine cyclohexyl ester
                              P2 = N-TFA-L-phenylalanyl-L-leucine cyclohexyl ester

REFERENCE

1. Parr, W., Pleterski, J., Yang, C., and Bayer, E., *J. Chromatogr. Sci.*, 9, 141, 1971.

Reproduced from the *Journal of Chromatographic Science* by permission of Preston Publications Inc.

Table GC 16
ENANTIOMERIC N-TRIFLUOROACETYL AND N-PENTAFLUOROPROPIONYL AMINO ACID ISOPROPYL ESTERS

Column packing		P1	P1	P2	P2
Temperature (°C)		100	100	100	100
Gas		He, 5 psi	He, 5 psi	He, 10 psi	He, 10 psi
Column					
Length (ft)		100	100	100	100
Diameter (inches I.D.)		0.02	0.02	0.02	0.02
Material		SS	SS	SS	SS
Detector		FID	FID	FID	FID
Amino acid	Enantiomer	$t_r^{a,c}$	$t_r^{b,d}$	$t_r^{a,c}$	$t_r^{b,d}$
Aspartic acid	D	0.313	0.350	0.290	0.325
	L	0.333	0.350	0.305	0.340
Methionine	D	0.494	0.477	0.476	0.482
	L	0.553	0.528	0.535	0.543
Phenylalanine	D	0.678	0.668	0.665	0.678
	L	0.761	0.752	0.751	0.782
Glutamic acid	D	0.914	0.911	0.911	0.919
	L	1.000	1.000	1.000	1.000
Tyrosine	D			1.948	2.335
	L			2.171	2.665
Lysine[f]	D			50.0[e]	26.5[e]
	L			55.5[e]	29.3[e]
Arginine	D				13.15[e]
	L				14.25[e]
Tryptophan[g]	D				17.65[e]
	L				18.70[e]

[a] N-trifluoroacetyl derivatives.
[b] N-pentafluoropropionyl derivatives.
[c] Relative to N-TFA-L-glutamic acid diisopropyl ester.
[d] Relative to N-PFP-L-glutamic acid diisopropyl ester.
[e] Retention time.
[f] 125°C, 12 psi.
[g] 130°C, 10 psi.

Column packing   P1 = N-TFA-L-valyl-L-valine cyclohexyl ester.
                            P2 = N-TFA-L-phenylalanyl-L-leucine cyclohexyl ester.

REFERENCE

1. Parr, W., Pleterski, J., Yang, C., and Bayer, E., *J. Chromatogr. Sci.*, 9, 141, 1971.

Reproduced from the *Journal of Chromatographic Science* by permission of Preston Publications Inc.

Table GC 17
DIASTEREOISOMERIC PENTAFLUOROPROPIONYL-DL-AMINO ACID (+)-3-METHYL-2-BUTYL ESTERS

Column packing	P1
Temperature	T1
Gas	H_2
Column	
Length (m)	25
Diameter	na
Material	G
Detector	FID

Amino acid (in order of emergence)	α^a
Alanine	1.09
Glycine	—
α-Aminobutyric acid	1.09
Threonine	1.03
Valine	1.08
Serine	1.05
Leucine	1.09
allo-Isoleucine	1.08
Isoleucine	1.09
Cysteine	1.07
Proline	1.06
Diaminopropionic acid	1.04
Methionine	1.05
Ornithine	1.06
Phenylalanine	1.04
Aspartic acid	1.04
Histidine	1.03
Lysine	1.05
Tyrosine	1.04
Glutamic acid	1.05
Arginine	1.05
Tryptophan	1.04

[a] The L-amino acid derivatives have longer retention times than the D-amino acid derivatives.

Column packing	P1	= SE-30
Temperature	T1	= 100°C, 140°C for diaminopropionic acid and the following amino acids, and 200°C for tryptophan.

REFERENCE

1. König, W. A., Rahn, W., and Eyem, J., *J. Chromatogr.*, 133, 141, 1977.

Table GC 18
ENANTIOMERS OF N(O,S)-PENTAFLUOROPROPIONYL AMINO ACID ISOPROPYL ESTERS

	P1		P2	
Column packing	P1		P2	
Temperature	T1		T2	
Gas	H_2		H_2	
Column				
Length (m)	20		20	
Diameter (mm)	0.3		0.3	
Form	Open tubular		Open tubular	
Material	Glass[a]		Glass[a]	
Detector	FID		FID	
Amino acid	$\alpha_{L/D}$	t_r	$\alpha_{L/D}$	t_r
Alanine	1.129	3.0	1.240	2.1
Valine	1.117	3.5	1.200	2.8
Threonine	1.069	4.5	1.125	2.5
Alloisoleucine	1.148	5.0	1.200	4.0
Isoleucine	1.158	6.0	1.192	4.4
Proline	1.005	7.0	1.025	6.3
Leucine	1.222	8.0	1.375	6.3
Serine	1.052	4.5	1.081	4.6
Cystine	1.060	5.0	1.118	4.1
Aspartic acid	1.031	9.0	1.027	6.3
Methionine	1.079	7.0	1.140	5.8
Glutamic acid	1.108	8.0	b	
Phenylalanine	1.086	8.0	1.132	8.5
Tyrosine	1.078	6.0	b	
Ornithine	1.081		b	
Lysine	1.065		b	
Tryptophan	1.044	16.0	b	

[a] AR glass pretreated with colloidal silicic acid or aerosil.
[b] This column bled severely and expired after a few hours.

Column packing	P1	= 0.15% N-propionyl-L-valine tert-butylamide polysiloxane, conditioned for 3 days at 170°C.
	P2	= 0.15% N-n-docosanoyl-L-valine tert-butylamide.
Temperature	T1	= Programed: 100°C to 170°C.
	T2	= Programed: 100°C to 140°C.

REFERENCE

1. Frank, H., Nicholson, G. J., and Bayer, E., *J. Chromatogr. Sci.,* 15, 174, 1977.

Reproduced from the *Journal of Chromatographic Science* by permission of Preston Publications Inc.

Table GC 19
AMINO ACID N-HEPTAFLUOROBUTYRYL ESTERS

Column packing	P1	P2	P3
Temperature	T1	T2	T3
Gas, Flow rate (ml/min)	N_2, 20	N_2	He, 4
Column			
Length (m)	3.50	6	60
Diameter (mm I.D.)	2	2	0.5
Material	G[a]	G	G[b]
Detector	FID	FID	FID
Literature	1	2	3

Amino acid	Retention Temperature (°C)	$r_{norleucine}$	t_r (sec)
Alanine	111.5[c]	0.499[d]	463[e]
Arginine	190.0	2.039	1305
Aspartic acid	170.0	1.391	994
S-Carboxymethyl cysteine			1344
Cysteine	143.0	1.154	1718
Cystine	230.0	2.568	
Glutamic acid	182.5	1.621	1131
Glycine	114.0	0.535	476
Histidine[f]	162.0	1.653	1433
Histidine[g]	194.5		
Hydroxyproline	150.0		872
Isoleucine	133.0	0.926	674
Alloisoleucine		0.913	
Kynurenine	218.0		1525
Leucine	131.0	0.905	658
Lysine	178.0	1.841	1194
Methionine	155.0	1.357	929
Norleucine			713
Ornithine	168.0		1071
Phenylalanine	165.0	1.549	1049
Pipecolic acid	147.0		
Proline	141.0	1.070	773
Serine	128.0	0.873	614
Threonine	126.0	0.837	598
Tryptophan	209.0		
Tyrosine	181.0	1.873	1219
Valine	124.0	0.750	583

[a] Pyrex glass.
[b] Glass silanized and deactivated.
[c] Isoamyl ester.
[d] n-propyl ester.
[e] Isobutyl ester.
[f] Diacyl derivative.
[g] Monoacyl derivative.

Column packing	P1 = 3% SE-30 on Gas-Chrom® Q.
	P2 = Open tubular column with dimethylsiloxane polymer chemically bound to it.
	P3 = SE-30.
Temperature	T1 = Programed: 70°C then increased at 40°C/min to 240°C.
	T2 = Programed: 70°C for 4 min then increased at 5°C/min up to 230°C.
	T3 = Programed; increased from 100°C to 250°C at 6°C/min and maintained at 250°C for 5 min.

Table GC 19 (continued)
AMINO ACID N-HEPTAFLUOROBUTYRYL ESTERS

REFERENCES

1. Zanetta, J. P. and Vincendon, G., *J. Chromatogr.*, 76, 91, 1973.
2. Jonsson, J., Eyem, J., and Sjoquist, J., *Anal. Biochem.*, 51, 204, 1973.
3. Pearce, R. J., *J. Chromatogr.*, 136, 113, 1977.

Table GC 20
N-HEPTAFLUOROBUTYRYL AMINO ACID ISOBUTYL ESTERS

Column packing	P1	P1
Temperature	T1	T2
Gas; Flow rate (mℓ/min)	N_2, 2	N_2, 2
Column		
Length (m)	25	25
Diameter (mm I.D.)	0.23	0.23
Material	G	G
Detector	FID	FID

Amino acid	t_r (min)	
Alanine	12.7	12.8
Glycine	13.2	13.3
α-Aminobutyric acid	15.2	15.7
β-Alanine	16.1	16.7
Valine	16.7	17.5
β-Aminoisobutyric	17.0	17.8
Threonine	17.4	18.4
Serine	18.1	19.2
Leucine	19.3	20.8
allo-Isoleucine	19.6	
Isoleucine	19.8	21.5
γ-Aminobutyric acid	21.3	23.4
Cycloleucine	22.2	24.6
Proline	23.1	25.8
4-Hydroxyproline	26.8	31.1
Methionine	28.5	33.4
Aspartic acid	30.7	36.7
Phenylalanine	32.4	39.1
Ornithine	33.6	41.1
Glutamic acid	35.3	43.4
Lysine	37.8	47.1
Tyrosine	38.4	47.9
Methionine sulfone	39.0	48.7
N^ϵ-monomethyllysine	39.6	49.7
Arginine	41.4	52.7
Histidine	15.2	15.2
Homoarginine	45.3	58.4
Lanthionine	48.8	63.5
Cystathionine	52.3	68.5
Cystine	54.2	71.3
Cys-S-S-homoCys	57.5	76.1
Homocystine	60.6	80.8

Column packing   P1 = OV-101

Table GC 20 (continued)
N-HEPTAFLUOROBUTYRYL AMINO ACID ISOBUTYL ESTERS

Temperature   T1 = Programed: 90°C, then 2°C/min increase.
               T2 = Programed: 90°C, then 3°C/min increase.

REFERENCE

1. Desgres, J., Boisson, D., and Padieu, P., *J. Chromatogr.*, 162, 133, 1979.

Reproduced by permission of Elsevier Scientific Publishing Company.

Table GC 21
ENANTIOMERIC DERIVATIVES OF AMINO ACIDS AND AMINES ON α-HYDROXYCARBOXYLIC ACID ESTERS

	P1	P2
Column packing	P1	P2
Temperature (°C)	100	100
Gas; Flow rate (ml/min)	He, 0.7	He, 0.7
Column		
Length (m)	40	40
Diameter (mm I.D.)	0.25	0.25
Material	G	G
Detector	na	na

Compound	t_r^d(min)	
L-alanine[a]	22.26	111.60
D-alanine	22.71	112.80
L-Leucine	67.07	
D-leucine	68.93	
L-valine	33.54	191.00
D-Valine	33.97	192.50
(−)-α-phenylethylamine[b]	82.34	
(+)-α-phenylethylamine	84.39	
(−)-α-phenylpropylamine	164.01	
(+)-α-phenylpropylamine	169.63	
(−)-α-(2,5-xylyl)ethylamine[c]	272.00	
(+)-α-(2,5-xylyl)ethylamine	281.20	

[a] Amino acids resolved as the N-trifluoroacetyl isopropyl esters.
[b] Resolved as the N-pentafluoropropyl derivative.
[c] Resolved as the N-trifluoroacetyl derivative.
[d] Measured from solvent peak.

Column packing   P1 = di-1-menthyl(+) tartrate.
                  P2 = di-dl-menthyl(−) malate.

REFERENCE

1. Oi, N., Kitahara, H., and Doi, T., *J. Chromatogr.*, 207, 252, 1981.

Table GC 22
L-α-CHLOROISOVALERYL DERIVATIVES OF DL-AMINO ACID METHYL ESTERS

Column packing	P1		P2		P3	
Gas	H_2		H_2		H_2	
Column						
Length (m)	33		25		14	
Diameter (mm I.D.)	0.25		0.25		0.25	
Material	G[a]		G		G	
DL-amino acid	α	T(°C)	α	T(°C)	α	T(°C)
Alanine	1.07	150	1.06	150	1.15	130
Aminobutyric acid	1.10	150	1.10	150	1.20	130
Valine	1.13	150	1.11	150	1.22	130
Isoleucine	1.10	150	1.09	150	1.20	130
Alloisoleucine	1.10	150	1.09	150	1.21	130
Leucine	1.07	150	1.06	150	1.19	130
Serine (O-TMS)	1.07	150				
Serine[b]			1.02	200		
Threonine (O-TMS)	1.10	150			1.18	130
Threonine[b]			1.07	200		
Proline	1.07	150	1.04	200	1.24	130
Aspartic acid	1.02	150	1.03	200	1.04	150
Phenylalanine			1.06	200	1.12	150
Glutamic acid			1.03	200	1.14	150
Methionine			1.03	200		
Ornithine			1.03	250		
Lysine			1.04	250		
Tryptophan			1.02	250		
N-methylvaline					1.03	130
N-methylisoleucine	1.08	150			1.06	130

[a] Pyrex glass.
[b] N,O-bis-L-α-chloroisovaleryl derivative.

Column packing   P1 = SE-30
    P2 = AmAc (LKB)
    P3 = N-TFA-L-Phe-L-Asp-bis-cyclohexyl ester. Before coating the capillary is treated with a 0.2% suspension of Silanox (Cabot) in CH_2Cl_2. Coating is performed with 0.1-0.2% solutions of the stationary phase in CH_2Cl_2.

REFERENCE

1. König, W. A., Stölting, K., and Kruse, K., *Chromatographia*, 10, 444, 1977.

Reproduced by permission of Friedr. Vieweg & Sohn Verlagsgesellschaft and the authors.

Table GC 23
N-ACETYL AMINO ACID ISOPROPYL ESTERS

Column packing	P1
Temperature	T1
Gas: Flow rate (ml/min)	He, 0.75
Column	
Length (m)	50
Diameter (mm I.D.)	0.27
Form	Capillary
Material	G
Detector	NPD[a]

Amino acid	$r_{norleucine}$
Alanine	0.610
Valine	0.750
Glycine	0.782
Isoleucine	0.900
Leucine	0.930
Norleucine	1.000
Proline	1.200
Threonine	1.270
Serine	1.420
Aspartic acid	1.630
Methionine	1.780
Cysteine	1.800
Phenylalanine	1.860
Hydroxyproline	1.960
Glutamic acid	1.990
Tyrosine	2.580
Ornithine	2.700
Histidine	2.760
Lysine	2.810
Arginine	2.860
Tryptophan	3.220
α-Aminoisobutyric acid	0.690
D-allo-Isoleucine	0.915
β-Amino-n-butyric acid	0.924
β-Amino-isobutyric acid	0.934
Pipecolic acid	0.962
β-Alanine	0.974
α-Amino-n-butyric acid	1.391
ε-Amino-n-butyric acid	1.406
Glycocyanomine	1.502
Methionine sulfoxide	1.584
Asparagine[b]	1.630
Selenomethionine	1.810
ε-Amino-n-caproic acid	1.901
Selenocysteine	1.972
Glutamine[c]	1.990
δ-Aminolevulinic acid	2.302
Aminoadipic acid	2.400
Sarcosine	2.411
Methionine sulfone	2.710
α,ε-Diaminopimelic acid	2.940
D-allo-Hydroxylysine	3.100
5-Hydroxylysine	3.990
allo-Cystathionine	4.012

Table GC 23 (continued)
N-ACETYL AMINO ACID ISOPROPYL ESTERS

- [a] Nitrogen-selective flame detector.
- [b] Derivatizes as aspartic acid.
- [c] Derivatizes as glutamic acid.

Column packing   P1 = 1:1 mixture of Carbowax® 20 M poly(ethylene glycol) and Silar 5 CP (Silicone AS1) phenyl (50%) cyanopropyl (50%) silicone.

Temperature   T1 = programed: 110°C to 190°C at 8°C/min, then held at 250°C for 10 min.

REFERENCE

1. Adams, R. F., Vandemark, F. L. and Schmidt, G. J., *J. Chromatogr. Sci.*, 15, 63, 1977.

Reproduced from the *Journal of Chromatographic Science* by permission of Preston Publications, Inc.

Table GC 24
DIASTEREOMERIC N-ACYL DL-ALANINE ESTERS

Column packing		P1	P2	P3	P4
Temperature (°C)		160	180	170	155
Gas, Flow rate (ml/min)		N_2,30	N_2,30	N_2,30	N_2,30
Column					
Length (m)		3	3	3	3
Diameter (mm I.D.)		3	3	3	3
Form		U	U	U	U
Material		G^a	G^a	G^a	G^a
Detector		FID	FID	FID	FID

Ester Group	N-acyl group	$r_{p,p'\text{-}DDT}$			
Methyl	d-Isoketopinyl	0.317	0.357	0.360	0.408
		0.307	0.356	0.359	0.403
Methyl	l-Dihydroteresantalinyl	0.178	0.198	0.130	0.172
		0.174	0.192	0.127	0.168
Methyl	l-Teresantalinyl	0.163	0.185	0.125	0.165
		0.168	0.191	0.131	0.182
n-Butyl	d-Isoketopinyl	0.953	0.960	0.607	0.781
		0.934	0.941	0.601	0.765
n-Butyl	l-Dihydroteresantalinyl	0.539	0.545	0.237	0.262
		0.523	0.524	0.230	0.256
n-Butyl	l-Teresantalinyl	0.497	0.490	0.223	0.256
		0.517	0.515	0.239	0.274

Column packing   P1 = 1.5% SE-30
                    P2 = 1.5% OV-1
                    P3 = 1.5% OV-17
                    P4 = 0.5% PEGA

REFERENCE

1. Nambara, T., Goto, J., Taguchi, K., and Iwata, T., *J. Chromatogr.*, 100, 180, 1974.

Reproduced by permission of Elsevier Scientific Publishing Company.

Table GC 25
DIASTEREOMERIC N-1-TERESANTALINYL DL-AMINO ACID ESTERS

Column packing	P1
Temperature	155°C
Gas; Flow rate (ml/min)	N_2, 30
Column	
Length (m)	3
Diameter (mm I.D.)	3
Form	U
Material	G[a]
Detector	FID

	Methyl ester		n-Butyl ester	
Amino acid	$r^b_{p,p'\text{-}DDT}$	α(L/D)	$r_{p,p'\text{-}DDT}$	α(L/D)
Alanine	0.164	1.10	0.256	1.07
Valine	0.191	1.05	0.329	1.03
Norvaline	0.242	1.08	0.401	1.05
Leucine	0.248	1.05	0.469	1.03
Isoleucine	0.250	1.03	0.474	1.04
Norleucine	0.311	1.05	0.596	1.04
Proline	0.648	1.20	1.07	1.04
Aspartic acid	0.922	1.05	2.38	1.01
Methionine	1.44	1.09	2.23	1.04
Ethionine	1.67	1.03	2.46	1.01
Glutamic acid	1.66	1.07	2.72	1.06
Phenylalanine	2.17	1.04	3.18	1.03

[a] Silanized glass.
[b] D enantiomer.

Column packing   P1 = 0.5% PEGA.

REFERENCE

1. Nambara, T., Goto, J., Taguchi, K., and Iwata, T., *J. Chromatogr.,* 100, 180, 1974.

Table GC 26
AMINO ACIDS, TRIMETHYLSILYL DERIVATIVES

Column packing	P1	P2	P3
Temperature	T1	T1	T2
Gas: Flow rate (ml/min)	A,30	A,50	N_2,17
Column			
Length (m)	1.83	1.83	1.83
Diameter (mm I.D.)	2	2	2
Material	SS	SS	Glass
Detector	FID	FID	FID
Literature	1	1	2

Amino acid		t_r	
Alanine	5.08	9.55	2.83
Arginine			12.07, 15.75, 18.39, 19.38[a]
Aspartic acid	15.65	19.40	11.21, 12.44[a]
Cysteine			13.72
Glutamic acid	17.67	20.13	14.31, 14.90[a]
Glycine	5.97	9.92	3.29, 6.84[a]
Homoserine	12.90		
Hydroxyproline	15.12	19.01	12.28
Isoleucine	9.79	14.41	6.61
Leucine	9.27	13.94	5.97
Lysine	21.84	23.75	19.79, 20.09[a]
Methionine	16.06	19.57	12.07, 13.33[a]
p-Methylphenylalanine	19.97		
Phenylalanine	18.23	21.30	15.24
Proline	10.39	13.17	5.66, 7.30[a]
Serine	11.58	16.06	8.19
Threonine	11.95	16.76	8.59
Tryptophan	31.20		32.11, 34.51[a]
Tyrosine	23.84	25.81	21.94
Valine	7.83	12.58	4.85

[a] Multiple peaks.

Column packing P1 = 3% OV-1 on Diatoport S, 80-100 mesh.
                       P2 = 3% OV-17 on Chromosorb® W HP, 80-100 mesh.
                       P3 = 10% OV-11 on Chromosorb® W, 100-120 mesh.
Temperature   T1 = Programed, 75°C for 3 min, then 6°C/min to 230°C, and held at 230°C for 3 min
                       T2 = Programed, 125°C for 3 min, then 5°C/min to 225°C.

REFERENCES

1. Pocklington, R., *Anal. Biochem.*, 45, 409, 1972.
2. Sarkar, S. K. and Malhotra, S. S., *J. Chromatogr.*, 199, 170, 1979.

Table GC 27
AMINO ACIDS, TRIMETHYLSILYL DERIVATIVES

Column packing	P1	P2
Temperature	T1	T1
Gas: Flow rate (ml/min)	He,80	He,80
Column		
Length (ft)	6	6
Diameter (inches O.D.)	0.25	0.25
Material	Pyrex glass	Pyrex glass
Detector	FID	FID

Table GC 27 (continued)
AMINO ACIDS, TRIMETHYLSILYL DERIVATIVES

Amino acid	I/100 (methylene units)	
Glycine	11.11	11.52
	13.18	13.21
Alanine	11.05	11.25
β-Alanine	11.90	12.30
	14.38	14.40
α-Aminobutyric acid	11.77	12.00
γ-Aminobutyric acid	15.46	15.57
α-Aminoisobutyric acid	11.48	11.60
β-Aminoisobutyric acid	12.16	12.48
	14.74	14.76
Valine	12.34	12.44
Leucine	12.84	12.91
Isoleucine	13.06	13.14
Sarcosine	11.43	11.73
Serine	13.80	14.10
Threonine	14.04	13.97
Proline	13.02	13.50
Hydroxyproline	15.46	14.63
Pipecolic acid	13.67	14.07
Phenylalanine	16.25	17.24
Tyrosine	19.52	20.19
Histidine	19.14	20.98
Tryptophan	22.15	23.77
Kynurenine	21.86	23.57
Cysteine	15.65	16.19
Cystine	23.12	23.95
Homocystine	25.63	26.38
Lanthionine	21.33	21.75
Cystathionine	22.35	22.84
Methionine	15.30	16.17
Cysteic acid	19.69	20.67
Djenkolic acid	24.58	25.60
Aspartic acid	15.41	15.88
Glutamic acid	16.29	16.89
Asparagine	16.87	18.00
Glutamine	17.73	18.87
Ornithine	18.53	18.21
Lysine	19.56	19.26
4-Hydroxylysine	18.90	18.55
	21.19	20.58
Citruline	18.36	17.69
Arginine	16.32	16.40

Column packing P1 = 3% OV-1 on Chromosorb® W HP, 80-100 mesh.
P2 = 3% OV-17 on Chromosorb® W HP, 80-100 mesh.
Temperature T1 = programed: 70°C to 325°C at 10°C/min.

REFERENCE

1. Butts, W. C., *Anal. Biochem.*, 46, 187, 1972.

Table GC 28
ALIPHATIC AMINES

Column packing	P1	P1
Temperature	T1	T2
Gas: Flow rate (ml/min)	N_2,50	N_2,50
Column		
Length (m)	2	2
Diameter (mm I.D.)	2	2
Material	Glass	Glass
Detector	FID or NP-FID[a]	FID or NP-FID

Amine	t_r (min)	
Methylamine	1.05 ± 0.005	6.68 ± 0.007
Dimethylamine	1.54 ± 0.005	7.58 ± 0.007
Ethylamine	1.83 ± 0.005	7.88 ± 0.007
Trimethylamine	2.26 ± 0.010	8.23 ± 0.005
Isopropylamine	3.49 ± 0.011	9.11 ± 0.005
n-Propylamine	4.83 ± 0.011	9.91 ± 0.005
Tert-butylamine	5.91 ± 0.017	10.49 ± 0.005
Diethylamine	7.03 ± 0.009	11.11 ± 0.005
Sec-butylamine	8.87 ± 0.014	12.04 ± 0.007
Isobutylamine	9.79 ± 0.018	12.53 ± 0.005
n-Butylamine	12.17 ± 0.027	13.65 ± 0.006

[a] Nitrogen-phosphorus flame ionization detector.

Column packing P1 = Ten grams of Chromosorb® 102 are added to 5% (v/v) trimethylchlorosilane (TMCS) in 50 ml of toluene. The mixture is placed under vacuum for 1 min, allowed to stand for 5 min filtered through a sintered glass funnel, washed with 100 ml of toluene followed by 50 ml of methanol, and dried at 80-90°C after a preliminary air-dry. The Chromosorb 102 TMCS is added to 25 ml of methanol containing 0.5 g of KOH. The mixture is dried under evaporator. The material is packed into the column and conditioned at 200°C for 3 days. During conditioning 10 µl of water is injected 30 to 40 times to ensure a stable column. Prior to use 10 µl of water is injected a few times at 170°C.

Temperature   T1 = 150°C.

T2 = Programed: 70°C for 4 min, then increased to 170°C at 30°C/min.

REFERENCE

1. Kuwata, K., Yamazaki, Y. and Uebori, M., *Anal. Chem.*, 52, 1980, 1980.

Reprinted with permission from *Anal. Chem.*, 52, 1980, 1980. Copyright (1980) American Chemical Society.

Table GC 29
LIGAND-EXCHANGE CHROMATOGRAPHY OF LOWER ALIPHATIC AMINES

Column packing	P1	P1	P2	P1	P1	P3	P3
Temperature (0°C)	60	60	60	60	60	60	60
Gas: Flow rate (ml/min)	N_2,20	N_2,20	N_2,20	N_2,20	N_2,30	N_2,20	N_2,30
NH_3 concentration in carrier gas, μmoles/ml	14.8	21.5	15.0	7.5	8.8	7.5	8.8
Column							
Length (m)	1	1	1	1	1	1	1
Diameter (mm I.D.)	4	4	4	4	4	4	4
Form	Spiral	Spiral	Spiral	Spiral	Spiral	Spiral	Spiral
Detector	FID	FID	FID	FID	FID	FID	FID

Amine				t_r			
Trimethylamine	0.5	0.4	0.4	0.7	0.4	0.4	0.2
Triethylamine	2.4	1.6	0.8	4.0	2.8	1.3	0.9
Dimethylamine	3.4	1.8	7.4	7.2	4.8	3.2	2.0
Diethylamine	5.7	2.6	4.0	11.6	8.4	5.8	3.8
Diisopropylamine	6.4	3.0	2.4				
Ethylamine	8.3	4.3	14.7				
Isopropylamine	11.2	5.0	8.4				
n-Propylamine	18.4	9.0	14.0				
sec-Butylamine	22.4	9.4	13.0				
Isobutylamine	30.1	12.9	14.5				
n-Butylamine	60.2	25.0	29.0				

Column packing P1 = ZP-1 (Cu^{2+})
               P2 = ZP-1 (NH_4^+)
               P3 = ZP-1 (Zn^{2+})

ZP-1 zirconium phosphate crystals, 60-80 mesh, are treated with 10% aqueous $CuCl_2$ or $ZnCl_2$ to convert them into the Cu^{2+} or Zn^{2+} form, and then with concentrated NH_4OH to form their ammonia complexes. The material is air dried, packed in the column, and conditioned by passing N_2 containing NH_3 and water vapor through the column.

REFERENCE

1. Fujimara, K. and Ando, T., *J. Chromatogr.*, 114, 15, 1975.

Table GC 30
DIAMINES

Column packing	P1
Temperature (°C)	170
Gas: Flow rate (ml/min)	He,20
Column	
Length (m)	3.0
Diameter (mm I.D.)	1.6
Material	SS
Detector	FID

Amine	I
Ethylenediamine	771.7
N-Methylethylenediamine	815.6
1,2-Diaminopropane	819.5

Table GC 30(continued)
DIAMINES

1,2-Diamino-2-methylpropane	843.7
sym-Dimethylethylenediamine	851.4
2,3-Diaminobutane	873.6
1,3-Diaminopropane	890.8
N,N-dimethyl-1,3-diaminopropane	919.0
N-Methyl-1,3-diaminopropane	928.0
1,3-Diaminobutane	929.6
1,4-Diaminobutane (putrescine)	999.3
1,5-Diaminopentane (cadaverine)	1100.8
Diethylenetriamine	1183.7
1,2-Diaminocyclohexane	1191.1

Column packing P1 = 28% Pennwalt 223 + 4% potassium hydroxide on Gas-Chrom® R, 80-100 mesh.

REFERENCE

1. Hansen, N. H., Kiens, K., and Nielsen, T., *J. Chromatogr. Sci.*, 9, 631, 1971.

Reproduced from the *Journal of Chromatographic Science*, by permission of Preston Publications, Inc.

Table GC 31
POLYAMINES

Column packing	P1	P2	P3
Temperature	T1	T1	T1
Gas: Flow rate (mℓ/min)	He,30	He,30	He,30
Column			
Length (inches)	12	12	18
Diameter (inches I.D.)	0.125	0.125	0.125
Material	SS	SS	SS
Detector	TC	TC[b]	TC

Polyamine	t_r		
Ethylenediamine	0.4	0.4	0.4
Piperazine	1.2	1.2	1.1
Triethylenediamine	2.2	2.0	1.8
Diethylenetriamine	2.2	2.0	1.9
1-(2-Aminoethyl)piperazine	3.6	3.2	2.6
Triethylenetetramine	a	7.2	5.3
Tetraethylenepentamine	a	a	a

[a] Failed to elute.
[b] TC = thermal conductivity.

Column packing P1 = Porapak® Q, 80-100 mesh.
                  P2 = 0.5% KOH on Porapak® Q, 80-100 mesh
                  P3 = Chromosorb® 103, 80-100 mesh.
Temperature T1 = Programed: 180°C to 245°C at 60°C/min.

REFERENCE

1. Casselman, A. A. and Bannard, R. A. B., *J. Chromatogr.*, 88, 33, 1974.

Reproduced by permission of the Elsevier Scientific Publishing Company.

Table GC 32
AMINES, LOWER ALIPHATIC AND SCHIFF BASES

Column packing	P1	P1	P2
Temperature	T1	T1	T2
Gas: Flow rate (mℓ/min)	N_2, 50	N_2, 50	N_2, 45
Column			
Length (m)	3	3	3
Diameter (mm I.D.)	3	3	3
Form	—	—	—
Material	Glass	Glass	Glass
Detector	TC[a]	TC	FID
Amine derivative	1	2	2
Literature	1	1	2
Amine	$r_{n\text{-}propylbenzene}$		$r_{ethylbenzene}$
Methylamine	0.15	1.14	2.32
Ethylamine	0.25	1.19	3.37
n-Propylamine	0.41	1.28	5.70
Isopropylamine	0.33	1.20	3.99
n-Butylamine	0.58	1.40	10.25
Isobutylamine	0.53	1.32	7.74
n-Amylamine	0.72	1.54	
Isoamylamine	0.68	1.47	
Dimethylamine	0.24		
Diethylamine	0.48		
Di-n-propylamine	0.74		
Trimethylamine	0.29		
Triethylamine	0.62		

[a] TC = thermal conductivity.

Column packing P1 = Tenax-GC, preconditioned at 280°C for 20 hr with a constant N_2 flow.
              P2 = 5% SE-30 on Shimalite W, 60-80 mesh, acid washed and silanized.
Temperature T1 = Programed: 100° for 1 min, 100° to 250° at 10°/min, 250° for 15 min.
              T2 = 100°.
Amine derivative 1. Free amine.
                    2. Schiff base derived from benzaldehyde.

REFERENCES

1. Hoshika, J., *Anal. Chem.*, 48, 1716, 1976.
2. Hoshika, J., *J. Chem.*, 115, 596, 1975.

Table GC 33
N-PERMETHYLATED C_{10}-C_{18} POLYAMINES

Column packing	P1	P1	P1	P2	P2	P2
Temperature (°C)	85	135	200	85	135	200
Gas: Flow rate (ml/min)	N_2,20	N_2,20	N_2,20	N_2,20	N_2,20	N_2,20
Column						
Length (m)	2	2	2	2	2	2
Diameter (mm)	2.5	2.5	2.5	2.5	2.5	2.5
Material	na	na	na	na	na	na
Detector	FID	FID	FID	FID	FID	FID

N-Permethylamine of:			I			
$H_2N(CH_2)_2NH_2$	974			821		
$H_2N(CH_2)_3NH_2$	1057			904		
$H_2N(CH_2)_4NH_2$	1152			1003		
$H_2N(CH_2)_5NH_2$	1261	1275		1108	1121	
$H_2N(CH_2)_6NH_2$		1375			1213	
$H_2N(CH_2)_7NH_2$		1474			1318	
$H_2N(CH_2)_8NH_2$		1573			1417	
N,N-Dimethyllaurylamine		1580			1512	
$H_2N(CH_2)_3NH(CH_2)_4NH_2$		1640			1447	
$H_2N(CH_2)_3NH(CH_2)_4NH(CH_2)_3NH_2$		2095	2143		1863	1886

Column packing P1 = 10% Carbowax® 20M on Chromosorb® W, 80-100 mesh.
P2 = 5% SE 52 on Chromosorb® W, 80-100 mesh.

REFERENCE

1. Giumanini, A. G., Chiavari, G., and Scarponi, F. L., *Anal. Chem.*, 48, 484, 1976.

Table GC 34
TETRADECYLAMINE, DERIVATIVES OF POSITIONAL ISOMERS

Column packing	P1
Temperature	T1
Gas: Flow rate (ml/min)	He, 6
Column	
Length (ft)	200
Diameter (inches)	0.02
Material	SS
Detector	FID

Amino position	Dimethylamine	Acetamide r^a	TFA derivative	TMS derivative
1	1.000	1.000	1.000	1.000
2	0.951	0.699	0.722	0.759
3	0.849	0.640	0.667	0.719
4	0.777	0.583	0.610	0.648
5	0.745	0.565	0.589	0.618
6	0.725	0.548	0.577	0.598
7	0.721	0.533	0.570	0.594

[a] Relative to terminal position = 1.00.

Table GC 34 (continued)
TETRADECYLAMINE, DERIVATIVES OF POSITIONAL ISOMERS

Column packing P1 = General Electric SF-96 silicone oil - 1000, modified with trioctadecylmethyl ammonium bromide. Available as GAS-QUAT-L from Lachat Chemical Company.

Temperature T1 = Programed: 100°C to 200°C at 10°C/min and then isothermal.

REFERENCE

1. Metcalfe, L. D. and Martin, R. J., *Anal. Chem.*, 44, 403, 1972.

Reprinted with permission from *Anal. Chem.*, 44, 403, 1972. Copyright (1972) American Chemical Society.

Table GC 35
AMINES, TRIFLUOROACETYL

Column packing	P1
Temperature	T1
Gas: Flow rate (mℓ/min)	He, 60
Column	
Length (cm)	140
Diameter (mm I.D.)	4
Material	G
Detector	FID

Amine	Elution temperature (°C)
Isobutylamine	110
Isopentylamine	118
1-Amino-2-propanol	124
Ethanolamine	130
3-Amino-1-propanol	144
3-Methylthiopropylamine	153
Histamine	160
Phenylethylamine	166
Tyramine	192
1,4-Butanediamine	205
1,5-Pentanediamine	209
Tryptamine	228
Indole	242
Skatole	244

Column packing P1 = 3% Poly A 103 on Gas Chrom® Q, 100-120 mesh. The column was conditioned at 280°C for 15 min with a He flow of 30 mℓ/min and 5 hr at 300°C with a flow of 45 mℓ/min.

Temperature T1 = Programed: 70°C for 3 min, then 3°C/min to 250°C. After the run the column temperature was lowered and held at 70°C for 10 min before the next injection.

REFERENCE

1. Cancalon, P. and Klingman, J. D., *J. Chromatogr. Sci.*, 10, 253, 1972.

Reproduced from the *Journal of Chromatographic Science* by permission of Preston Publications Inc.

Table GC 36
AMINES, N-TRIFLUOROACETYL DERIVATIVES

Column packing	P1	P2
Temperature	T1	T2
Gas: Flow rate (ml/min)	He, 3	He, 30
Column		
Length (m)	46	1.80
Diameter (cm)	I.D., 0.05	O.D., 0.32
Form	Capillary column	Packed column
Material	SS	SS
Detector	FID	FID

N-Trifluoroacetyl derivative of:	t_r	
Dimethylamine	13.4	—
Diethylamine	15.3	4.9
tert-Butylamine	16.1	6.8
tert-Amylamine	19.6	8.6
Isopropylamine	22.1	9.6
N-Methyl-n-butylamine	22.9	12.4
sec-Butylamine	25.2	13.6
Ethylamine	28.8	15.6
Isobutylamine	30.4	17.3
n-Propylamine	31.4	18.5
Methylamine	31.7	18.6
n-Butylamine	36.4	21.1
Isoamylamine	38.9	22.6
n-Amylamine	41.7	23.7
n-Hexylamine	47.2	26.4
Ammonia	50.9	28.8

Column packing P1 = FFAP (Carbowax® 20 M terminated with 2-nitroterephthalic acid).
P2 = 10% FFAP on Chromosorb® W, 80-100 mesh.
T1 = Programed; 80°C for 8 min then 80°C to 150°C at 2°/min.
T2 = Programed: 50°C for 2 min then 50°C to 200°C at 5°/min.

1. Jungclaus, G. A., *J. Chromatogr.*, 139, 174, 1977.

Reproduced by permission of Elsevier Scientific Publishing Company.

Table GC 37
N-TFA SECONDARY AMINES AND N-TFA-AMINO ACID ISOPROPYL ESTER ENANTIOMERS

Column packing	P1		P2		P3		P4		P5		P6	
Gas	H_2		H_2		H_2		H_2		H_2		H_2	
Column												
(Length (m)	40		40		40		44.5		26.5		39	
Diameter (mm I.D.)	0.2		0.2		0.2		0.25		0.25		0.25	
Material	G^a		G^a		G^a		G^b		G^b		G^b	
Literature	1		1		1		2		2		2	
Racemate	α^c	T(°C)	α	T(°C)	α	T(°C)	α	T(°C)	α	T(°C)	α	T(°C)
R,S-Alanine	1.053	70	1.016	100	1.020	90	1.037	90	1.038	80		
R,S-Valine	1.075	70	1.017	100	1.020	90	1.031	90	1.044	80		
R,S-2-Aminopentane					1.014	90	1.010	100				
R,S-2-Aminohexane							1.012	100				
R,S-2-Amino-5-methylhexane							1.013	100				
R,S-2-Aminoheptane					1.022	90	1.013	100			1.023	120
R,S-2-Amino-6-methylheptane					1.024	90	1.013	100				
R,S-2-Aminooctane	1.048	80			1.024	90	1.014	100				
R,S-α-Phenylethylamine	1.090	90			1.055	110	1.033	110	1.039	100	1.107	120

^{a} Borosilicate glass.
^{b} Pyrex glass.
^{c} The S-enantiomers are retarded and have longer retention times.

Column packing P1 = (S)-mandelic acid-(S)-α-phenylethylamide
P2 = O-benzyloxycarbonyl-(S)-mandelic acid-(S)-α-phenylethylamide
P3 = N-((S)-mandeloyl)-(S)-valine cyclohexyl ester
P4 = O-benzyloxycarbonyl-(S)-mandelic acid tert-butylamide
P5 = O-lauroyl-(S)-mandelic acid tert-butylamide
P6 = 2-(S)-phenylbutyric acid-(S)-α-phenylethylamide

Note: Before coating, the capillary was treated with a 0.2% suspension of Silanox (Cabot) in CH_2Cl_2. Coating was performed with 0.1—0.2% solutions of the stationary phase in CH_2Cl_2.

REFERENCES

1. König, W. A., Sievers, S., and Schulze, U., *Angewandte Chemie*, 92, 935, 1980.
2. König, W. A. and Sievers, S., *J. Chromatogr.*, 200, 189, 1980.

Table GC 38
N-TRIFLUOROACETYL-L-ALANYLAMIDES OF CHIRAL AMINES

Column packing	P1
Temperature	150°C
Gas, pressure	He, 0.5 bar
Column	
Length (m)	30
Diameter (mm I.D.)	0.25
Form	Capillary
Material	G[a]
Detector	FID

N-TFA-L-alanylamide of	t_r(min)
(−)-2-Aminopentane	6.25
(+)-2-Aminopentane	6.70
(±)-2-Amino-3-methylpentane	7.55
	7.95
	8.45
	8.80
(±)-2-Aminohexane	8.55
	9.30
(±)-2-Amino-5-methylhexane	10.20
	11.20
(±)-2-Aminoheptane	12.25
	13.50
(±)-2-Amino-6-methylheptane	15.00
	16.70
(±)-2-Aminooctane	18.25
	20.40
(+)-α-Phenylethylamine	42.25
(−)-α-Phenylethylamine	46.70
(±)-Amphetamine[b]	7.71
	8.50

[a] Borosilicate glass.
[b] 200°C.

Column packing   P1 = column treated with 0.2% suspension of Silanox (Cabot Corporation) in CH_2Cl_2 before coating with 0.2% OV-17 in CH_2Cl_2.

REFERENCE

1. Kruse, W., Francke, W., and König, W. A., *J. Chromatogr.*, 170, 423, 1979.

Reproduced by permission of Elsevier Scientific Publishing Company.

Table GC 39
ENANTIOMERIC AMINES AS DIASTEREOMERS

Column packing	P1
Temperature	170°C
Gas: Flow rate (ml/min)	He,60
Column	
Length (ft)	6
Diameter (inches I.D.)	0.125
Form	U
Material	Glass[a]
Detector	FID

Compound	α
N-TFA L-valyl dl-α-methylbenzylamine	1.212
N-TFA L-alanyl dl-α-methylbenzylamine	1.214
N-TFA L-leucyl dl-α-methylbenzylamine	1.200
N-TFA L-prolyl dl-α-methylbenzylamine	1.212
N-PFP L-prolyl dl-α-methylbenzylamine	1.192
N-HFB L-prolyl dl-α-methylbenzylamine	1.221
N-TFA L-prolyl dl-α-methylphenethylamine	1.075
N-PFP L-prolyl dl-α-methylphenethylamine	1.068
N-HFB L-prolyl dl-α-methylphenethylamine	1.107
N-HFB L-leucyl dl-α-methylphenethylamine	NS[b]
N-HFB L-valyl dl-α-methylphenethylamine	1.123
N-TEA L-prolyl dl-p-chloro-α-methylphenethylamine	1.176
N-PFP L-prolyl dl-p-chloro-α-methylphenethylamine	1.175
N-HFB L-prolyl dl-p-chloro-α-methylphenethylamine	1.188
N-TFA L-prolyl dl-α-ethylphenethylamine	1.075
N-PFP L-prolyl dl-α-ethylphenethylamine	1.175
N-HFB L-prolyl dl-α-ethylphenethylamine	1.117
N-TFA L-alanyl dl-α-ethylphenethylamine	1.051
N-TFA L-leucyl dl-α-ethylphenethylamine	NS[b]
N-TFA L-prolyl dl-1-methyl-3-phenylpropylamine	NS[b]
N-PFP L-prolyl dl-1-methyl-3-phenylpropylamine	NS[b]
N-HFB L-prolyl dl-1-methyl-3-phenylpropylamine	NS[b]
N-TFA L-alanyl dl-1-methyl-3-phenylpropylamine	1.073
N-TFA L-leucyl dl-1-methyl-3-phenylpropylamine	1.139

[a] Glass, silanized with 1% hexamethyldisilazane in toluene.
[b] NS = no separation observed.

Column packing   P1 = 5% SE-30 on Anakrom AB, 70-80 mesh.

REFERENCE

1. Souter, R. W., *J. Chromatogr.*, 108, 265, 1975.

Reproduced by permission of Elsevier Scientific Publishing Company.

Table GC 40
ENANTIOMERIC AMINES AS DIASTEREOMERS

Column packing	P1
Gas: Flow rate (mℓ/min)	He,60
Column	
Length (ft)	6
Diameter (inches I.D.)	0.125
Form	U
Material	Glass[a]
Detector	FID

Compound	α[b]
140°C	
N-TFA L-prolyl dl-1-methylhexylamine	1.190
N-TFA L-valyl dl-1-methylhexylamine	1.135
N-TFA L-leucyl dl-1-methylhexylamine	1.099
N-TFA L-alanyl dl-1-methylhexylamine	1.135
170°C	
N-TFA L-prolyl dl-1-methylhexylamine	1.188
N-TFA L-valyl dl-1-methylhexylamine	1.122
N-TFA L-leucyl dl-1-methylhexylamine	1.051
N-TFA L-alanyl dl-l-methylhexylamine	1.117
N-TFA L-prolyl dl-α-methylbenzylamine	1.420
N-PFP L-prolyl dl-α-methylbenzylamine	1.380
N-HFB L-prolyl dl-α-methylbenzylamine	1.428
N-TFA 1-valyl dl-α-methylbenzylamine	1.397
N-TFA L-leucyl dl-α-methylbenzylamine	1.389
N-TFA L-alanyl dl-α-methylbenzylamine	1.461
N-TFA L-prolyl dl-α-methylphenethylamine	1.340
N-TFA L-valyl dl-α-methylphenethylamine	1.183
N-TFA L-leucyl dl-α-methylphenethylamine	1.112
N-TFA L-alanyl dl-α-methylphenethylamine	1.187
200°C	
N-TFA L-prolyl dl-α-methylbenzylamine	1.326
N-PFP L-prolyl dl-α-methylbenzylamine	1.298
N-HFB L-prolyl dl-α-methylbenzylamine	1.304
N-TFA L-valyl dl-α-methylbenzylamine	1.327
N-TFA L-leucyl dl-α-methylbenzylamine	1.315
N-TFA L-alanyl dl-α-methylbenzylamine	1.322
N-TFA L-prolyl dl-α-methylphenethylamine	1.259
N-PFP L-prolyl dl-α-methylphenethylamine	1.214
N-HFB L-prolyl dl-α-methylphenethylamine	1.206
N-TFA L-valyl dl-α-methylphenethylamine	1.163
N-TFA L-leucyl dl-α-methylphenethylamine	1.095
N-TFA L-alanyl dl-α-methylphenethylamine	1.138
N-TFA L-prolyl dl-α-methyl-3,4-(methylenedioxy)-phenethylamine	1.331
F-TFA L-valyl dl-α-methyl-3,4-(methylenedioxy)-phenethylamine	1.154
N-TFA L-leucyl dl-α-methyl-3,4-(methylenedioxy)-phenethylamine	1.076
N-TFA L-alanyl dl-α-methyl-3,4-(methylenedioxy)-phenethylamine	1.163
N-TFA L-prolyl dl-p-chloro-α-methylphenethylamine	1.309
N-TFA L-valyl dl-p-chloro-α-methylphenethylamine	1.102
N-TFA L-leucyl dl-p-chloro-α-methylphenethylamine	1.036
N-TFA L-alanyl dl-p-chloro-α-methylphenethylamine	1.136
N-TFA L-prolyl dl-α-ethylphenethylamine	1.252
N-TFA L-leucyl dl-α-ethylphenethylamine	1.078
F-TFA L-alanyl dl-α-ethylphenethylamine	1.120
N-TFA L-prolyl dl-1-methyl-3-phenylpropylamine	1.096
N-TFA L-leucyl dl-1-methyl-3-phenylpropylamine	1.153
N-TFA L-alanyl dl-1-methyl-3-phenylpropylamine	1.177
N-TFA L-prolyl dl-0,α-dimethylphenethylamine	1.294

Table GC 40 (continued)
ENANTIOMERIC AMINES AS DIASTEREOMERS

N-TFA L-leucyl dl-0,α-dimethylphenethylamine	1.082
N-TFA L-alanyl dl-0,α-dimethylphenethylamine	1.152

^a Glass, silanized with 1% hexamethyldisilazane in toluene.
^b The (+) amine eluted first regardless of the L-amino acid reagent used in the formation of the diastereomers.

Column packing   P1 = 5% DEGS (diethyleneglycol succinate) on Anakrom AB, 70-80 mesh.

REFERENCE

1. Souter, R. W., *J. Chromatogr.*, 108, 265, 1975.

Reproduced by permission of Elsevier Scientific Publishing Company.

Table GC 41
ENANTIOMERIC AMINE DERIVATIVES

Column packing	P1	P2
Temperature	115.6°C	115.6°C
Gas	He	He
Column		
Length (cm)	91	91
Diameter (mm I.D.)	3.2	3.2
Material	Borosilicate glass	Borosilicate glass
Detector	FID	FID

Amine derivative	$\alpha_{S/R}$	
(R,S)-N-TFA-2-aminoheptane	1.04	1.02
(R,S)-N-TFA-2-aminooctane	1.05	1.00
(R,S)-N-TFA-2-aminoethyl cyclohexane	1.07	1.03
(R,S)-N-TFA-2-aminoethylbenzene	1.10	1.04

Column packing   P1 = Carbonyl-*bis*-((S)-valine isopropyl ester).
                 P2 = Carbonyl-1-((S)-valine isopropyl ester)-1'-(glycine isopropyl ester)

REFERENCE

1. Lochmüller, C. H. and Hinshaw, J. V., Jr., *J. Chromotogr.*, 178, 411, 1979.

Table GC 42
ENANTIOMERIC AMINE DERIVATIVES

Column packing	P1	P2	P3	P4
Temperature (°C)	100	100	100	100
Gas	He	He	He	He
Column				
Length (m)	40	40	40	40
Diameter (mm I.D.)	0.23	0.23	0.23	0.23
Material	G	G	G	G
Detector	FID	FID	FID	FID
Compound			$\alpha_{R,S}$	
N-TFA-2-aminoheptane	1.015	1.007	1.023	1.034
N-TFA-2-aminooctane	1.018			1.033
N-TFA-2-aminoethylcyclohexane	1.016	1.007	1.022	1.036
N-TFA-2-aminoethylbenzene	1.029	1.032	1.036	1.065
N-PFP-2-aminoethylbenzene	1.030	1.047	1.027	1.050
N-HFB-2-aminoethylbenzene	1.031	1.032	1.026	1.053

Column packing   P1, P2, P3, and P4 are chiral polysiloxanes formed when copolymers of dimethyl and methyl-γ-aminopropyl silane are modified by reaction to include amine bonded moieties, creating a "urea"-type phase and a "peptide"-type phase of molecular weight about 1500. P1 is a 2-4% urea type phase, P2 a 32% urea type phase, P3 a 2-4% peptide type phase and P4 a 32% peptide type phase. For column packings P1 and P2 the columns were treated with barium carbonate, pre-coated with a non-extractable layer of Carbowax® 20M, and then dynamically coated with the copolymers. For copolymer P3 a capillary was baked out at 300°C with helium gas flow, prepared with a nonextractable layer of Carbowax® 20M and then coated statically with the copolymer. The columns were conditioned with carrier gas flow (oxygen-free helium) at 160°C for 3 hr prior to use.

REFERENCE

1. Lochmüller, C. H. and Hinshaw, J. V. Jr., *J. Chromatogr.*, 202, 363, 1980.

Table GC 43
ENANTIOMERIC AMINE DERIVATIVES

Column packing	P1	P1	P2	P2	P3	P3	P4	P5	P5
Temperature	T1	T2	T3	T4	T5	T6	T7	T8	T9
Gas	He	He	He	He	He	He	He	He	He
Column									
Length (ft)	6	6	6	6	6	6	6	6	6
Diameter (inches)	0.125	0.125	0.125	0.125	0.125	0.125	0.125	0.125	0.125
Material	G	G	G	G	G	G	G	G	G
Detector	FID	FID	FID	FID	FID	FID	FID	FID	FID

Compound					α				
N-TFA(dl)2-aminooctane			1.036	N.R.[a]					
N-TFA(dl)2-aminoethylbenzene	1.062	1.067	1.173	1.092			1.102	1.119	1.073
N-PFP(dl)2-aminoethylbenzene	1.067	1.072	1.257	1.097			1.118	1.658	1.080
N-HFB(dl)2-aminoethylbenzene	1.063	1.067	1.336	1.098			1.099	1.399	1.079
N-TFA(dl)2-amino-3-phenylpropane			N.R.	N.R.	N.R.	N.R.	N.R.		N.R.
N-TFA(dl)2-amino-4-phenylbutane			1.108	N.R.	1.192	N.R.	N.R.		1.050

[a] N.R. = no resolution.

Column packing   P1 = 5.0% carbonyl-bis-(L-valine methyl ester) on Chromosorb® G AW DMCS, 100-120 mesh.
    P2 = 5.0% carbonyl-bis-(L-valine ethyl ester) on Chromosorb® G AW DMCS, 100-120 mesh.
    P3 = 5.0% carbonyl-bis-(L-valine isopropyl ester) on Chromosorb® G AW DMCS, 100-120 mesh.
    P4 = 5.0% carbonyl-bis-(L-valine tert-butyl ester) on Chromosorb® G AW DMCS, 100-120 mesh.
    P5 = 5.0% carbonyl-bis-(D-leucine isopropyl ester) on Chromosorb® G AW DMCS, 100-120 mesh.

Temperature   T1 = 390.19°K (smectic phase)
    T2 = 415.70°K (isotropic liquid phase)
    T3 = 385.02°K (smectic phase)
    T4 = 398.04°K (isotropic liquid phase)
    T5 = 377.28°K (smectic phase)
    T6 = 396.96°K (isotropic liquid phase)
    T7 = 415.61°K
    T8 = 361.67°K (smectic phase)
    T9 = 395.63°K (isotropic liquid phase)

REFERENCE

1. Lochmüller, C. H. and Souter, R. W., *J. Chromatogr.*, 88, 41, 1974.

Table GC 44
ENANTIOMERIC SECONDARY AMINE DERIVATIVES

Column packing	P1	P1
Temperature	T1	T2
Gas: Flow rate (mℓ/min)*	$N_2, 1$	$N_2, 1$
Column		
Length (m)	40	40
Diameter (mm)	0.25	0.25
Form	Open tubular	Open tubular
Material	Borosilicate glass	Borosilicate glass
Detector	FID	FID

Compound	K_1[a]	K_2	K_1	K_2
N-TFA-2-aminohexane	0.90	1.18	12.23	12.82
N-PFP-2-aminohexane	0.96	1.47	9.85	10.29
N-HFB-2-aminohexane	1.31	2.14	11.43	12.02
N-TFA-3-aminohexane	1.13	1.22	11.68	11.91
N-TFA-2-aminoheptane	1.74	2.35	20.48	21.47
N-TFA-2-amino-3,3-dimethylbutane	0.40	0.44	4.66	4.87
N-TFA-3-methylcyclohexamine	3.17	4.01	33.36	34.25
N-TFA-α-methylbenzylamine	6.12	11.72	73.44	82.98

[a] K_1 and K_2 are the respective capacity ratios of the enantiomers.

Column packing   P1 = 15% ureide of L-valine isopropyl ester.
Temperature   T1 = 100°C (solid).
     T2 = 120°C (liquid).

REFERENCE

1. Corbin, J. A. and Rogers, L. B., *Anal. Chem.*, 42, 974, 1970.

Reprinted with permission from *Anal. Chem.*, 42, 974, 1970. Copyright (1970) American Chemical Society.

Table GC 45
ENANTIOMERS OF AMINE AND AMINO ACID DERIVATIVES

Column packing	P1		P1		P2		P2	
Temperature (°C)	100		120		100		120	
Gas: Flow rate (mℓ/min)	$N_2, 2$		$N_2, 2$		$N_2, 2$		$N_2, 2$	
Column								
Length (m)	40		40		40		40	
Diameter (mm)	0.25		0.25		0.25		0.25	
Form	Open tubular		Open tubular		Open tubular		Open tubular	
Material	G[a]		G[a]		G[a]		G[a]	
Detector	FID		FID		FID		FID	

Compound	K_1[b]	K_2	K_1	K_2	K_1	K_2	K_1	K_2
N-TFA-DL-alanine isopropyl ester	0.44	0.55	7.93	8.26	10.34	11.53	4.70	5.11
N-PFP-DL-alanine-isopropyl ester	0.42	0.61	6.70	6.97	8.67	9.64	3.91	4.24
N-TFA-DL-2-aminooctane	1.97	2.72	23.72	25.01	25.95[c]		10.75[d]	

Table GC 45 (continued)
ENANTIOMERS OF AMINE AND AMINO ACID DERIVATIVES

[a] Borosilicate glass.
[b] K_1 and K_2 are the respective capacity ratios of the enantiomers. The emergence time of methane, which was assumed to be nonretained, was used in calculating K.
[c] Single, flat-topped peak suggestive of a partial separation.
[d] Single, broad peak.

Column packing. P1 = ureide of L-valine isopropyl ester.
                 P2 = N-TFA-L-valyl-L-valine cyclohexyl ester.

REFERENCE

1. Corbin, J. A. and Rogers, L. B., *Anal. Chem.*, 42, 1786, 1970.

Reprinted with permission from *Anal. Chem.*, 42, 1786, 1970. Copyright (1970) American Chemical Society.

Table GC 46
ENANTIOMERS OF CHIRAL N-TRIFLUOROACETYL AMINES

Column packing	P1
Gas: Flow rate (ml/min)	He, 3
Column	
Length (m)	50
Diameter (mm I.D.)	0.5
Form	Capillary
Material	SS
Detector	FID

Amine	Enantiomer	t_r(min)	Temp(°C)
2-Aminoheptane	L-	11.72	110
	D-	12.06	
2-Aminooctane	L-	23.82	110
	D-	24.32	
2-Aminononane	L-	42.86	110
	D-	44.00	
2-Aminodecane	L-	88.10	110
	D-	90.56	
3-Aminocyclohexane	I	7.96	120
	II	7.96	
	I	35.50	80
	II	36.20	
	I	70.10	65
	II	71.60	
α-Phenylethylamine	R	16.68	130
	S	17.26	
α-(1-Naphthyl)ethylamine	R	42.80	180
	S	44.20	

Column packing   P1 = N-docosanoyl-L-valine-2-(2-methyl)-n-heptadecylamide.

REFERENCE

1. Charles, R. and Gil-Av, E., *J. Chromatogr.*, 195, 317, 1980.

Reproduced by permission of Elsevier Scientific Publishing Company.

Table GC 47
BIOGENIC AMINES, TRIMETHYLSILYL DERIVATIVES

Column packing	P1	P1
Temperature	167°C	190°C
Gas; Flow rate (mℓ/min)	N_2,75	N_2,75
Column		
Length (ft)	6	6
Diameter (mm I.D.)	3	3
Form	U	U
Material	G[a]	G[a]
Detector	FID	FID

Amine	t_r (mins)	
Dopamine	3.07	2.58
Norepinephrine	4.58	4.01
Epinephrine	5.67	4.84
Tryptamine		2.95
Histamine		2.89
3-Methoxytyramine	2.27	2.07
Normetanephrine	3.60	3.30
Metanephrine	4.31	3.90

[a] Glass silanized with 5% dimethyldichlorosilane in toluene.

Column packing   P1 = 3% OV-1 on Gas Chrom® Q, 100-120 mesh. The packed column is conditioned at 295°C for 3 hr with no gas flow, then at 275°C for 24 hr with a N_2 flow rate of 20 mℓ/min.

REFERENCE

1. Cashaw, J. L., Walsh, M. J., Yamanaka, K., and Davis, V. E., *J. Chromatogr. Sci.*, 9, 98, 1971.

Table GC 48
BIOLOGICALLY SIGNIFICANT AROMATIC AMINES, TRIMETHYLSILYL DERIVATIVES

Column packing	P1	P2
Temperature	T1	T1
Gas; Flow rate (mℓ/min)	He,80	He,80
Column		
Length (ft)	6	6
Diameter (inches O.D.)	0.25	0.25
Material	Pyrex glass	Pyrex glass
Detector	FID	FID

Compound	I/100	
Phenylethylamine	15.68	16.31
Tyramine	19.22	19.59
Dopamine	21.04	21.12
3-Methoxytyramine	20.48	21.16
3,4-Dimethoxyphenylethylamine	19.78	21.38
Epinephrine	20.00	19.80
Metanephrine	19.38	19.82
Norepinephrine	19.49	19.52
Normetanephrine	18.97	19.56
Phenylalanine	16.25	17.24

Table GC 48 (continued)
BIOLOGICALLY SIGNIFICANT AROMATIC AMINES, TRIMETHYLSILYL DERIVATIVES

N-Acetylphenylalanine	17.90	19.33
Tyrosine	19.52	20.19
3,4-Dihydroxyphenylalanine (DOPA)	21.24	21.63

Column packing P1 = 3% OV-1 on Chromosorb® W, HP, 80-100 mesh
  P2 = 3% OV-17 on Chromosorb® W, HP, 80-100 mesh.
  T1 = Programed 70°C to 325°C at 10°C/min.

REFERENCE

1. Butts, W. C., *Anal. Biochem.*, 46, 195, 1972.

Table GC 49
SECONDARY AMINES AS SULFONAMIDES

	P1	P2
Column packing	P1	P2
Temperature	T1	T2
Gas: Flow rate (ml/min)	N_2, 40	N_2, 40
Column		
Length (m)	1	3
Diameter (mm I.D.)	3	3
Material	G	G
Detector	FPD[a]	FPD

Amine	t_r(min)	
Dimethylamine	2.83	2.42
Diethylamine	4.20	3.20
Diisopropylamine	5.41	3.91
Di-n-propylamine	6.18	4.38
Pyrrolidine	6.18	4.74
Morpholine	7.10	5.20
Diisobutylamine	7.68	5.20
Di-n-butylamine	9.23	6.41

[a] Flame photometric detector.

Column packing P1 = 3.5% SE-30 on acid-washed and DMCS-treated Chromosorb® W, 60-80 mesh.
  P2 = 3% OV-1 on acid-washed and DMCS-treated Chromosorb® W, 60-80 mesh.
  T1 = Programed: 140 to 200°C at 5°C/min.
  T2 = Programed: 200 to 240°C at 4°C/min.

REFERENCE

1. Hamano, T., Mitsuhashi, Y., and Matsuki, Y., *J. Chromatogr.*, 190, 462, 1980.

Reproduced by permission of the Elsevier Scientific Publishing Company.

Table GC 50
ENANTIOMERIC AMINE AND AMINO ACID METHYL ESTER DERIVATIVES

Column packing	P1	P1
Gas: Flow rate (mℓ/min)	N_2, 40	N_2, 40
Column		
Length (m)	5	5
Diameter (mm I.D.)	3	3
Material	Silanized glass	Silanized glass
Detector	FID	FID

	Drimanamide		Chrysanthemamide	
Compound	I	Temp(°C)	I	Temp(°C)
(R)-(+)-α-Methylbenzylamine	2535	187	1990	171
(S)-(−)-α-Methylbenzylamine	2515	187	1990	171
(S)-(+)-Amphetamine	2600	200	2025	160
(R)-(−)-Amphetamine	2620	200	2035	160
(R)-(+)-N,α-dimethylbenzylamine	2630	200	2035	171
(S)-(−)-N,α-dimethylbenzylamine	2630	200	2035	171
(S)-(+)-Methamphetamine			2095	186
(R)-(−)-Methamphetamine			2110	186
(R)-(−)-Valine methyl ester	2335	190	1800	143
(S)-(+)-Valine methyl ester	2340	190	1815	143
(R)-(+)-Proline methyl ester	2465	220	1925	143
(S)-(−)-Proline methyl ester	2470	220	1930	143
(R)-(−)-Phenylglycine methyl ester	2750	220	2170	189
(S)-(+)-Phenylglycine methyl ester	2750	220	2170	189
(R)-(+)-Phenylalanine methyl ester	2810	218	2240	189
(S)-(−)-Phenylalanine methyl ester	2820	218	2255	189

Column packing   P1 = 1% SE-30 on Gas-Chrom® Q, 100-120 mesh.

REFERENCE

1. Brooks, C. J. W., Gilbert, M. T., and Gilbert, J. D., *Anal. Chem.*, 45, 896, 1973.

Reprinted with permission from *Anal. Chem.*, 45, 896, 1973. Copyright (1973) American Chemical Society.

Amino Acids and Amines

Table GC 51
DERIVATIVES OF PHENYLETHYLAMINES AND RELATED AMINES

Column packing	P1	P2	P2	P3	P3	P4	P4	P5	P6	P2
Temperature (°C)	250	220	230	230	250	230	250	230	255	285
Gas; Flow rate (ml/min)	N₂,32	N₂,44	He,127	He,127	He,127	He,127	He,127	He,127	CH₄,12	CH₄,12
Column										
Length (ft)	6	3	3	6	6	6	6	6	5	5
Diameter (mm I.D.)	2	2	2	0.125ª	0.125ª	0.125ª	0.125ª	0.125ª	2	2
Form				U	U	U	U	U	U	U
Material	G^b	G^b	G^b	G	G	G	G	G	G^b	G^b
Detector	EC^c	EC	EC	EC	EC	EC	EC	EC	MS	MS
Literature	1	1	1	2	2	2	2	2	3	3

Amine					t_r(min)					
Benzylamine	0.89^c	1.38^c	0.98^c	4.2^d	2.2^d	3.7^d	1.8^d	2.1^d		
Amphetamine	0.98	1.18	0.89							
Phenylethylamine	1.18	1.97	1.38	5.6	2.9	5.0	2.4	2.9	3.10^d	5.54^d
Phenylpropanolamine	1.38	1.38	0.98							
Phenylethanolamine	1.58	1.97	1.38	5.4	2.6	4.3	2.0	2.9	4.05	5.06
1-Methylhistamine	1.77	4.92	3.35							
o-Tyramine	1.87	2.76	1.77	7.4	3.7	6.0	2.8	4.0		
p-Chlorophenylethylamine	2.17	4.13	2.76	12.4	6.0	11.2	4.9	7.0		
m-Tyramine	2.36	3.94	2.56	10.0	4.8	8.1	3.7	4.5	5.95	8.70
Histamine	2.56	9.15	5.91							
p-Tyramine	2.56	4.63	2.95	12.4	5.8	9.7	4.3	5.6	6.85	10.04
Metaraminol	2.56	2.46	1.77							
m-Octopamine	2.76	3.64	2.36							
p-Octopamine	3.25	4.33	2.76	10.7	4.9	7.6	3.3	5.6	8.50	8.62
3-Methoxytyramine	3.94	7.28	4.92							
Dopamine	4.53	6.99	4.53							
Normetanephrine	4.53	6.00	3.94	14.8	6.6	10.9	5.2	6.6		
Norepinephrine	5.12	5.91	3.74	18.8	8.6	15.8	6.5	7.6		
Tryptamine	5.71			27.8	14.6	24.2		20.4		
Serotonin	10.83									
p-Methoxyphenylethylamine				12.5	6.0	11.2	4.9	5.9		

Table GC 51 (continued)
DERIVATIVES OF PHENYLETHYLAMINES AND RELATED AMINES

4-Hydroxy-3-methoxybenzylamine	14.8	6.8	12.0	5.2	6.5
3,4-Dimethoxyphenylethylamine	22.8	10.1	20.8	8.4	9.4

^a Inches.
^b Silanized glass.
^c N-DNT, O-TMS amines.
^d N-DNP, O-TMS amines.
^e ⁶³Ni electron capture detector.

Column packing   P1 = 3% OV-1 on Supelcoport, 80-100 mesh.
    P2 = 3% SP-2250 on Supelcoport, 80-100 mesh.
    P3 = 1% OV-17 on Gas-Chrom® Q, 80-100 mesh.
    P4 = 1% OV-25 on Gas-Chrom® Q, 80-100 mesh.
    P5 = 1% OV-210 on Gas-Chrom® Q, 80-100 mesh.
    P6 = 3% OV-101.

REFERENCES

1. Doshi, P. S. and Edwards, D. J., *J. Chromatogr.*, 176 359, 1979.
2. Edwards, D. J. and Blau, K., *Anal. Biochem.*, 45, 387, 1972.
3. Edwards, D. J., Doshi, P. S., and Hanin, I., *Anal. Biochem.*, 96, 308, 1979.

Section I.II.
LIQUID CHROMATOGRAPHY TABLES*

Whenever possible, tables are arranged according to classes of chemical compounds. This was not always possible when different types of compound were chromatographed under the same experimental conditions. The reader is referred to the compound index for specific compounds which may appear in different tables.

---

* Tables LC 1-12, 18, 23-25, and 27-32 were prepared by A. J. Smith.

Table LC 1
AUTOMATIC ION-EXCHANGE CHROMATOGRAPHY OF AMINO ACIDS WITH SODIUM BUFFERS

Peak	Elution time (min)	Peak	Elution time (min)
Glutathione (performic oxidized)	17.9	Isoleucine	51.2
		N^2-Acetyllysine	51.8
Cysteic acid	18.3	Leucine	52.6
Phosphoserine	18.5	Prolylglycine	54.5
O^4-Sulfotyrosine	19.3	Norleucine	54.6
Glutathione (reduced)	24.4	Threonylmethionine	56.6
Methionine sulphoxides	25.2	Tyrosine	58.1
S-Carboxymethylcysteine	25.5	β-Alanine	58.2
4-Hydroxyproline	26.8	$O^{4'}$-Acetyltyrosine	58.4
Aspartic acid	27.3	Phenylalanine	61.5
Methionine sulphone	27.8	Aminoethanol	63.0
Threonine	29.7	Ammonia	66.0
Asparagine	30.2	S-Aminoethylcysteine	69.1
Glutamine	30.4	Ornithine	69.4
Serine	30.8	N^6-Trimethyllysine	69.6
Homoserine	33.2	N^6-Dimethyllysine	73.1
Glutamic acid	33.8	Lysine	74.1
Citrulline	35.0	Homoserine lactone	74.4
Proline	36.9	N^6-Methyllysine	74.7
Cysteine	37.9	Histidine	76.3
Glycine	42.0	N^{π}-Methylhistidine	76.7
Glycylaspartic acid	42.3	Buffer change	80.1
Alanine	43.2	Leucyltyrosine	81.7
2-Aminobutyric acid	44.4	Arginylglutamic acid	81.8
Cystine	45.1	Glycyltryptophan	82.1
Glucosamine	45.5	Diiodotyrosine	82.8
Valine	46.0	N^G-Dimethylarginine	83.4
Serylglycine	46.2	Tryptophan	84.9
Galactosamine	47.5	N^G-Methylarginine	86.4
Methionine	48.1	Arginine	86.9
Glycylglycine	48.3	Lysyllysine	87.5
allo-Isoleucine	49.8		

Conditions: Technicon TSM Amino Acid Analyzer
Column = 21.5 × 0.5 cm. Technicon C-3 Chromobeads
Buffers =
1. pH 2.98 sodium citrate, 0.25 N-Na⁺, 1.5 mM ascorbic acid
2. pH 3.60 sodium citrate, 0.68 N-Na⁺, 1.5 mM ascorbic acid
3. pH 4.00 sodium citrate, 1.02 N-Na⁺, 1.3 mM ascorbic acid
4. pH 9.50 sodium citrate, 0.71 N-Na⁺, 2.0 mM ascorbic acid

Buffers were applied for the following times, buffer 1, 13.5 min; buffer 2, 28 min; buffer 3, 15 min; buffer 4, 10.5 min.
Samples loaded in 0.067 M sodium citrate buffer, pH 2.0 Ammonia removed by filtration column 6.5 × 0.9 cm of Durrum DC-3 or Fisher Rexyn 101, 40-100 mesh resin.
Flow rates = Buffer 30 ml/hr. Ninhydrin 72 ml/hr.
Temperature = 60°C

REFERENCE

1. Niece, R. L., *J. Chromatogr.*, 103, 25, 1975.

Reproduced by permission of the Elsevier Scientific Publishing Company.

Table LC 2
AUTOMATIC ION-EXCHANGE CHROMATOGRAPHY OF AMINO ACIDS WITH LITHIUM BUFFERS ON A SINGLE COLUMN SYSTEM

Amino acid	Retention time (min-sec)	Amino acid	Retention time (min-sec)
Phosphoserine	6-17	L-Leucine	119-15
Taurine	9-45	L-Norleucine	123-50
Phosphoethanolamine	11-09	L-Tyrosine	131-07
Urea	14-23	L-Phenylalanine	138-13
L-Aspartic acid	27-18	β-Alanine	149-36
Hydroxy-L-proline	29-31	DL-β-Aminoisobutyric acid	161-02
L-Threonine	33-02	γ-Aminobutyric acid	173-47
L-Serine	34-41	L-Homocystine	180-24
L-Asparagine	37-34	Ethanolamine	196-59
L-Glutamic acid	39-29	L-Tryptophan	201-43
L-Glutamine	42-27	Ammonia	205-06
Sarcosine	47-13	D-allo-δ-Hydroxylysine	216-48
α-Aminoadipic acid	53-42	L-allol-δ-Hydroxylysine	219-04
L-Proline	59-08	Creatinine	226-56
Glycine	62-27	L-Ornithine	236-10
L-Alanine	67-39	L-Lysine	245-57
L-Citrulline	71-16	L-Histidine	253-00
L-α-Amino-n-butyric acid	77-29	L-1-Methylhistidine	263-00
L-Valine	87-08	L-3-Methylhistidine	268-00
Cysteine	106-49	L-Carnosine	279-45
L-Methionine	109-41	L-α-Aminoguanidino-propionic acid	289-59
D-Cystathionine	112-08	L-Arginine	321-53
L-Cystathionine	112-40		
L-Isoleucine	115-10		

Conditions: Durrum D-500 amino acid analyzer
Column = 48 × 0.175 cm. Durrum DC-4a resin
Buffers = 1. 0.244 M lithium citrate, pH 2.88
2. 0.356 M lithium citrate, pH 3.60
3. 0.668 M lithium citrate, pH 4.17
4. 0.644 M lithium citrate, pH 5.28
5. 1.193 M lithium citrate, pH 6.5
Flow rate = Buffer 6.3 mℓ/hr.
Temperature = 41,45,67°C.
The buffers are used in sequence. Buffer changes are made at 90, 160, 208, and 290 min from the time of sample injection. Temperature changes are made at 60 and 120 min.

REFERENCE

1. Lee, P. L. Y., Biochem. Med., 10, 107, 1974.

Reproduced by permission of Academic Press Inc.

Table LC 3
AUTOMATIC ION-EXCHANGE CHROMATOGRAPHY OF AMINO ACIDS WITH LITHIUM BUFFERS ON A TWO COLUMN SYSTEM

Compound	Elution time (min)	Compound	Elution time (min)
Cysteic acid	18	Methionine	315
Phosphoserine	21	allo-Isoleucine	317
Glycerophosphoethanolamine	26	Cystathionine	321
Taurine	30	Isoleucine	329
Phosphoethanolamine	33	Leucine	336
Urea	43	Norleucine	346
Aspartic acid	79	Tyrosine	358
Hydroxyproline	88	Phenylalanine	382
Threonine	101	β-Alanine	420
Serine	106	β-Aminoisobutyric acid	426
Asparagine	115	γ-Aminobutyric acid	447
Glutamic acid	122	Ethanolamine	473
Glutamine	130	Ammonia	496
Sarcosine	144	Ornithine	511
α-Aminoadipic acid	167	Lysine	520
Proline	177	Tryptophan	538
Glycine	183	Histidine	564
Alanine	200	1-Methylhistidine	567
Citrulline	213	Carnosine	576
α-Aminobutyric acid	232	Homocarnosine	581
Valine	265	3-Methylhistidine	597
Cystine	310	Arginine	685

Conditions: Beckman model 120B amino acid analyzer.
Column = 55 × 0.9 cm. Beckman AA-15 resin (both columns identical).
Glass ammonia filtration column with Pierce Hi-Rez Type DC-3 resin.
Buffers = Lithium citrate buffers pH 2.85, 3.80, and 4.30.
Flow rate = Buffer 60 mℓ/hr. Ninhydrin 30 mℓ/hr.
Temperature = 37,55°C.

The column temperature is maintained at 37°C for 150 min and then increased to 55°C. The optimal time for buffer changes from the first to the second and the second to the third are 245 min and 390 min, respectively. Total run time is 11½ hr.

REFERENCE

1. Melancon, S. B. and Tayco, J., *J. Chromatogr.*, 63, 404, 1971.

Reproduced by permission of the Elsevier Scientific Publishing Company.

Table LC 4
AUTOMATIC ION-EXCHANGE CHROMATOGRAPHY OF AMINO ACIDS IN BIOLOGICAL MATERIALS

Ninhydrin positive compound	Retention time (min)	Ninhydrin positive compound	Retention time (min)
Acidic and neutral amino acid analysis		Valine	219
Cysteic acid	17	Cystine	231
Homocysteic acid	17	Norvaline	231
Phosphoserine	18	Pipecolic acid	232
Taurine	25	Homocitrulline	234
Phosphoethanolamine	28	Cystathionine	234
Laevulinic acid	28*	Methionine	236
Urea	36	Lanthionine	237
Aspartic acid	65	Dihydroxyphenylalanine	239
S-Carboxymethylcysteine	65	Isoleucine	244
Glutathione (reduced)	66	Leucine	248
Hydroxyproline	69	Norleucine	255
Methionine sulfone	73	β-(2-Thienyl)-DL-alanine	259
Methionine sulfoxides	75,78	Tyrosine	264
Threonine	80	Phenylalanine	284
Serine	83	β-Alanine	295
Asparagine	90	Homocystine	302
Glutamic acid	93	β-Aminoisobutyric acid	315
Glutathione (oxidised)	97	Basic amino acid analysis	
Glutamine	98	5-Hydroxylysine	100
Homoserine	102	allo-Hydroxylysine	104
Sarcosine	111	δ-Aminolaevulinic acid	105
α-Aminoadipic acid	120	γ-Aminobutyric acid	120
β-(2-Thienyl)-DL-serine	125	Ornithine	126
Proline	138	Ethanolamine	131
Glycine	146	Ammonia	140
Hippuric acid	146	Kynurenine	155
Citrulline	153	1-Methylhistidine	157
Alanine	160	Lysine	175
Glucosamine	181	N-ε-Methyllysine	176
α-Aminobutyric acid	188	Histidine	188

Table LC 4 (continued)
AUTOMATIC ION-EXCHANGE CHROMATOGRAPHY OF AMINO ACIDS IN BIOLOGICAL MATERIALS

Ninhydrin positive compound	Retention time (min)	Ninhydrin positive compound	Retention time (min)
Mannosamine	190	3-Methylhistidine	199
Galactosamine	204	Creatinine	220
Creatine	221	Canavanine	239
Anserine	224	N^G, N^G-Dimethylarginine	247
Tryptophan	230	$N^G, N^{G'}$-Dimethylarginine	255
α-Amino-β-guanidino-propionic acid	232	S-β-(4-Pyridylethyl)cysteine	268
Monoiodotyrosine	233	N^G-Monomethylarginine	288
ε-Aminocaproic acid	234	Arginine	300
Carnosine	236	Diiodotyrosine	318
Homocarnosine	236	Homocysteine thiolactone	322

Conditions: Hitachi model KLA-5 amino acid analyzer.

Column = Acidic, neutral - 40 × 0.9 cm.
Basic - 25 × 0.9 cm.
Hitachi Custom Sulfonated Resin 2618 (10% cross-link, 11.5 μ).

Buffers = Acidic, neutral - 1. 0.25 N lithium citrate, pH 2.90
2. 0.25 N lithium citrate, pH 4.30
Basic - 1. 0.38 N sodium citrate, pH 4.10
2. 0.50 N sodium citrate, pH 6.09

Flow rates = Buffer 60 mℓ/hr., Ninhydrin 30 mℓ/hr.
Temperature = 43,57°C

The two lithium citrate buffers are used for acidic and neutral amino acids and the two sodium citrate buffers for basic amino acid analysis. For acidic and neutral amino acids, a change from buffer 1 to buffer 2 is made after 3 hr 14 min; the temperature of 43°C is increased to 57°C after 1 hr 14 min and reduced to 43°C after 2 hr 40 min. For basic amino acids a change from buffer 1 to buffer 2 is made after 3 hr 12 min and the temperature used is 43°C throughout.

REFERENCE

1. Murayama, K. and Shindo, N., *J. Chromatogr.*, 143, 137, 1977.

Reproduced by permission of the Elsevier Scientific Publishing Company.

Table LC 5
ION-EXCHANGE CHROMATOGRAPHY OF AMINO ACIDS, AMINO SUGARS AND AMINO ALCOHOLS RELATED TO PEPTIDOGLYCANS

Compound	Elution time (min)
Muramicitol	40
Muramic acid	62
Glutamic acid	65
Glycine	90
Alanine	95
2,6-Diaminopimelic acid	100
Isoglutamine	110
5-Hydroxy-4-aminopentanoic acid	115
Glucosamine	125
Glucosaminitol	135
Ammonia	167
Alaninol	170

Conditions: LKB® BC-200 amino acid analyzer.
Column = 54 × 0.9 cm. Aminex A-6 resin.
Buffers = 1. 0.2 M sodium citrate, pH 3.17
2. 0.35 M sodium citrate, pH 5.28
3. 1.2 M sodium citrate, pH 6.45
Samples loaded in 0.2 M sodium citrate, pH 2.2.
Buffer changes are sequenced as follows: buffer 1 to buffer 2 after 64 min and buffer 2 to buffer 3 after 100 min.
Flow rates = Buffer 60 ml/hr, Ninhydrin 30 ml/hr.
Temperature = 55°C.

REFERENCE

1. Hadzija, O. and Keglevic, D., *J. Chromatogr.*, 138, 458, 1977.

Table LC 6
ION-EXCHANGE CHROMATOGRAPHY OF SOME UNCOMMON DIAMINO ACIDS

Compound	Relative peak elution time pH 5.28	Relative peak elution time pH 4.55	Color value (C = H × W) per nmole at 570 nm.
Lysine	1.0	1.0	0.142
2,5-DAH (*threo* & *erythro*)	1.05		
Histidine	1.10		0.144
threo-3,5-DAH	1.27	1.81	0.0475[a]
erythro-3,5-DAH	1.30	1.90	0.0482[a]
β-Lysine	1.40		0.139
2,4-DAP	1.43		0.125
Ammonia	1.48		
Arginine	2.14		0.150

[a] At 440 nm.

Conditions: Beckman model 120C amino acid analyzer
Column = 54 × 0.9 cm. Beckman AA-15 resin
13 × 0.9 cm. Beckman PA-35 resin

Table LC 6 (continued)
ION-EXCHANGE CHROMATOGRAPHY OF SOME UNCOMMON DIAMINO ACIDS

	Relative peak elution time		
Compound	pH 5.28	pH 4.55	Color value (C = H × W) per nmole at 570 nm.

Buffers =
20 × 0.9 cm. Beckman PA-35 resin
1. 0.20 N sodium citrate, pH 4.24
2. 0.35 N sodium citrate, pH 4.55
3. 0.35 N sodium citrate, pH 5.28

The pH 4.24 and pH 5.28 buffers are made according to Benson, J. V. and Patterson, J. A., *Anal. Chem.*, 37, 1108, 1964, the pH 4.55 buffer is made by acidifying the pH 5.28 buffer with HCl. The actual peak elution times of lysine with the pH 5.28 buffer on the 13 cm and 54 cm columns are approximately 47 and 176 min, respectively. With the pH 4.55 buffer the peak elution time of lysine is approximately 218 min on the 54 cm column.

Flow rate = Buffer 68 mℓ/hr
Temperature = 55.8°C

Abbreviations:
DAH = diaminohexanoic acid.
DAP = diaminopentanoic acid.

REFERENCE

1. Herbst, M. M., Baltimore, B. G., Bozler, G., and Barker, H. A., *Anal. Biochem.*, 58, 322, 1974.

Reproduced by permission of Academic Press Inc.

Table LC 7
ION-EXCHANGE CHROMATOGRAPHY OF GLUTAMYL DERIVATIVES OF β-LYSINE AND LACTAMS OF 3,5-DIAMINOHEXANOIC ACID

Diamino acid isomer	Peak elution times		Relative color values
	Absolute (min)	Relative	
D-β-Lysine	88	1.00	1.00
L-β-Lysine	128	1.46	0.93
3,5-DAH[a]			
D-*erythro*	85	1.00	1.00
L-*erythro*	91	1.08	0.98
threo-B	102	1.20	1.00
threo-A	109	1.29	1.00

[a] Lactam preparation procedure - 1.0 mg of dry 3,5-DAH·2HCl was dissolved in 1.0 mℓ. of 2N HCl in absolute methanol by stirring vigorously for 5 min at room temperature (23°C) and then rapidly evaporating the solution to dryness under vacuum in a 40°C bath. The residue was dissolved in enough 0.45 M sodium borate buffer, pH 10.2 to give a 10 mM solution and a 0.1 mℓ aliquot was used for reaction with glutamyl-NCA; approximately 94% of the material is esterified by this method.

Conditions:
Beckman model 120C amino acid analyzer
Column = 54 × 0.9 cm Beckman AA-15 resin
Buffer = 0.35 N sodium citrate, pH 4.55
Flow rate = Buffer 68 mℓ/hr
Temperature = 55.8°C

Abbreviations:
DAH = diaminohexanoic acid
NCA = N-carboxy-anhydride

REFERENCE

1. Herbst, M. M., Baltimore, B. G., Bozler, G., and Barker, H. A., *Anal. Biochem.*, 58, 322, 1974.

Reproduced by permission of Academic Press Inc.

Table LC8
ION-EXCHANGE CHROMATOGRAPHY OF METHYLATED BASIC AMINO ACIDS AND RELATED COMPOUNDS

Compounds	pH =	Elution time (min.) 5.657	5.734	Observed constants C[a]	Operational color yields[b]
Tyrosine		151	148	114.0	0.89
Phenylalanine		165	162	115.4	0.91
δ-Hydroxy-DL-lysine		315	312	151.2	1.19
δ-allo-Hydroxy-DL-lysine		330	328	149.5	1.17
DL-Ornithine		405	401	151.3	1.19
Lysine		433	427	127.1	1.00
ε-N-Monomethyllysine		496	489	153.0	1.20
ε-N,ε-N-Dimethyllysine		536	525	148.4	1.16
ε-N,ε-N,ε-N-Trimethyllysine		567	552	111.6	0.88
Histidine		629	576	117.4	0.92
Ammonia		660	682	105.9	0.83
3-Methylhistidine		696	653	129.9	1.02
1-Methylhistidine		725	707	112.7	0.88
N^G,N^G-Dimethylarginine		1226	1178	24.2	0.19
N^G,N'^G-Dimethylarginine		1271	1235	25.7	0.20
Arginine		1534	1460	122.2	0.96

[a] Values for C are the observed constants per micromole of ninhydrin positive compound (C in H × W/C = μmole).
[b] Relative to norleucine at 1.00 at 570 nm.

Conditions: Beckman model 120B amino acid analyzer
Column = 60 × 0.9 cm. Durrum DC-6a resin (11 μm)
Buffer = 0.35 N sodium citrate, pH 5.734 or 5.657 containing 0.1 ml/l. octanoic acid
Flow rate = Buffer 30 ml/hr., ninhydrin 15 ml/hr
Temperature = 28°C

REFERENCES

1. Zarkadas, C. G., Can. J. Biochem., 53, 96, 1975.

Table LC 9
CHROMATOGRAPHY OF METHYLATED AMINO ACIDS ON DIFFERENT ION-EXCHANGERS

		Time of elution of peak maximum (min.)		
Amino acid	pH	Beckman UR (50°C)	Aminex A-5 (37°C)	Ostion KS LG0802 (37°C)
Lysine	5.79	240	—	61
	5.38	263	—	74
	5.32	291	—	76
	5.28	309	103	90
	5.16	335	113	92
	4.26	—	120	129
Monomethyllysine	5.79	265	—	73
	5.38	285	—	80
	5.32	300	—	89
	5.28	317	112	102
	5.16	341	122	106
	4.26	—	127	135—140
Dimethyllysine	5.79	282	—	77
	5.38	298	—	84
	5.32	307	—	81
	5.28	325	118	95
	5.16	345	126	98
	4.26	—	130	135—140
Trimethyllysine	5.79	300	—	80
	5.38	310	—	87
	5.32	314	—	85
	5.28	333	—	95
	5.16	345	—	98
	4.26	—	100	135—140
Histidine	5.79	323	—	86
	5.38	328	—	95
	5.32	328	—	109
	5.28	343	—	124
	5.16	352	—	125
	4.26	—	145	135—140
1-Methylhistidine	5.79	340	—	91
	5.38	342	—	104
	5.32	345	—	118
	5.28	349	125	131
	5.16	355	135	133
	4.26	—	136	135—140
3-Methylhistidine	5.79	360	—	96
	5.38	355	—	108
	5.32	355	—	125
	5.28	356	138	139
	5.16	360	152	141
	4.26	—	160	135—140
Arginine	5.79	—	—	120
	5.38	—	—	131
	5.32	—	—	145
	5.28	—	214	147
	5.16	—	255	152
	4.26	—	265	155-160
Asym.-dimethylarginine	5.79	382	—	125
	5.38	385	—	136
	5.32	—	—	149
	5.28	—	165	152

Table LC 9 (continued)
CHROMATOGRAPHY OF METHYLATED AMINO ACIDS ON DIFFERENT ION-EXCHANGERS

		Time of elution of peak maximum (min.)		
Amino acid	pH	Beckman UR (50°C)	Aminex A-5 (37°C)	Ostion KS LG0802 (37°C)
	5.16	—	202	154
	4.26	—	205	155-160
Sym.-dimethylarginine	5.79	388	—	130
	5.38	400	—	144
	5.32	—	—	156
	5.28	—	182	155
	5.16	—	214	154
	4.26	—	216	155—160

Conditions: Amino acid analyzer based on Technicon components.
Column = 30 × 0.9 cm.
Buffers = 0.35 N citrate buffers of pH 4.26, 5.16, 5.28, 5.38 and 5.79.
Flow rates = Buffer 60 ml/hr., ninhydrin 30 ml/hr.
Temperature = Constant in range 37 - 55°C.

REFERENCE

1. Helm, R., Vancikova, O., Macek, K., and Deyl, Z., *J. Chromatogr.*, 133, 390, 1977.

Reproduced by permission of Elsevier Scientific Publishing Co.

Table LC 10
CATION-EXCHANGE CHROMATOGRAPHY OF IODOAMINO ACIDS

Compound	Approx. elution volume (ml)	Peak area measured (cm²/nmole)	Relative peak area per mole	Relative peak area per iodine atom
Monoiodotyrosine	37	1.52	1.00	1.00
Diiodotyrosine	50	3.82	2.50	1.25
3-Monoiodothyronine	80	1.53	1.00	1.00
3'-Monoiodothyronine	96	0.50	0.33	0.33
3,5-Diiodothyronine	80	2.56	1.68	0.84
3,3'-Diiodothyronine	113	2.32	1.53	0.76
3',5'-Diiodothyronine	88	2.04	1.34	0.67
3,3',5-Triiodothyronine	106	2.94	1.93	0.64
3,3',5'-Triiodothyronine	96	3.27	2.15	0.72
Thyroxine	91	5.72	3.75	0.94
Monoiodohistidine	—	0.005	<0.01	<0.01
Diiodohistidine	27	0.07	0.04	0.02
Iodide	7[a]	2.96	1.94	1.94

[a] Iodide was not retarded in this chromatogram.

Conditions: Column = 15 × 1 cm. Cation-exchange resin AG50W-X4 (30-35 μ)
Column equilibrated with 0.04 M ammonium acetate buffer pH 4.7 containing 30% (v/v) ethanol at 50°C and a gradient of increasing pH was prepared from 100 ml of the starting buffer and 100 ml of 0.65 N ammonium hydroxide. Effluent continuously analyzed by the iodine-catalyzed ceric ammonium sulfate-arsenious acid reaction with a Technicon Auto-Analyzer.

REFERENCE

1. Sorimachi, K. and Ui, N., *Anal. Biochem.*, 67, 157, 1975.

Reproduced by permission of Academic Press Inc.

Table LC 11
ION-EXCHANGE CHROMATOGRAPHY OF O-SULFATE ESTERS OF HYDROXY AMINO ACIDS

Amino acid	Elution time (min.)	O-Sulfate ester	Elution time (min.)
Hydroxyproline	44.5	Hydroxyproline O-SO$_4$	17.0
Serine	56.5	Serine O-SO$_4$	18.0
Threonine	52.5	Threonine O-SO$_4$	18.0
Tyrosine	160.0	Tyrosine O-SO$_4$	21.5
Cysteic acid	18.5		

Conditions: Beckman amino acid analyzer.
Ion-exchanger = Dowex 1-X10, 200-400 mesh; alternatively UR-30 resin.
Buffers = 1. sodium citrate, pH 3.25.
2. sodium citrate, pH 4.30.
The buffer of pH 3.25 is changed to the buffer of pH 4.30 at 85 min.
Flow rates = Buffer 68 mℓ/hr., ninhydrin 34 mℓ/hr.
Temperature = 56°C.

REFERENCE

1. Dziewiatkowski, D. D., Riolo, R. L., and Hascall, V., *Anal. Biochem.*, 50, 442, 1972.

Reproduced by permission of Academic Press Inc.

Table LC 12
AUTOMATIC ION-EXCHANGE CHROMATOGRAPHY OF SULFUR CONTAINING AMINO ACIDS

Compound	Elution time (min)	Peak area[a] (per nmol.)
DL-Cysteic acid	5;49	39360
S-Carboxymethyl cysteine	26;26	41261
DL-Methionine sulfoxide	32;05	44072
DL-Methionine sulfone	31;05	34655
Glutathione-NEM[b]	27;45	29531
Cysteine-NEM[b]	47;15	30056
	60;15	23402
L-Cystine	104;52	30192
L-Cysteine-D-penicillamine mixed disulfide	106;26	38377
Methionine	107;16	31027
D-Penicillamine disulfide	109;34	58960
Cystathionine	110;34	33957
L-Homocysteine-L-cysteine Mixed disulfide	128;36	58194
L-Homocystine	173;41	32686

[a] Peak areas are expressed as a summation of digitized photometric output by the computer.
[b] Glutathione-NEM and cysteine-NEM refer to the *N*-ethylmaleimide condensation products with the thiol groups of these substances (viz. S-(1-ethyl-2,5-dioxopyrrolidin-3-yl-L-cysteine).

Conditions: Durrum D-500 amino acid analyzer.
Column = 48 × 0.175 cm. Durrum DC-4a resin
Buffers = 1. 0.244 N lithium citrate, pH 2.88
2. 0.356 N lithium citrate, pH 3.60
3. 0.668 N lithium citrate, pH 4.17
4. 0.644 N lithium citrate, pH 5.28
5. 1.193 N lithium citrate, pH 6.50
Flow rate = Buffer 6.3 mℓ/hr.
Temperature = 41, 45, 67°C.

REFERENCE

1. Bowie, L., Crawhall, J. C., Gochman, N., Johnson, K., and Schneider, J. A., *Clin. Chim. Acta*, 68, 349, 1976.

Table LC 13
LIGAND EXCHANGE CHROMATOGRAPHY OF AMINO ACID ENANTIOMERS ON POLYSTYRENE RESIN CONTAINING L-HYDROXYPROLINE

Packing	P1		P1		P1	
Column						
Length (cm)	140		140		140	
Diameter (mm I.D.)	7.8		7.8		7.8	
Solvent	S1		S2		S3	
Flow rate (mℓ/hr)	10		10		10	
Temperature	rt		rt		rt	
Detection	UV (206 nm)		UV (206 nm)		UV (206 nm)	
Amino acid			V_r			
	L	D	L	D	L	D
Glycine	6.44					
β-Alanine	0.28					
Alanine	5.82	6.04				
Aminobutyric acid	6.48	7.95				
Norvaline	11.2	19.9				
Norleucine	21.4	47.4				
Valine	7.27	11.8				
Isovaline	6.8	8.5				
Leucine	14.2	24.2				
Isoleucine	11.1	20.9				
Serine	3.47	4.48			42.5	
Threonine	3.47	5.27				
Allothreonine	2.65	3.85				
Homoserine	5.32	6.65				
Methionine	11.7	14.3				
Asparagine	4.60	5.37				
Glutamine	2.46	3.70				
Phenylglycine	6.15	13.6				
Phenylalanine	33.8	97.6				
α-Phenyl-α-alanine	11.9	12.5				
Tyrosine	8.95	19.8				
Phenylserine	22.6	41.4				
β-Phenyl-β-alanine	1.25	2.23				
Proline	14.6	57.8				
Hydroxyproline	9.18	29.1				
Allohydroxyproline	29.4	17.7				
Azetidine carboxylic acid	14.0	31.5				
Ornithine	34.4		2.0	2.0		
Lysine			2.5	3.04		
Histidine			14.6	5.22		
Tryptophan			20.7	36.5		
Aspartic acid					11.5	11.5
Glutamic acid					2.2	1.8
Iminodiacetic acid					32.6	

Packing   P1 = An asymmetric resin prepared by aminating a chloromethylated polystyrene containing 11 mol% of cross-links of diphenylmethane structure with L-hydroxyproline methyl ester. The resin particles were of irregular shape and had an average size in the swollen state of 100 μm. On treatment with excess copper-ammonia solution the resin was saturated with copper (II) ions to an extent of 92% of the theoretical capacity calculated for fixed complexes containing two fixed ligands per copper (II) ion.

Solvent   S1 = 0.1 M NH$_4$OH containing 1.2×10^{-5} M CuSO$_4$.
S2 = 1.5 M NH$_4$OH containing 2.0×10^{-4} M CuSO$_4$.
S3 = 0.025 M Na(NH$_4$)$_2$PO$_4$, pH 8.3 containing 2.5×10^{-5} M CuSO$_4$.

REFERENCE

1. Davankov, V. A. and Zolotarev, Yu. A., *J. Chromatogr.*, 155, 285, 1978.

Table LC 14
LIGAND EXCHANGE CHROMATOGRAPHY OF AMINO ACID ENANTIOMERS ON POLYSTYRENE RESIN CONTAINING L-PROLINE

Packing	P1	P1	P1	P1
Column				
Length (cm)	14	14	14	14
Diameter (mm I.D.)	7.8	7.8	7.8	7.8
Solvent	S1	S2	S3	S4
Flow rate (mℓ/hr)	10	10	10	10
Temperature	rt	rt	rt	rt
Detection	UV	UV	UV	UV

Amino acid	V_r							
	L	D	L	D	L	D	L	D
Glycine	5.0						36	
Alanine	6.75	7.25						
Aminobutyric acid	7.20	8.50						
Norvaline	14.2	18.2	4.10	5.75				
Norleucine	25.5	39.4	12.0	18.5				
Valine	9.0	11.6						
Serine	4.0	4.35						
Threonine	4.0	5.5						
Allothreonine	3.25	5.0						
Asparagine	4.25	5.0						
Glutamine	3.75	4.5						
Proline	17.0	70	6.25	25.0				
Hydroxyproline	9.9	38.2	3.50	13.5				
Allohydroxyproline	43.5	18.8						
Phenylglycine	11.3	18.8						
Leucine			13.0	16.5				
Isoleucine			7.0	10.5				
Phenylalanine			31.5	51.5	6.0	9.25		
Tyrosine			2.65	6.5				
Methionine			6.25	6.5				
Lysine					2.5	2.75		
Ornithine					2.5	2.5		
Histidine					15.5	5.75		
Tryptophan					5.5	7.8		
Aspartic acid							4.25	3.75
Glutamic acid							2.0	1.25
Iminodiacetic acid							7.5	

Packing   P1 = An asymmetric resin prepared by aminating a chloromethylated polystyrene matrix containing 11 mol% of cross-links of diphenylmethane structure with proline methyl ester. The resin particles were of irregular shape, their size in the swollen state being about 100 μm. The resin was charged with copper (II) ions from a copper-ammonia solution until it contained 80% of the theoretical amount of copper corresponding to the formation of fixed complexes containing two fixed ligands per copper ion.

Solvent   S1 = 0.1 M NH$_4$OH containing 1.2×10^{-5} M CuSO$_4$.
S2 = 0.3 M NH$_4$OH containing 3.8×10^{-5} M CuSO$_4$.
S3 = 1.5 M NH$_4$OH containing 2.0×10^{-4} M CuSO$_4$.
S4 = 0.017 M (NH$_4$)$_3$PO$_4$, pH 8.8, containing 2.5×10^{-5} M CuSO$_4$.

REFERENCE

1. Davankov, V. A. and Zolotarev, Yu. A., *J. Chromatogr.*, 155, 295, 1978.

Table LC 15
LIGAND EXCHANGE CHROMATOGRAPHY OF AMINO ACID ENANTIOMERS ON POLYSTYRENE RESIN CONTAINING L-AZETIDINE CARBOXYLIC ACID

	P1	P1	P1	P1
Packing Column				
Length (cm)	14	14	14	14
Diameter (mm I.D.)	7.8	7.8	7.8	7.8
Solvent	S1	S2	S3	S4
Flow rate (mℓ/hr)	10	10	10	10
Temperature	rt	rt	rt	rt
Detection	UV	UV	UV	UV

Amino acid	V_r							
	L	D	L	D	L	D	L	D
Glycine		7.2						
Alanine	11.2	11.9						
Aminobutyric acid	15.0	19.2	1.85	2.4				
Valine	24.0	41.0	3.1	5.4				
Norvaline	52	64	9.12	11.4				
Tyrosine	9.6	19.0						
Methionine			7.2	9.3				
Proline			7.5	18.6				
Hydroxyproline			3.6	8.1				
Allohydroxyproline			8.3	5.7				
Leucine			18.2	22.5				
Isoleucine			15.1	25.5				
Norleucine			25.2	35.4	3.10	4.28		
Phenylglycine			4.8	6.6				
Ornithine					2.1	2.1		
Lysine					1.8	1.91		
Histidine					27.6	15.3		
Tryptophan					33.2	37.4		
Phenylalanine					7.25	13.5		
Serine							6.0	12.9
Threonine							13.7	10.7
Asparagine							13.8	9.6
Glutamine							17.4	21.8
Aspartic acid							3.05	2.7
Glutamic acid							9.0	7.2
Iminodiacetic acid								2.4

Packing   P1 = An asymmetric resin prepared by aminating a chloromethylated polystyrene matrix containing 11 mol% of cross-links of diphenylmethane structure by azetidine carboxylic acid methyl ester. The resin particles were of irregular shape and their size in the swollen form was about 100 μm. The resins were charged with copper (II) ions from a copper-ammonia solution, until they contained 80% of the theoretical amount of copper corresponding to the formation of fixed complexes containing two fixed ligands per copper ion.

Solvent   S1 = 0.1 M NH$_4$OH containing 1.2 × 10^{-5} M CuSO$_4$.
S2 = 0.3 M NH$_4$OH containing 3.8 × 10^{-5} M CuSO$_4$.
S3 = 1.5 M NH$_4$OH containing 2.0 × 10^{-4} M CuSO$_4$.
S4 = 0.017 M (NH$_4$)$_3$PO$_4$, pH 8.8, containing 2.5 × 10^{-5} M CuSO$_4$

REFERENCE

1. Davankov, V. A. and Zolotarev, Yu. A., *J. Chromatogr.*, 155, 295, 1978.

Table LC 16
LIGAND EXCHANGE CHROMATOGRAPHY OF AMINO ACID ENANTIOMERS ON POLYSTYRENE RESIN CONTAINING L-ALLOHYDROXYPROLINE

Packing	P1	P1	P1	P1
Column				
Length (cm)	14	14	14	14
Diameter (mm I.D.)	7.8	7.8	7.8	7.8
Solvent	S1	S2	S3	S4
Flow rate (ml/hr)	10	10	10	10
Temperature	rt	rt	rt	rt
Detection	UV	UV	UV	UV

Amino acid	\multicolumn{8}{c}{V_r}
	L	D	L	D	L	D	L	D
Glycine		5.55						
Alanine	8.9	9.2						
Aminobutyric acid	9.6	11.0	4.1	6.0				
Norvaline	13.4	19.0						
Norleucine	22.9	33.4						
Valine	8.6	13.6						
Leucine	21.6	33.7	8.8	13.8	1.3	2.0		
Isoleucine	16.2	28.2	7.03	12.2				
Serine	4.38	5.25						
Threonine	4.82	7.15						
Asparagine	4.38	5.25						
Glutamine	3.94	5.52						
Phenylglycine	9.35	16.7						
Allohydroxyproline	18.7	27.8						
Hydroxyproline	21.3	34.1						
Proline	52.5	96.0	20.5	38.1	3.15	5.75		
Phenylalanine			15.2	47.2				
Tyrosine			3.21	7.58				
Methionine			8.62	13.1				
Lysine					2.25	3.0		
Ornithine					1.0	1.2		
Histidine					6.8	9.1		
Tryptophan					63.0	68.8		
Aspartic acid							9.8	6.8
Glutamic acid							16.0	11.0
Iminodiacetic acid							8.0	

Packing P1 = The asymmetric resin was prepared by aminating chloromethylated polystyrene containing 11 mol% cross-links of diphenylmethane structure using L-allohydroxyproline methyl ester. The resin particles have an irregular shape with an average size of about 100 μm when swollen. The resin was charged with copper ions from a copper-ammonia solution, resulting in quantitative formation of complexes containing two fixed ligands per copper ion.

Solvent S1 = 0.1 M NH$_4$OH containing $1.2 \times 10^{-5} M$ CuSO$_4$.
S2 = 0.3 M NH$_4$OH containing $3.8 \times 10^{-5} M$ CuSO$_4$.
S3 = 1.5 M NH$_4$OH containing $2.0 \times 10^{-4} M$ CuSO$_4$.
S4 = 0.017 M (NH$_4$)$_3$PO$_4$ pH 8.8, containing $2.0 \times 10^{-5} M$ CuSO$_4$.

REFERENCE

1. Davankov, V. A. and Zolotarev, Yu. A., *J. Chromatogr.*, 155, 303, 1978.

Table LC 17
LIGAND EXCHANGE CHROMATOGRAPHY OF AMINO ACID ENANTIOMERS ON L-PHENYLALANINE MODIFIED POLYACRYLAMIDE LOADED WITH COPPER (II)

Column packing

	P1
Column	
Length (mm)	300
Diameter (mm I.D.)	9
Material	G
Solvent	S1
Flow rate (mℓ/hr)	30
Detection	UV

Amino acid	K_L	K_D
Alanine	1.68	2.31
Asparagine	3.04	4.12
Aspartic acid	1.02	1.34
Glutamic acid	1.13	1.50
Glutamine	1.47	2.20
allo-Hydroxyproline	5.25	6.59
Isoleucine	1.74	2.79
Leucine	2.29	3.26
Lysine	6.85	9.34
Methionine	3.15	6.98
Norleucine	2.33	3.68
Ornithine	2.84	3.78
Phenylalanine	3.61	4.85
Phenylglycine	1.51	2.54
Serine	2.04	2.67
Threonine	2.52	3.36
Tryptophan	8.95	12.70
Tyrosine	5.17	7.58
Valine	1.53	2.35

Column packing   P1 = Bio-Gel® P-4 polyacrylamide beads (Serva, Heidelberg, GFR) treated with formaldehyde and L-phenylalanine. The packing contained 1.4 mmol of L-phenylalanine groups per gram and the water uptake was 300%. Before introduction into the column, sample was treated with excess copper (II)-ammonia solution and subsequently with a solution of KCl in 1.0 N NH$_4$OH to achieve the desired content of Cu(II)ions (60%).

Solvent   S1 = 2% ammonium phosphate, pH 9.2.

REFERENCE

1. Zolotarev, Yu. A., Myasoedov, N. F., Penkina, V. I., Dostovalov, I. N., Petrenik, O. V., and Davankov, V. A., *J. Chromatogr.*, 206, 231, 1981.

Table LC 18
REVERSE-PHASE HIGH PERFORMANCE LIQUID CHROMATOGRAPHY OF FREE AMINO ACIDS

Hydrophobic[a] amino acids	Elution time (min.)	Hydrophilic[b] amino acids	Elution time (min.)
Lysine	2.55	Glutamic acid	1.9
All other hydrophilics	2.7-2.9	Aspartic acid	2.0
Histidine	3.3	Lysine	2.2
Proline	3.4	Serine, threonine	2.7
Valine	4.1	Alanine, glycine	2.8
Methionine	5.1	Histidine	3.15
Leucine, isoleucine	6.5		
Tyrosine	11.1		
Phenylalanine	18.0		
Tryptophan	54.0		

[a] Column, μ-Bondapak-C_{18}; eluant, H_2O containing 0.1% H_3PO_4.
[b] Column, μ-Bondapak-alkylphenyl; eluant, H_2O.

Conditions: Column = 30 × 0.4 cm. (10 μm resin).
Flow rate = 1.5 ml/min., maintained by a pressure of 2000 psi.
Temperature = room temperature (22°C).
Waters Associates HPLC system with Cecil UV-monitor.

REFERENCE

1. Hancock, W. S., Bishop, C. A., and Hearn, M. T., *Anal. Biochem.*, 92, 170, 1979

Reproduced by permission of Academic Press Inc.

Table LC 19
AROMATIC AMINO ACIDS ON β-CYCLODEXTRIN POLYURETHANE RESINS

Column packing	P1	P1	P2	P2
Column				
Length (cm)	35	35	35	35
Diameter (mm I.D.)	5.2	5.2	5.2	5.2
Material	G	G	G	G
Solvent	S1	S2	S1	S2
Flow rate (ml/hr)	20	20	20	20
Detection	D1	D1	D1	D1

Amino acid	$r_{phenylglycine}$			
β-(3,4-Dihydroxyphenyl)alanine	1.15		1.12	
Kynurenine	7.84	7.00	10.52	5.81
Phenylalanine	25.82[a]	11.03[a]	20.24[a]	8.94[a]
Phenylglycine	1.00	1.00	1.00	1.00
Tryptophan	6.28	5.43	7.77	5.08
Tyrosine	2.23	1.91	2.96	1.87

[a] Initial retention time is shown because the peak is too broad.

Column packing P1 = β-H6XDI-P-6.0-M
P2 = β-XDI-P-5.8-A
The resins are prepared by reaction of cyclodextrin with diisocyanates in solvent, followed by addition of precipitant. β = cyclodextrin. H6XDI = 1,3-bis(isocyanatomethyl) cyclohexane. XDI = 1,3-bis(isocyanatomethyl) benzene. The resin particles have a size of 149-177 μ.

Solvent S1 = phosphate buffer, pH 5.5
S2 = phosphate buffer, pH 8.2
Detection D1 = UV at 254 and/or 280 nm

REFERENCE

1. Mizobuchi, Y., Tanaka, M. and Shono, T., *J. Chromatogr.*, 208, 35, 1981.

Table LC 20
SEPARATION OF AMINO ACID ENANTIOMERS BY HPLC USING CHIRAL ELUANTS

Packing	P1
Column	
Length (cm)	12
Diameter (cm I.D.)	0.2
Solvent	S1
Flow rate (mℓ/hr)	10
Temperature (°C)	75
Detection	D1

	t'_R(min)[a]	
Amino acid	L	D
(Cysteic acid)	(0)	(0)
Alanine	17.2	17.2
Alloisoleucine	23.5	25.5
Allothreonine	10.0	11.4
α-Amino-n-butyric acid	16.2	16.6
Asparagine	14.0	14.0
Aspartic acid	2.2	2.2
3,4-Dihydroxyphenylalanine	22.3	28.5
Ethionine	29.0	30.3
p-Fluorophenylalanine	64.0	75.6
Glutamic acid	3.3	3.3
Glutamine	12.9	12.9
Isoleucine	21.5	23.7
Leucine	28.3	28.6
Methionine	23.8	24.6
Norleucine	30.8	32.3
Norvaline	21.5	22.3
Phenylalanine	48.7	55.0
Serine	12.9	13.4
Threonine	13.0	13.7
Tyrosine	30.8	39.4
m-Tyrosine	34.1	41.1
o-Tyrosine	36.7	43.2
Valine	16.3	17.7

[a] t'_R = adjusted retention time.

Packing P1 = Durrum DC-4a resin, 5 μ.
Solvent S1 = 0.1 N acetate with 8×10^{-3} M CuSO$_4$ and 16×10^{-3} M L-proline.
Detection D1 = Fluorometry after reaction with o-phthalaldehyde.

REFERENCE

1. Hare, P. E. and Gil-Av, E., *Science*, 204, 1226, 1979.

Table LC 21
RESOLUTION OF AMINO ACID ENANTIOMERS ON A REVERSED PHASE COLUMN USING MIXED CHELATE COMPLEXATION

Packing			P1	P1	P1
Column					
Length (cm)			10	10	10
Diameter (mm I.D.)			4	4	4
Material			SS	SS	SS
Solvent			S1	S2	S3
Flow rate (mℓ/min)			1.0	1.0	1.0
Temperature (°C)			30	30	30
Detection			D1	D1	D1

Amino Acid			K		
Alanine	L		2.40	0.78	0.12
	D		3.24	1.10	0.12
Asparagine	D		2.70	0.60	0.18
	L		2.93	0.60	0.18
Aspartic acid	D		1.44	0.10	0.05
	L		2.00	0.29	0.05
Glutamic acid	D		2.20	0.28	0.10
	L		2.63	0.58	0.10
Glutamine	D		3.52	0.68	0.20
	L		3.52	0.68	0.20
Isoleucine	L			9.40	2.60
	D			19.74	4.40
Leucine	L			16.00	3.82
	D			26.60	5.64
Norleucine	L			16.80	4.10
	D			31.50	6.60
Norvaline	L		12.00	5.20	1.22
	D		25.00	9.24	1.96
Phenylalanine	L			37.40	9.50
	D			71.20	14.70
Phenylglycine	L		22.81	9.20	2.60
	D		52.00	21.80	5.00
Tryptophan	L				13.20
	D				20.80
Tyrosine	L		10.51	4.50	1.20
	D		21.86	8.00	1.90
Valine	L		8.25	3.70	1.00
	D		16.90	6.40	1.54

Packing   P1 = Chemically bonded *n*-octylsilyl silica gel Develosil C8, 5 μm particle size (Nomura Chemical), unreacted accessible silanol groups being capped with trimethylsilane.

Solvent   S1, S2, S3 consist of acetonitrile in an aqueous solution containing 1 mM N-(*p*-toluenesulfonyl)-L-phenylalanine (TosPhe) and 0.5 mM CuSO$_4 \cdot$ 5H$_2$O, pH 6.0.

       S1 = 0% acetonitrile
       S2 = 10% acetonitrile
       S3 = 15% acetonitrile

Detection   D1 = Reaction with *o*-phthalaldehyde reagent containing EDTA·2 Na (2.5 g/ℓ) and measurement of fluorescence intensity at 455 nm, excitation being at 340 nm.

REFERENCE

1. Nimura, N., Suzuki, T., Kasahara, Y., and Kinoshita, T., *Anal. Chem.*, 53, 1380, 1981.

Table LC 22
LIGAND EXCHANGE CHROMATOGRAPHY OF AMINO ACID ENANTIOMERS

Packing	P1
Column	
Length (cm)	25
Diameter (mm I.D.)	4.6
Material	SS
Solvent	S1
Flow rate (mℓ/min)	2
Temperature (°C)	50
Detection	D1

Amino acid	K_D[a]	K_L[a]
Alanine	1.3	1.3
Arginine	2.4	3.0
Asparagine	3.0	4.1
Aspartic acid	3.2	4.1
Citrulline [b]	2.1	1.6
Cystine	5.4	5.4
DOPA	3.4	11.2
Ethionine[b]	4.0	4.6
Glutamic acid	2.0	2.0
Glutamine	1.0	1.0
Histidine	6.7	12.1
5-Hydroxytryptophan	7.4	25.2
Isoleucine[b]	2.7	3.6
Leucine	3.1	3.1
Lysine[b]	1.8	2.2
Methionine[b]	3.0	3.4
Norleucine[b]	3.1	3.7
Norvaline[b]	2.0	2.2
Ornithine	0.6	0.75
Phenylalanine	3.2	9.4
Phenylserine	6.2	10.2
Proline	2.4	1.4
Serine	2.0	3.2
Threonine	2.0	3.2
Tryptophan	7.8	27.4
o-Tyrosine	2.8	9.0
m-Tyrosine	2.2	6.0
p-Tyrosine	3.3	10.2
Valine	2.5	3.8

[a] K_D and K_L are the capacity ratios of the D and L enantiomers, respectively.
[b] Temperature 80°C.

Packing P1 = Chemically bonded chiral phase prepared by bonding 3-glycidoxypropyltrimethoxysilane to silica, followed by reaction with L-proline. The phase was loaded with CU(II).
Solvent S1 = 0.05 M KH$_2$PO$_4$, pH 4.6.
Detection D1 = UV and polarimeter.

REFERENCE

1. **Gubitz, G., Jellenz, W., and Santi, W.**, *J. Chromatogr.*, 203, 377, 1981.

Table LC 23
CHROMATOGRAPHY OF DANSYL AMINO ACIDS ON POLYAMIDE COLUMNS

Amino acid	Benzene-acetic acid (9:1)		Benzene-acetic acid (6:4)	
	V_e [a]	V_e/V_{pro} [b]	V_e	V_e/V_{pro}
Leucine	138	1.53	80	1.02
Valine	124	1.38	82.5	1.05
Histidine	143	1.64	—	—
Methionine	152	1.69	81	1.04
Alanine	163	1.81	92	1.08
Lysine	196	2.18	108	1.38
Glycine	765	8.39	117	1.50
Tryptophan	500	5.55	134	1.71
Threonine	865	9.60	142	1.82
Serine	2540	28.21	485	6.21
Cysteine	—	—	980	12.57
Arginine	—	—	384	4.92
Hydroxyproline	700	7.76	—	—
Proline	90	1.00	78	1.00
Phenylalanine	155	1.72	85	1.09
Glutamine	370	4.12	126	1.61
Asparagine	1350	15.00	370	4.74

[a] V_e = elution volume.
[b] V_e/V_{pro} = elution volume relative to proline.

Conditions: Column = 100 × 1 cm filled with Woelm polyamide (15 g)
Temperature = 35°C
Fluorescence detection; 340 nm excitation, 500 nm emission. Gradient elution may be employed to provide an improved separation: - 0-300 min, gradient of benzene/benzene-acetic acid (9:1), 200 ml of each solvent; 300-1100 min, benzene-acetic acid (9:1); and 1100-2500 min, benzene-acetic acid (6:4).

REFERENCE

1. Deyl, Z. and Rosmus, J., J. Chromatogr., 69, 129, 1972.

Reproduced by permission of the Elsevier Scientific Publishing Company.

Table LC 24
REVERSE-PHASE HIGH PERFORMANCE LIQUID CHROMATOGRAPHY OF DANSYL AMINO ACIDS

	Capacity factor in solvent system (all aqueous, 1% acetic acid)			
Amino acid derivative	Acetonitrile %			30% Methanol
	50	40	30	
Dns[a]-Arginine	0.1	—	—	0.1
Dns-NH_2	0.4	0.4	0.4	0.4
Dns-Isoleucine	1.1	1.4	1.8	—
Dns-Tryptophan	1.2	1.3	—	—
Dns-Leucine	1.4	—	—	3.8
Dns-Tyrosine	1.7, 2.0	1.7, 2.0	2.0, 2.6	2.6
Dns-Glycine	3.1	3.4	3.6	6.4
Dns-Threonine	3.2	3.2	—	—
Dns-Hydroxyproline	3.3	3.4	—	—
Dns-Proline	3.4	3.4	—	4.9
Dns-Phenylalanine	3.5	—	—	—
Dns-Methionine	4.0	—	—	—
Dns-Alanine	4.8	—	—	—
Dns-Serine	4.9	—	—	6.6
Dns-Glutamic acid	19.4	—	—	16.5
Dns-Aspartic acid	23.2	—	—	—
Dns-OH	28.6	—	—	11.2

[a] 5-dimethylamino-1-naphthalene sulfonyl derivative.

Conditions: Waters Associates model ALC201 HPLC system with UV-detector.
Column = 25 × 0.46 cm. Whatman Partisil® PAC (10 μm).
Flow rate = 10 mℓ/min.

REFERENCE

1. Hsu, K-T. and Currie, B. L., *J. Chromatogr.*, 166, 555, 1978.

Reproduced by permission of the Elsevier Scientific Publishing Company.

Table LC 25
REVERSE-PHASE HIGH PERFORMANCE LIQUID CHROMATOGRAPHY OF OPTICAL ISOMERS OF DANSYL AMINO ACIDS USING METAL CHELATE ELUANTS

Dansyl amino acid	R^a = ethyl		R = isopropyl		R = isobutyl	
	k'^b	α^c	k'	α	k'	α
α-Alanine	3.4	1.00	3.65	0.79	6.5	0.86
	3.4		4.6		7.6	
α-Aminobutyric acid	2.5	1.00	2.85	0.88	4.8	0.89
	2.5		3.25		5.4	
Norleucine	8.25	0.96	10.0	0.86	18.0	0.92
	8.60		11.6		19.5	
Leucine	6.15	0.94	7.3	0.84	14.2	1.08
	6.50		8.7		13.1	
Threonine	2.55	1.19	2.7	0.71	5.1	0.82
	2.15		3.8		6.2	
Serine	4.0	1.25	4.1	0.67	7.8	0.74
	3.2		6.1		10.5	
Aspartic acid	1.9	1.00	2.9	1.16	5.4	1.00
	1.9		2.5		5.4	

[a] R = alkyl substituent on metal chelate in mobile phase of the bonded reversed phase column.
[b] k' = capacity factor.
[c] $\alpha = k'_L/k'_D$.

Conditions: Waters Associates HPLC system with UV-detector
Column = 15 × 0.46 cm. Shandon Southern Hypersil silica-C_8 (5 μm)
Flow rate = 2 mℓ/min.
Temperature = 30°C.
Eluant = 0.65 mM L-2-R-dien-Zn(II); 0.17 M-NH₄Ac to pH 9.0 with aqueous ammonia; acetonitrile-H_2O (35:65).

Weighed amounts of the dien and metal salt are added to the appropriate amount of an aqueous ammonium acetate solution. The pH is adjusted with aqueous NH_3 and the appropriate amount of acetonitrile added. The mobile phase is then degassed with He.

REFERENCES

1. LePage, J. N., Davies, G., Seitz, D. E., and Karger, B. L., Anal. Chem., 51, 433, 1979.

Table LC 26
REVERSED PHASE SEPARATION OF ENANTIOMERS OF DANSYL AMINO ACIDS USING CHIRAL METAL CHELATE ADDITIVES

Column packing	P1		P2	
Column				
Length (cm)	15		15	
Diameter (mm)	4.6		4.6	
Solvent	S1		S2	
Flow rate (ml/min)	2.0		2.0	
Temperature (°C)	30		25	
Detection	UV		UV	
Literature	1		2	
Dansyl amino acid	K_L	K_D	K_L	K_D
Alanine	2.5	5.6	13.3	7.3
Arginine	0.6	1.0	3.1	2.5
Asparagine	2.25	4.3	14.7	12.0
Aspartic acid	4.8	3.6	6.6	9.6
α-n-Butyric acid	2.5	4.85	9.6	7.8
Citrulline			3.7	3.1
Cysteic acid	4.15	6.8	16.5	18.0
Glutamic acid	0.7	1.45	24.0	19.8
Glutamine	0.8	0.95	4.1	3.1
Histidine	69.3	101.5	18.2	10.4
Isoleucine	5.0	6.0	16.80	15.56
Alloisoleucine	4.9	5.4		
Leucine	6.2	12.8	23.5	19.3
Lysine	37.1	55.3	9.5	7.9
Methionine	4.4	8.5	18.2	12.3
Methionine sulfone	1.5	2.6	45.1	25.7
Norleucine	7.35	14.5	26.4	20.6
Norvaline	4.2	8.2	16.2	13.0
Ornithine	33.25	55.2	9.4	7.8
Phenylalanine	14.4	26.3	10.3	4.6
Phenylglycine	8.9	14.8		
Proline	2.0	2.0	37.3	36.3
Serine	3.2	7.8	29.6	8.5
Threonine	2.5	5.6	5.9	3.9
Tryptophan	18.0	32.6	11.2	4.5
Tyrosine	19.25	33.4	32.0	18.7
Valine	3.0	4.5	52.9	50.3

Column packing P1 = 5 μm Hypersil bonded with n-octyl groups.
P2 = 5 μm C_8 Hypersil.
Solvent S1 = 0.8 mM C_3-C_8-dien-Zn(II), 0.19 M ammonium acetate, pH 9, acetonitrile-water (35:65) v/v). (C_3-C_8-dien = L-2-ethyl-4-octyl-diethyltriamine).
S2 = 4 mM L-prolyl-n-octylamide-Ni(II), 0.088 M ammonium acetate, pH 9, methanol-water(60:40).

REFERENCES

1. Lindner, W., LePage, J. N., Davies, G., Seitz, D. E., and Karger, B. L., *J. Chromatogr.*, 185, 323, 1979.
2. Tapuhi, Y., Miller, N., and Karger, B. L., *J. Chromatogr.*, 205, 325, 1981.

Table LC 27
REVERSE PHASE HIGH PERFORMANCE LIQUID CHROMATOGRAPHY OF FLUORESCAMINE-AMINO ACID REACTION PRODUCTS

	Retention volume (ml)		
Amino acid reactant	Acid alcohol derivative	Lactone derivative	Acid alcohol/lactone peak area ratio
2-Aminobutyric acid	6.0	10.2	2.5
3-Aminobutyric acid	5.4	8.4	0.64
4-Aminobutyric acid	4.5	—	No lactone formed
Alanine	5.0	8.6	2.84
Valine	6.8	16.2	5.4
Leucine	8.7	21.0	3.7
Isoleucine	8.9	22.0	4.8
α-Aminoisobutyric acid	4.2	7.8	12.3
L-Alanyl-L-alanine	28.0	—	No lactone formed
L-Alanine-*tert.*-butyl ester	4.5	—	No lactone formed

Conditions: Waters Associates model ALC202 HPLC system with fluorometric detector.
Column = 30 × 0.4 cm. μ-Bondapak ODS-C_{18} (10 μm).
Flow rate = 1.5 ml/min.
Pressure = 3100 psi.
Temperature = ambient.
Eluant = solvent of water (3.0 × 10^{-3} moles of camphorsulphonic acid) adjusted to pH3.5 with tetrapropylammonium hydroxide and methanol.

REFERENCE

1. McHugh, W., Sandmann, R. A., Haney, W. G., Sood, S. P., and Wittmer, D. P., *J. Chromatogr.*, 124, 376, 1976.

Reproduced by permission of the Elsevier Scientific Publishing Company.

Table LC 28
REVERSE-PHASE HIGH PERFORMANCE LIQUID CHROMATOGRAPHY OF PHENYLTHIOHYDANTOIN AMINO ACIDS ON A TRIPEPTIDE BONDED STATIONARY PHASE

PTH-amino acid	Capacity ratios (k') with various mobile phases					
	1% Citric acid in H_2O, pH 2.5	5% Methanol pH 2.5	10% Methanol pH 2.5	1% Sodium citrate in H_2O, pH 7.4	Dist. H_2O pH 5.5	Methanol
PTH-L-histidine monohydrochloride	0.13	0.13	0.10	10.8	4.88	0.33
PTH-L-arginine[b]	0.25	0.13	0.10	5.86	0.00	0.00
PTH-D,L-threonine[b]	4.86	3.80	3.10	5.75	—	0.50
PTH-D,L-serine	6.11	4.36	3.95	4.35	—[a]	0.70
PTH-L-asparagine	6.88	5.34	4.05	4.65	12.1	0.75
PTH-S-methyl-L-cysteine	7.21	5.13	4.15	3.80	—	—[a]
PTH-L-glutamine[b]	8.40	6.05	4.75	5.23	—[a]	0.70
PTH-glycine	8.75	7.10	5.75	7.44	—	0.75
PTH-D,L-alanine[b]	12.1	9.38	7.28	7.56	—	0.63
PTH-L-hydroxyproline	13.8	9.38	8.75	9.58	—	0.50
PTH-L-methionine sulphone	14.0	9.80	8.00	7.98	—[a]	0.88
PTH-L-glutamic acid[b]	24.6	16.6	14.3	15.1	—[a]	—[a]
PTH-D,L-valine	28.0	18.5	14.2	29.8	—[a]	0.50
PTH-L-proline	34.6	24.5	17.3	22.5	—[a]	0.63
PTH-D,L-aspartic acid[b]	36.2	25.0	22.2	9.10	—[a]	—[a]
PTH-D,L-methionine	39.0	27.0	19.7	28.7	—[a]	0.63
PTH-L-tyrosine[b]	46.2	31.3	23.3	39.8	—[a]	0.60
PTH-isoleucine	49.5	31.6	22.9	40.1	—[a]	0.38
PTH-L-leucine	54.8	33.0	25.5	62.1	—[a]	0.38
PTH-norleucine[b]	65.7	42.4	31.8	54.6	—	0.50
PTH-D,L-phenylalanine[b]	85.9	58.5	44.0	107.0	—[a]	0.63
PTH-(S-carboxymethyl)-L-cysteine	143.0	85.1	68.8	13.9	—	—
PTH-D,L-tryptophan[b]	198.0	145.0	103.0	224.0	—	0.75
PTH-(ß-phenylthiocarbamyl)-L-lysine	385.0	217.0	162.0	285.0	—	0.93
PTH-L-cysteic acid (K salt)	—[a]	—[a]	—[a]	9.22	—	—[a]

[a] No peak observed in a reasonable length of time.

[b] Indicates PTH derivatives that could be easily separated with 5% methanol.

Conditions: Waters Associates HPLC system with UV-detector.
Column = 25 × 0.21 cm. L-Val-L-Phe-L-Val bonded on Partisil-10® (0.26 mmoles peptide bonded per gram Partisil-10®).

REFERENCE

1. Fong, G. W-K. and Grushka, E., *J. Chromatogr.*, 142, 299, 1977.

Reproduced by permission of Elsevier Scientific Publishing Co.

Table LC 29
ISOCRATIC REVERSE-PHASE HIGH PERFORMANCE LIQUID CHROMATOGRAPHY OF PHENYLTHIOHYDANTOIN AMINO ACIDS

Compound	t_R(min)[a]	k'[b]	Coefficient of variation (%)[c]
PTH-Cysteic acid (K salt)	3.30	0.04	±3.9
PTH-Aspartic acid	4.60	0.45	±2.4
PTH-(S-Carboxymethyl)cysteine	5.20	0.64	±0.9
PTH-Asparagine	6.00	0.89	±1.9
PTH-Serine	6.80	1.14	±1.8
PTH-Glutamine	7.00	1.20	±1.0
PTH-Glutamic acid	7.30	1.30	±0.8
PTH-Threonine	7.60	1.39	±1.0
PTH-Glycine	8.60	1.71	±1.5
PTH-Histidine monohydrochloride	10.1	2.18	±2.2
PTH-Methionine sulphone	10.7	2.37	±0.4
PTH-Alanine	12.4	2.90	±0.3
PTH-Arginine	15.0	3.72	±2.5
PTH-Tyrosine	15.6	3.91	±0.8
PTH-Methionine	27.8	7.75	±2.3
PTH-Proline	27.8	7.75	±0.8
PTH-Valine	28.6	8.00	±2.3
PTH-Tryptophan	46.6	13.7	±3.8
PTH-Isoleucine	49.8	14.7	±1.8
	51.6[d]	15.2[d]	±2.9
PTH-Phenylalanine	53.6	15.9	±3.6
PTH-Leucine	57.6	17.1	±2.8
PTH-N-Phenylthiocarbamoyllysine	71.0	21.3	±7.2

[a] t_R = retention time.
[b] k' = capacity ratio.
[c] Calculated from the ratio t_R (PTH-amino acid) to t_R (internal standard).
[d] Produced two peaks in the chromatogram.

Conditions: Waters Associates HPLC system with UV-detector.
Column = Two 15 × 0.39 cm stainless steel tubes in tandem connected by low dead-volume swagelok fitting. Merck LiChrosorb® RP8 (5 μm). Flow rate = 0.9 mℓ/min
Pressure = 3000 psi.
Temperature = 50°C
Eluant = 25:75 (v:v) Acetonitrile:sodium acetate pH 4.6 (adjusted with conc. acetic acid)

REFERENCE

1. Abrahamsson, M., Gröningsson, K. and Castensson, S., *J. Chromatogr.*, 154, 313, 1978.

Table LC 30
GRADIENT REVERSE-PHASE HIGH PERFORMANCE LIQUID CHROMATOGRAPHY OF PHENYLTHIOHYDANTOIN AMINO ACIDS

Compound	Mean of T_R (min)[a]	Mean of T_R/T_R THR.	C.V.[b] of T_R (±%)	C.V. of T_R/T_R THR.
PTH-Cysteic acid[c]	4.09		0.7	
	4.48		1.1	
PTH-Aspartic acid[c]	5.53		1.2	
	5.75		1.5	
PTH-(S-Carboxymethyl)-cysteine[c]	8.06		2.9	
	8.45		2.9	
PTH-Glutamic acid[c]	10.16		2.8	
	10.77		2.7	
PTH-Asparagine	11.44		1.7	
PTH-Serine	12.99	0.932	1.8	0.42
PTH-Glutamine	13.44	0.964	1.8	0.22
PTH-Threonine	13.93	(1.00)	1.7	(0.00)
PTH-Glycine	14.34	1.029	1.6	0.17
PTH-Histidine	15.32		1.4	
PTH-Methionine-SO$_2$	16.40		0.7	
PTH-Alanine	18.15		0.9	
PTH-Tyrosine	21.35		0.6	
PTH-Arginine	22.29		0.9	
PTH-(S-Methyl)-cysteine	22.44		0.5	
PTH-Proline	23.27		0.6	
PTH-Methionine	24.26		0.6	
PTH-Valine	24.67		0.4	
PTH-Tryptophan	27.44		0.4	
PTH-Phenylalanine	28.02		0.4	
PTH-(ε-φ-Thiocarbamyl)-lysine	29.07		0.6	
PTH-Isoleucine	29.50		0.6	
PTH-Leucine	30.07		0.6	
PTH-Norleucine	30.61		0.6	

Note: Results from 27 repetitive injections of the 24 PTH-amino acids indicated.

[a] T_R = retention time.
[b] C.V. = coefficient of variation.
[c] These compounds were detected as multiple peaks.

Conditions: Altex model 332 HPLC system with UV-detector.
        Column = 25 × 0.46 cm. (pre-column 4.5 × 0.46 cm.). Altex Ultrasphere-ODS
        Flow rate = 1.3 ml/min
        Temperature = 45°C
        Eluant = A- 5% THF in 4.2 mM acetic acid adjusted to pH 5.16 with NaOH B - 10% THF in acetonitrile

Elution is started with solvent A, while the concentration of solvent B is increased from 0 to 40% during 20 min.

REFERENCE

1. Somack, R., *Anal. Biochem.*, 104, 464, 1980.

Reproduced by permission of Academic Press Inc.

Table LC 31
REVERSE-PHASE HIGH PERFORMANCE LIQUID CHROMATOGRAPHY OF PHENYLTHIOHYDANTOIN AMINO ACIDS WITH VARIOUS GRADIENT MOBILE PHASES

PTH-amino acid	t_R (min)[a] system a	PTH-amino acid	t_R (min)[a] system b	PTH-amino acid	t_R (min)[a] system c
1. Asparagine	7.5	1. Aspartic acid	6.7	1. Aspartic acid	4.9
2. Aspartic acid	8.1	2. Asparagine	7.6	2. Glutamic acid	5.6
3. Serine	8.7	3. Serine	8.45	3. Asparagine	6.1
4. Threonine	9.3	4. Histidine	8.45	4. Serine	6.8
5. Histidine	10.0	5. Threonine	8.9	5. Glutamine	7.1
6. Glutamine	10.4	6. Glycine	9.4	6. Threonine	7.9
7. Glycine	10.6	7. Glutamine	10.5	7. Glycine	9.0
8. Glutamic acid	13.5	8. Glutamic acid	11.1	8. Alanine	11.7
9. Arginine	14.2	9. Arginine	11.3	9. Tyrosine	12.5
10. Alanine	15.1	10. Alanine	13.0	10. Histidine	14.5
11. Tyrosine	19.1	11. Tyrosine	15.4	11. Methionine	15.5
12. Valine	22.1	12. Proline	18.2	12. Valine	15.8
13. Proline	22.1	13. Valine	19.2	13. Proline	16.3
14. Methionine	22.1	14. Methionine	19.2	14. Tryptophan	16.8
15. Tryptophan	24.6	15. Tryptophan	20.8	15. Lysine	18.2
16. Isoleucine	24.8	16. Phenylalanine	23.3	16. Phenylalanine	18.4
17. Phenylalanine	24.8	17. Isoleucine	23.6	17. Isoleucine	19.0
18. Leucine	25.2	18. Leucine	24.4	18. Leucine	19.6
19. Lysine	25.7	19. Lysine	25.4	19. Arginine	21.0

[a] t_R = retention time.

Conditions: Waters Associates HPLC system

System a: Column = 30 × 0.4 cm. Waters Associates μ-Bondapak C_{18}
　　Flow rate = 2.0 mℓ/min.
　　Temperature = room temperature.
　　Eluant = A- 90% 0.02 N sodium acetate pH 3.7, 10% acetonitrile
　　　　　　B- 10% acetate buffer pH 3.7, 90% acetonitrile
　　　　　　5% Buffer B (initial conditions) to 90% Buffer B (final conditions). Curve 7 of the gradient programmer

System b: Column = 30 × 0.4 cm. Waters Associates μ-Bondapak C_{18}
　　Flow rate = 2.0 mℓ/min.
　　Temperature = room temperature
　　Eluant = A- 900 mℓ water and 2.5 mℓ glacial acetic acid are combined, the pH adjusted to 4.4 with 1 N-NaOH, then 100 mℓ methanol and 0.05 mℓ acetone are added.
　　　　　　B- 100mℓ water, 900 mℓ methanol and 0.25 mℓ glacial acetic acid.
　　　　　　5% Buffer B (initial conditions) to 45% Buffer B (final conditions). Curve 6 of the gradient programmer

System c: Column = 25 × 0.45 cm. DuPont Zorbax ODS
　　Flow rate = 0.8 mℓ/min.
　　Temperature = 62°C
　　Eluant = A- 99% 0.01 N sodium acetate pH 4.8, 1% acetonitrile
　　　　　　B- 100% acetonitrile
　　　　　　24% Buffer B (initial conditions) to 44% Buffer B (final conditions). Curve 6 of the gradient programmer

REFERENCE

1. Harris, J. U., Robinson, D. and Johnson, A. J., *Anal. Biochem.*, 105, 239, 1980.

Reproduced by permission of Academic Press Inc.

Table LC 32
SEPARATION OF L- AND D-AMINO ACIDS AS DIASTERIOMERIC METHOXY-α-METHYL-1-NAPHTHALENE-ACETYLAMINO ACID METHYL ESTER DERIVATIVES BY HIGH PERFORMANCE LIQUID CHROMATOGRAPHY

Amino acid	k' D	k' L	α	R	Mobile phase
Alanine	1.91	3.10	1.62	5.84	A
Valine	1.76	1.06	1.66	9.17	A
Norvaline	2.17	1.22	1.78	5.90	A
Leucine	1.83	0.78	2.35	6.14	A
Norleucine	1.92	0.96	2.00	6.00	A
Isoleucine	1.53	0.88	1.74	4.26	A
Proline	1.42	2.38	1.68	2.48	A
Phenylglycine	2.56	1.32	1.94	6.22	A
Phenylalanine	2.41	1.52	1.59	4.35	A
Serine	2.52	7.32	2.90	8.63	C
Threonine	1.20	1.85	1.55	2.98	D
Tyrosine	2.02	1.74	1.16	1.02	B
Dihydroxyphenylalanine	2.32	2.04	1.14	0.67	C
Cysteine	2.08	0.98	2.12	2.80	C
Methionine	1.60	1.24	1.29	1.61	B
Aspartic acid	5.61	7.00	1.25	2.52	A
Glutamic acid	6.46	6.81	1.05	0.62	A
Ornithine	2.00	3.52	1.76	3.33	D
Lysine	1.60	2.20	1.38	1.97	D
Histidine	1.36	1.96	1.45	1.47	D

^{a} t_o = 5.0 min. k' and α values refer to capacity ratio and separation factor for a pair of diastereomers, respectively.

$$R = \text{resolution value} = \frac{2(t_{R2} - t_{R1})}{(W_1 + W_2)}$$

where t_{R1} and t_{R2} are retention times and W_1 and W_2 are the bases of triangles derived from the peaks.

Conditions: Waters Associates model ALC/GPC 202 R401 HPLC system with UV-detector.
Column = 12 × 0.25 in. Waters Associates μ-Porasil®.
Temperature = Ambient.
Flow rate = 0.7 mℓ/min.
Eluant = Cyclohexane-ethyl acetate, (A) 4:1, (B) 2:1, (C) 1:1, (D) 2:3.

REFERENCE

1. Goto, J., Hasegawa, M., Shimada, K. and Nambara, T., *J. Chromatogr.*, 152, 413, 1978.

Table LC 33
THIOUREA DERIVATIVES[a] OF ENANTIOMERIC AMINO ACID ETHYL ESTERS

Packing	P1	P1
Column		
Length (cm)	25	25
Diameter (mm I.D.)	0.4	0.4
Material	SS	SS
Solvent	S1	S2
Flow rate (mℓ/min)	0.4	0.4
Temperature	rt	rt
Detection	UV	UV
Amino acid	K	
L-Serine	1.70	0.95
D-Serine	1.50	0.83
L-Alanine	2.69	1.50
D-Alanine	2.93	1.62
L-Proline	4.28	1.58
D-Proline	4.67	1.61
L-Aspartic acid	5.38	2.67
D-Aspartic acid	5.71	2.83
L-Glutamic acid	5.90	2.85
D-Glutamic acid	6.26	2.97
L-Tyrosine		2.57
D-Tyrosine		2.90
L-Valine		3.50
D-Valine		4.13
L-Phenylglycine		4.31
D-Phenylglycine		4.81
L-Tryptophan		5.66
D-Tryptophan		6.64
L-Isoleucine		5.67
D-Isoleucine		6.71
L-Leucine		5.70
D-Leucine		6.73
L-Phenylalanine		6.85
D-Phenylalanine		8.65

[a] Derivatives formed by reaction of the amino acid ethyl ester with 2,3,4,6-tetra-O-acetyl-β-D-glucopyranosyl isothiocyanate (GITC).

Packing   P1 = LiChrosorb® RP-18, 5 μm.
Solvent   S1 = 50% aqueous methanol.
            S2 = 60% aqueous methanol.

REFERENCE

1. Nimura, N., Ogura, H., and Kinoshita, T., *J. Chromatogr.*, 202, 375, 1980.

Table LC 34
DIASTEREOMERS OF N-d-10-CAMPHORSULFONYL AMINO ACID p-NITROBENZYL ESTERS

Packing		P1	P1	P2
Column				
Length		25 (cm)	25 (cm)	25 (cm)
Diameter		2.2 (mm I.D.)	2.2 (mm I.D.)	2.0 (mm)
Material		SS	SS	
Solvent		S1	S2	S3
Flow rate		0.4 (mℓ/min)	0.4 (mℓ/min)	0.5 (mℓ/min)
Temperature		rt	rt	rt
Detector		UV	UV	UV
Literature		1	1	2
Amino acid	Enantiomer	t_r (min)		
Leucine	L	3.9	2.7	7.6
	D	4.4	2.8	8.6
Isoleucine	L	4.4	2.9	7.0
	D	5.0	3.1	7.6
Phenylalanine	L	6.2	3.3	11.0
	D	8.5	4.1	12.4
Methionine	L	7.4	3.6	
	D	10.0	4.6	
Alanine	L	7.2	3.7	17.0
	D	9.3	4.4	22.3
Glutamic acid	L	12.8	4.2	
	D	16.8	5.2	
Tryptophan	L	29.2	9.0	
	D	49.6	14.9	
Tyrosine	L	33.2	11.6	
	D	47.2	16.2	

Packing P1 = Varian MicroPak-NH$_2$, average particle size 10 μm
       P2 = Varian MicroPak Si-5
Solvent S1 = Isooctane-dichloromethane-isopropanol (79:16:5)
       S2 = Isooctane-dichloromethane-isopropanol (63:32:5)
       S3 = Isooctane-isopropanol (98.5:1.5)

REFERENCES

1. Furukawa, H., Mori, Y., Takeuchi, Y., and Ito, K., *J. Chromatogr.*, 136, 428, 1977.
2. Furukawa, H., Sakakibara, E., Kamei, A., and Ito, K., *Chem. Pharm. Bull.*, 23, 1625, 1975.

Table LC 35
ALIPHATIC AMINES

Packing	P1
Column	
Length (mm)	550
Diameter (mm I.D.)	9
Material	Glass
Solvent	Water
Temperature	50°C
Detection	Flow coulometric

Amine	V_R
Methylamine	17.3
Ethylamine	17.4
Diethylamine	18.0
Triethylamine	21.5
n-Propylamine	19.3
Isopropylamine	18.3
n-Butylamine	23.6
n-Amylamine	30.0
n-Hexylamine	49.5
n-Heptylamine	76.0
n-Octylamine	—

Packing P1 = Hitachi hydroxide-form anion exchange resin, particle size 18 ± 2 μm, 8% cross-linking.

REFERENCE

1. Tanaka, K., Ishizuka, T., and Sunahara, H., *J. Chromatogr.*, 172, 484, 1979.

Table LC 36
AMINES

Packing	P1
Column	
Length (mm)	250
Diameter (mm)	9
Material	na
Solvent	S1
Flow rate	460 mℓ/hr
Temperature	rt
Detection	Conductimetric

Amine	Concentration (ppm)	t_r
Monomethylamine	8	6.5
Monomethylamine	80	6.3
Monomethylamine	800	5.2
Dimethylamine	8	8.2
Dimethylamine	80	7.7
Dimethylamine	800	6.3
Trimethylamine	8	10.3
Trimethylamine	80	9.9
Trimethylamine	400	8.7
Tetramethylammonium Br	10	12.0
Tetramethylammonium Br	100	12.3
Tetramethylammonium Br	1000	9.8
Monoethylamine	14	7.2
Monoethylamine	140	6.8
Diethylamine	20	10.8
Diethylamine	200	10.0
Triethylamine	40	15.5
Triethylamine	400	14.0
Tetraethylammonium Br	100	27.5
Tetraethylammonium Br	1000	22.0
n-Butylamine	20	14.2
n-Butylamine	200	12.7
Cyclohexylamine	200	23.6
Cyclohexylamine	1000	19.5
Tri-n-butylamine		30
Tetra-n-butylammonium Br		30
Monoethanolamine	20	5.3
Diethanolamine	20	5.5
Triethanolamine	200	5.9
Monoisopropanolamine	20	5.6
Diisopropanolamine	40	6.2
Triisopropanolamine	400	6.6
Ammonia	1	5.1
Ammonia	10	5.1
Ammonia	100	4.9

Packing P1 = Separation column: sulfonated styrene-divinylbenzene copolymer, 0.024 m equiv/g, 180-325 mesh. Stripper column: Dowex 1 × 8 (OH$^-$), 200-400 mesh.

Solvent S1 = 0.01 N HCl.

REFERENCE

1. Small, H., Stevens, T. S., and Brown, W. C., *Anal. Chem.*, 47, 1805, 1975.

Reprinted with permission from *Anal. Chem.*, 47, 1805, 1975. Copyright (1975) American Chemical Society.

Table LC 37
HIGH PERFORMANCE ION PAIR PARTITION CHROMATOGRAPHY OF BIOGENIC AMINES AND THEIR METABOLITES

Packing	P1	P1	P1	P2	P3
Column					
Length (cm)	25	25	25	25	25
Diameter (mm I.D.)	3	3	3	3	3
Material	SS	SS	SS	SS	SS
Solvent	S1	S2	S3	S4	S5
Flow rate	na	na	na	na	na
Temperature	25 ± 1°C	25 ± 1°C	25 ± 1°C	25 ± 1°C	25 ± 1°C
Detection	UV	UV	UV	UV	UV
Amine			$r_{tyramine}$		
Phenylethylamine	0.3	0.4	0.5	0.6	0.2
Tyramine	1.0	1.0	1.0	1.0	1.0
3-Methoxytyramine	1.5	1.2	3.1	1.6	0.9
Dopamine	2.9	2.2	2.7	2.0	3.4
Normetanephrine	3.2	3.0	6.9	3.3	2.8
Metanephrine	3.1	2.8	12.3	4.0	1.7
Noradrenaline	6.5	6.1	7.2	4.8	11.5
Adrenaline	6.2	5.4	12.8	5.3	6.8

Packing   P1 = Porasil® E, 54-75 μ
            P2 = silica gel, 4 μ
            P3 = silica gel, 10 μ
Solvent (mobile phase)   S1 = butanol-hexane (1:1)
                           S2 = ethyl acetate-hexane (9:1)
                           S3 = tributyl phosphate-hexane (1:3)
                           S4 = ethyl acetate-tributyl phosphate-hexane (7.25:1:1.75)
                           S5 = butanol-methylene chloride (2:3)
Aqueous phase = 0.1 M HClO$_4$/0.9 M NaClO$_4$; for S5 0.2 M HClO$_4$/0.8 M NaClO$_4$

REFERENCE

1. Persson, B-A. and Karger, B. L., *J. Chromatogr. Sci.*, 12, 521, 1974.

Reproduced from the Journal of Chromatographic Science by permission of Preston Publications Inc.

Table LC 38
LIGAND-EXCHANGE CHROMATOGRAPHY OF DIAMINES AND POLYAMINES

Packing Column	P1	P2	P3	P4	P5	P6	P7	P8
Length (cm)	13	13	13	13	24	24	36	36
Diameter (cm)	0.9	0.9	0.9	0.9	0.63	0.63	0.63	0.63
Material	G	G	G	G	G	G	G	G
Solvent	S1	S2	S3	S4	S5	S6	S7	S8
Flow rate (mℓ/hr)	20	20	20	20	20	20	50	50
Temperature	rt	rt	rt	rt	rt	rt	rt	rt
Detection	RI	RI	RI	RI	RI	RI	RI	RI

Amine	V_e (as multiples of the bulk column volume)							
1,3-Diaminopropane	14.3	2.0	5.4	1.9	12.2	2.1	5.6	1.5
1,4-Diaminobutane	2.9	2.9	1.7	3.3		3.6	2.6	2.3
1,5-Diaminopentane	3.4	4.4	2.3	4.8		7.4	3.3	2.5
1,6-Diaminohexane	4.4	5.6	2.4	6.4	8.4	14.4	3.5	2.8
Spermidine	4.2	3.6	1.9	4.3	3.9	4.6	2.3	1.7
Spermine	8.9	4.2	2.0	5.5	6.4	5.1	1.9	1.3
Histamine	4.4	1.4		1.2	18.6		4.9	2.0
Lysine	1.1	0.8	1.6		1.0		1.0	
Histidine	2.0	2.1	1.5		1.0		1.5	
Arginine	3.1	1.3	4.0	1.2	5.8	1.3	6.0	1.2
n-Butylamine	1.9	2.0	1.4	1.9		4.3	1.2	1.0
Ammonia, M	1.1	1.7	1.1	1.6	7.1	7.6	3.8	4.9

Packing P1 = Cellex CM-Cu
       P2 = Cellex CM-Zn
       P3 = Cellex CM-Ni
       P4 = Cellex CM-NH$_4$
       P5 = Aminex Q-150S-Cu
       P6 = Aminex Q-150S-Zn
       P7 = Bio-Rex 70-Cu
       P8 = Bio-Rex 70-Zn
Solvent S1 = 0.97 M NH$_3$, 10^{-4} M CuSO$_4$
       S2 = 0.97 M NH$_3$, 10^{-4} M ZnSO$_4$
       S3 = 0.97 M NH$_3$, 10^{-4} M NiSO$_4$
       S4 = 0.97 M NH$_3$, 10^{-4} M (NH$_4$)$_2$SO$_4$
       S5 = 5.5 M NH$_3$, 10^{-4} M CuSO$_4$
       S6 = 5.5 M NH$_3$, 10^{-4} M ZnSO$_4$
       S7 = 5.3 M NH$_3$, 10^{-4} M CuSO$_4$
       S8 = 5.3 M NH$_3$, 10^{-4} M ZnSO$_4$

REFERENCE

1. Navratil, J. D. and Walton, H. F., *Anal. Chem.*, 47, 2443, 1975.

Reprinted with permission from *Anal. Chem.*, 47, 2443, 1975. Copyright (1975) American Chemical Society.

Table LC 39
LIGAND-EXCHANGE CHROMATOGRAPHY OF DIAMINES AND POLYAMINES

Packing	P1	P1	P1	P1
Column				
Length (cm)	24	24	24	24
Diameter (cm)	0.63	0.63	0.63	0.63
Material	G	G	G	G
Solvent				
NH_3, M	5.1	5.4	5.5	5.6
Zn, influent, ppm	50	46	110	115
Zn, effluent, ppm	40	60	80	112
Flow rate, (mℓ/hr)	20	20	20	20
Temperature (°C)	20	55	55	55
Detection	RI	RI	RI	RI

Amine		V_e(mℓ)		
1,2-Diaminopropane	60.0	45.0		
1,3-Diaminopropane	16.5	16.6	15.0	
1,4-Diaminobutane	21.0	19.3	15.7	16.5
1,5-Diaminopentane	40.0	33.7	27.7	27.7
1,6-Diaminohexane		60.0		
Spermidine	24.0	26.3	21.0	20.2
Spermine	30.0	32.5	24.5	23.1
Histamine			35.5	

Packing P1 = Aminex A-7-Zn. Raising the temperature increased the stripping of zinc from the column, and more zinc sulfate had to be added to the influent.

REFERENCE

1. Navratil, J. D. and Walton, H. F., *Anal. Chem.*, 47, 2443, 1975.

Reprinted with permission from *Anal. Chem.*, 47, 2443, 1975. Copyright (1975) American Chemical Society.

Table LC 40
AUTOMATED ION-EXCHANGE CHROMATOGRAPHY OF DIAMINES, POLYAMINES, AND BASIC AMINO ACIDS

Packing	P1
Column	
Length (cm)	9
Diameter (cm)	0.9
Material	G
Solvent	S1
Flow rate (mℓ/hr)	70
Temperature	T1
Detection	D1

Compound	t_r (min)
A. Buffer 1	
Hydroxylysine	31
Ornithine	34
Histidine	35
Lysine	36
N-ε-Monomethyllysine	41
N-ε-Dimethyllysine	44
N,N'-bis-(carboxyethyl)-1,4-diaminobutane	45
Trimethyllysine	47
N-Carboxyethyl-1,4-diaminobutane	47.5
1-Methylhistidine	49
NH$_3$	57
Monoacetyl-1,3-diaminopropane	59
α-Amino-β-guanidinopropionic acid	72
Monoacetyldiaminobutane	73
N^G,N^G-Dimethylarginine	90
N^G,N'^G-Dimethylarginine	94
α-Amino-γ-guanidinobutyric acid	94
Monocarbamyl-1,4-diaminobutane	97
Arginine	110
Monoacetyl-1,5-diaminopentane	111
B. After buffer change	
Homoarginine	31.5
S-Carboxyamidomethylglutathionylspermidine	37.5
Glutathionylspermidine (reduced)	48
1,4-Diamino-2-hydroxybutane	61
Norepinephrine	62.5
S-(N-Ethylsuccinimido)-glutathionylspermidine	65.5
1,3-Diaminopropane	68
N^1-Monoacetylspermidine	75
N^8-Monoacetylspermidine	77
1,4-Diaminobutane	81
Histamine	105
1,5-Diaminopentane	111
Glutathionylspermidine (disulfide)	128
N-(3-Aminopropyl)-1,3-diaminopropane	141
Spermidine	160
Agmatine	198
Cystamine	224
N,N'-bis(3-aminopropyl)-1,3-diaminopropane	229
Spermine	263

Packing P1 = Beckman, PA-35 resin
Solvent S1 = Buffer I. Sodium citrate, 0.35 M Na$^+$, pH 6.32

Table LC 40 (continued)
AUTOMATED ION-EXCHANGE CHROMATOGRAPHY OF DIAMINES, POLYAMINES, AND BASIC AMINO ACIDS

	Buffer II. Sodium citrate, 0.35 M Na⁺-NaCl, 1 M Na⁺ mixture, pH 5.81
	Buffer IIIA. Sodium citrate 0.35 M Na⁺-NaCl, 2 M Na⁺ mixture, pH 5.64
	Buffer IIIB. Potassium citrate, 0.35 M K⁺-KCl, 2 M K⁺ mixture
	Buffer IIIC. is prepared by mixing 80 parts of buffer IIIA and 20 parts of buffer IIIB.
	Buffer I was used for 95-120 min the elution then being switched automatically to a three-vessel exponential gradient. The three flasks were connected in series, so that the output of flask 3 flowed into flask 2, the output of flask 2 flowed into flask 1 and the output of flask 1 flowed into the column. Flasks 1 and 2 were 50 ml round-bottom flasks with air-tight connections; each contained 60 ml of buffer II and was stirred by a magnetic stirrer.
	Flask 3 was an open flask containing 500 ml of buffer IIIC.
	Regeneration of the column was carried out with 25 ml of 0.2 N NaOH-0.1% disodium ethylenediamine tetraacetate, followed by 140 ml of buffer I
Temperature T1 =	30°C while using buffer I, 54°C after the buffer change
Detection D1 =	Ninhydrin colorimetry

REFERENCE

1. Tabor, H., Tabor, C. W., and Irreverre, F., *Anal. Biochem.*, 55, 461, 1973.

Reproduced by permission of Academic Press, Inc.

Table LC 41
POLYAMINES

Packing	P1	P2	P3	P4
Column				
Length (cm)	24	9	11.5	11.5
Diameter (cm)	0.60	0.40	0.45	1.75
Material	na	na	na	SS
Solvent	S1	S2	S3	S4
Flow rate (ml/hr)	50	25	26	27
Temperature	T1	T2	T3	T4
Detection	D1	D2	D3	D2
Literature	1	2	3	4
Amine		t_r		
Agmatine	219	87.7	81.6	
Aminopropylcadaverine			86.8	
Ammonia		67	12.1	30.5
Arginine		86	14.2	41.1
Cadaverine	176	59.6	63.5	41.36

Table LC 41 (continued)
POLYAMINES

Carbamylputrescine			32.1	
1,7-Diaminoheptane			114.8	
1,3-Diaminopropane			55.8	31.49
Ethanolamine			27.7	
Hexamethylenediamine		81.7		
Histamine	122	53.8	59.3	38.30
Histidine			14.7	
Lysine	57	7.6	19.2	
1-Methylhistamine				34.35
Phenylethylamine	208	109.3		
Putrescine	140	51.1	57.4	34.19
Spermidine		75.2	73.8	59.38
Spermine		116.5	107.7	70.00
Sym-homospermidine			76.3	
Sym-norspermidine			70.7	
Sym-norspermine			99.8	
Tryptamine	281			
Tyramine	196	99.1		

Packing
- P1 = Chromobeads A (Technicon)
- P2 = DC 6A resin (Durrum)
- P3 = DC 4A resin (Durrum)
- P4 = Na⁺ form high performance cation-exchange resin with 16% cross-linkage

Solvent
- S1 = Sodium citrate-NaCl buffers containing Brij solution 10.0 mℓ/ℓ. pH 4.10 buffer changed to pH 7.50 buffer after 10 min; pH 8.20 buffer after the elution of lysine (57 min) and pH 11.50 buffer containing 100 mℓ propanol-2/1 after the elution of putrescine (140 min)
- S2 = pH 5.65 buffer, 0.2 N sodium citrate, 1.0 N NaCl for 35 min, then pH 5.65 buffer, 0.2 N sodium citrate, 2.6 N NaCl for 83 min
- S3 = pH 5.60 0.20 N sodium citrate, 0.30 N NaCl, 4% ethanol, then pH 5.65 0.20 N sodium citrate, 2.50 N NaCl, 6% ethanol
- S4 = pH 5.85 buffers, 0.20 N sodium citrate for 26 min, 0.20 N sodium citrate + 1.7 N NaCl for 30 min and 0.20 N sodium citrate + 3.30 N NaCl for 19 min
- T1 = 65°C
- T2 = 60°C
- T3 = 61°C for 48 min and then 78°C
- T4 = 78°C

Detection
- D1 = Ninhydrin colorimetry
- D2 = o-Phthalaldehyde fluorimetry

REFERENCES

1. Dierick, N. A., Vervaeke, I. F., Decuypere, J. A., and Henderickx, H. K., *J. Chromatogr.*, 129, 403, 1976.
2. Villanueva, V. R., Adlakha, R. C., and Cantera-Soler, A. M., *J. Chromatogr.*, 139, 381, 1977.
3. Adlakha, R. C. and Villanueva, V. R., *J. Chromatogr.*, 187, 442, 1980.
4. Perini, F., Sadow, J. B., and Hixson, C. V., *Anal. Biochem.*, 94, 431, 1979.

Table LC 42
AMINES, m-TOLUOYL DERIVATIVES

Packing	P1
Column	
Length (mm)	300
Diameter (mm)	4
Material	SS
Solvent	S1
Flow rate	0.8 ml/min
Temperature	rt
Detection	UV

Amine	t_r
Ammonia	1.00
Ethylenediamine	1.82
Diethylamine	2.02
1,4-Butanediamine	2.07
Piperazine	2.11
1,5-Pentanediamine	2.43
Benzylamine	2.68
Diethylenetriamine	2.73
Aniline	3.02
Monoethanolamine	3.70
1,7-Heptanediamine	4.14
p-Phenylenediamine	4.77
Triethylenetetramine	4.89
1,8-Octanediamine	5.89

Packing P1 = μBondapak C_{18}, particle size 10 μm
Solvent S1 = Water-acetonitrile, 2:1.5

REFERENCE

1. Chen, C. M. and Farquharson, R. A., *J. Chromatogr.*, 178, 358, 1979.

Reproduced by permission of the Elsevier Scientific Publishing Company.

Table LC 43
FLUORESCAMINE DERIVATIVES OF AMINES

Packing	P1
Column	
Length (cm)	15
Diameter (cm I.D.)	0.4
Material	SS
Solvent	S1
Flow rate (ml/min)	1.0
Temperature	30 ± 0.5°C
Detection	Fluorescence

Compound	r^a
Guanidine	Not detectable

Table LC 43 (continued)
FLUORESCAMINE DERIVATIVES OF AMINES

Agmatine	Not detectable		
Histidine	0.05		
Aspartic acid	0.07		
Glutamic acid	0.07		
Threonine	0.07		
Tyrosine	0.07		
Citrulline	0.07		
Glutathione	0.07		
Carnosine	0.07		
Glycine	0.08		
Valine	0.08		
Histamine	0.08		
Urea	0.08		
DOPA	0.09		
Serotonin	0.09		
Noradrenalin	0.25		
Alanine	0.33	0.10^b	
Dopamine	0.35		
Tyramine	0.47		
Methionine	0.54	0.09^b	
Tryptophan	0.64		
Spermidine	0.71		
Ornithine	0.76	0.06^b	
Spermine	0.77		
1,3-Diaminopropane	0.80	0.59^b	
Leucine	0.81		
Angiotensin II	0.82	0.54^b	
Lysine	0.82	0.67	0.08^b
Isoleucine	0.82	0.35	0.10^b
Putrescine	0.83		
Phenylalanine	0.88	0.37^b	
Cadaverine	0.89		
1,6-Hexanediamine	1.00		

^{a} Relative to retention time of fluorescamine derivative of 1,6-hexanediamine taken as 1.00.

^{b} Two or three peaks were observed on the chromatograms.

Packing P1 = LiChrosorb® RP-18, 5 μm.

Solvent S1 = Methanol linear gradient elution. Two solutions are required. A salt solution is first prepared by dissolving 13.37 g of NH_4Cl, 15.76 g of sodium benzene sulfonate and 1 g of acetic acid in about 400 ml of water. The pH is brought to 4.0 with 1 M NaOH and the solution is diluted to 500 ml. Solution A is prepared by mixing 100 ml of salt solution, 175 ml of water and 225 ml of methanol; solution B by mixing 100 ml of salt solution with 400 ml of methanol. Both solutions are degassed before use. Linear gradient elution with a methanol concentration between 45 and 80% is performed.

REFERENCE

1. Kai, M., Ogata, T., Haraguchi, K., and Ohkura, Y., J. Chromatogr., 163, 151, 1979.

Reproduced by permission of the Elsevier Scientific Publishing Copany.

Table LC 44
BIOGENIC AMINES AS
o-PHTHALALDEHYDE DERIVATIVES

Packing	P1
Column	
Length (mm)	300
Diameter (mm)	4
Material	—
Solvent	S1
Flow rate, (mℓ/min)	1.5
Temperature (°C)	30
Detection	D1

Amine	t_r (min)
Histamine	13.21
Norepinephrine	30.41
Normetanephrine	51.76
Dopamine	69.33
Serotonin	77.46
Tyramine	82.97
Octopamine[a]	43.13

[a] Internal standard.

Packing P1 = μBondapak phenyl
Solvent   S1 = NaH_2PO_4, 50 mmole/ℓ, pH 5.10 containing 32% methanol for the first elution step (50 min) and 45% methanol for the second elution step
Detection D1 = Fluorometry

REFERENCE

1. Davis, T. P., Gehrke, C. W., Gehrke, C. W., Jr., Cunningham, T. D., Kuo, K. C., Gerhardt, K. O., Johnson, H. D., and Williams C. H., *J. Chromatogr.*, 162, 293, 1979.

Reproduced by permission of the Elsevier Scientific Publishing Company.

Section I.III.

PAPER CHROMATOGRAPHY TABLES*

Wherever possible, tables are arranged according to classes of chemical compounds. This was not always possible when different types of compound were chromatographed under the same experimental conditions. The reader is referred to the compound index for specific compounds which may appear in different tables.

---

* Tables PC 2, 3, 5, and 6 were prepared by Z. Deyl.

Table PC 1
AMINO ACIDS

Paper	P1	P1	P1	P1	P1	P1
Solvent	S1	S2	S3	S4	S5	S6
Technique	T1	T1	T1	T1	T1	T1
Detection	D1	D1	D1	D1	D1	D1

Amino acid			$R_F \times 100$			
Alanine	27	25	27	52	48	45
Aspartic acid	10	9	11	31	26	26
Glutamic acid	15	11	12	44	29	28
Glycine	22	24	20	43	40	41
Histidine	16	17	18	40	36	33
Hydroxyproline	27	23	27	50	47	42
Leucine	53	48	50	76	69	67
Lysine	14	14	15	37	37	31
Proline	36	37	35	57	52	60
Tryptophan	48	41	47	66	64	57
Valine	40	36	38	64	61	65

Paper P1 = Whatman 3MM.
Solvent S1 = Acetonitrile - 0.1 M ammonium acetate (70:30) pH 4.0.
S2 = Acetonitrile - 0.1 M ammonium acetate (70:30) pH 7.2
S3 = Acetonitrile - 0.1 M ammonium acetate (70:30) pH 9.0.
S4 = Acetonitrile - 0.1 M ammonium acetate (60:40) pH 4.0.
S5 = Acetonitrile - 0.1 M ammonium acetate (60:40) pH 7.2.
S6 = Acetonitrile - 0.1 M ammonium acetate (60:40) pH 9.2. The buffer pH was adjusted by the addition of acetic acid or concentrated ammonium hydroxide.
Technique T1 = Ascending.
Detection D1 = Spraying with 0.2% ninhydrin in 95 mℓ of n-butanol and 5 mℓ of 10% aqueous acetic acid and heating at 110°C.

REFERENCE

1. Heimer, E. P., *J. Chem. Education*, 49, 547, 1972.

Table PC 2
ION EXCHANGE PC OF RACEMIC DL-AMINO ACIDS

Paper	P1
Solvent	S1
Detection	D1
Literature	1

Compound	$R_F \times 100$
L-Arg	0
L-Lys	0
DL-Asp	0.52 (D-)
	0.78 (L-)
L-Cys	87
DL-Met	30 (L-)
	75 (D-)
DL-Ser	0 (L-)
	55 (D-)
L-Ser	0
L-Tyr	7 (D-)
DL-Ala	58 (L-)
DL-Leu	0 (D-)

Table PC 2 (continued)
ION EXCHANGE PC OF RACEMIC DL-AMINO ACIDS

DL-Ile	95 (L-)
	0 (D-)
DL-Phe	70 (L-)
	0 (D-)
	85 (L-)
Gly	0

Paper P1 = Alginate-silica gel paper.
Solvent S1 = Pyridine-water-amylalcohol-$Na_2HPO_4 \cdot 7 H_2O$ (0.5% aqueous solution) (7:7:2:6).
Detection D1 = Ninhydrin.

REFERENCE

1. El Din Awad, A. M. and El Din Awad, O. M., *J. Chromatogr.*, 93, 393, 1974.

Table PC 3
METHYLATED AMINO ACIDS

Paper	L1	L1
Solvent	S1	S2
Detection	D1	D1
Literature	1	1
Compound	$R_F \times 100$	
N^E-Trimethyllysine	14	18
N^G-Monomethylarginine	27	26
$N^G N^G$-Dimethylarginine	32	28
$N^G N'^G$-Dimethylarginine	38	33

Paper   P1 = Whatman 1 or equivalent.
Solvent S1 = Pyridine-acetone-3 M ammonia (10:6:5).
        S2 = 1-butanol-acetic acid-water (4:1:12).
Detection D1 = Ninhydrin.

REFERENCE

1. Tomita, T. and Nakamura, K., *Hoppe-Seyler's Z. Physiol. Chem.*, 358, 413, 1977.

Reproduced by permission of Verlag Walter de Gruyter & Co.

Table PC 4
SULFUR-CONTAINING AMINO ACIDS

Paper	P1
Solvent	S1
Technique	T1
Detection	D1

Amino acid	$R_{cystathionine}$
L-cystathionine sulfone	0.0-0.35
L-cystathionine sulfoxide	0.39
L-cysteine	3.36
L-cystine	0.14-0.86
L-homocysteine	4.43
L-homocysteine thiolactone	4.21
L-homocystine	2.28
L-homolanthionine	1.90
L-homoserine	3.00
L-methionine	4.10

Paper   P1 = Whatman No. 1.
Solvent   S1 = 2-propanol — 88% formic acid — 10 mM 2-mercaptoethanol (7:1:2).
Technique   T1 = Descending.
Detection   D1 = Ninhydrin.

REFERENCE

1. Datko, A. H., Mudd, S. H., and Giovanelli, J., *Anal. Biochem.*, 62, 531, 1974.

Table PC 5
SELENOTAURINES

Paper	P1	P1	P1
Solvent	S1	S2	S3
Detection	D1	D1	D1
Literature	1	1	1

Compound	$R_F \times 100$		
Selenohypotaurine	78	23	14
Seleno homohypotaurine	91	20	27
Selenotaurine	52	11	21
Selenohomotaurine	52	11	12
Hypotaurine	68	20	19
Homohypotaurine	75	22	15
Taurine	49	16	30
Homotaurine	55	16	26
Thiotaurine	49	25	57
Homothiotaurine	55	22	49

Paper   P1 = Whatman 1.
Solvent   S1 = Water saturated phenol in the presence of ammonia vapor.
           S2 = *n*-butanol-acetic acid-water (4:1:5), upper phase.
           S3 = 2,4,6-trimethylpyridine-2,6-dimethyl pyridine 1:1, water saturated.
Detection   D1 = Iodoplatinate.

REFERENCE

1. De Marco, C., Cossu, P., Dernini, S., and Rinaldi, A., *J. Chromatogr.*, 129, 369, 1976.

Reproduced by permission of The Elsevier Scientific Publishing Company.

Table PC 6
AROMATIC AMINES, NITRO DERIVATIVES

Paper	P1
Solvent	S1
Technique	T1
Detection	D1
Literature	1

Compound	$R_F \times 100$
2,4-Dinitroaniline	0
N-Methyl-2,4-dinitroaniline	4
N-Ethyl-2,4-dinitroaniline	20.8
N-n-Propyl-2,4-dinitroaniline	50.0
N-n-Butyl-2,4-dinitroaniline	60.0
N-Isopropyl-2,4-dinitroaniline	49.5
N-Isobutyl-2,4-dinitroaniline	60.5
N-Cyclohexyl-2,4-dinitroaniline	64.0
N-Allyl-2,4-dinitroaniline	24.0
N-Phenyl-2,4-dinitroaniline	52.4
N,N-Dimethyl-2,4-dinitroaniline	15.0
N,N-Diethyl-2,4-dinitroaniline	64.0
N,N-Di-n-propyl-2,4-dinitroaniline	81.0
N,N-Di-isopropyl-2,4-dinitroaniline	7.0
N-(2,4-Dinitrophenyl)piperidine	64.0
2,6-Dinitroaniline	14.9
N-Methyl-2,6-dinitroaniline	32.1
N-Ethyl-2,6-dinitroaniline	68.2
N-n-Propyl-2,6-dinitroaniline	80.3
N-n-Butyl-2,6-dinitroaniline	87.3
N-Cyclohexyl-2,6-dinitroaniline	87.5
N-Allyl-2,6-dinitroaniline	75.0
N,N-Dimethyl-2,6-dinitroaniline	74.4
N,N-Diethyl-2,6-dinitroaniline	83.6
N,N-Di-n-propyl-2,6-dinitroaniline	33.5
N-(2,6-Dinitrophenyl)piperidine	87.8
2,4,6-Trinitroaniline	0
N-Methyl-2,4,6-trinitroaniline	4.0
N-Ethyl-2,4,6-trinitroaniline	25.5
N-n-Propyl-2,4,6-trinitroaniline	54.0
N-n-Butyl-2,4,6-trinitroaniline	72.0
N-Cyclohexyl-2,4,6-trinitroaniline	87.0
N-Allyl-2,4,6-trinitroaniline	33.0
N,N-Dimethyl-2,4,6-trinitroaniline	29.0
N,N-Diethyl-2,4,6-trinitroaniline	78.0
N,N-Di-n-propyl-2,4,6-trinitroaniline	91.0
N,N-Di-n-butyl-2,4,6-trinitroaniline	92.0
N-(2,4,6-Trinitrophenyl)piperidine	81.2

Paper   P1 = Whatman No. 2 impregnated with formamide-methanol (5:2).
Technique   T1 = Descending, 30 min at 21 ± 0.3°C.
Detection   D1 = UV light.

REFERENCE

1. Zemanova, E. and Zeman, S., *J. Chromatogr.*, 154, 33, 1978.

Section I.IV.

THIN-LAYER CHROMATOGRAPHY TABLES*

Wherever possible, tables are arranged according to classes of chemical compounds. This was not always possible when different types of compound were chromatographed under the same experimental conditions. The reader is referred to the compound index for specific compounds which may appear in different tables.

---

* Tables TLC 1-32 were prepared by Z. Deyl.

Table TLC 1
AMINO ACIDS, HYDROXYPROLINE, AND RELATED COMPOUNDS

Layer	L1	L1	L1	L1	L1
Solvent	S1	S2	S3	S4	S5
Detection	D1	D1	D1	D1	D1
Literature	1	1	1	1	1
Compound			$R_F \times 100$		
L-Hydroxyproline	54	42	58	37	20
D-allo-Hydroxyproline	45	34	51	31	14
4-Hydroxy-1-pyrrolin-2-carboxylic acid	54			48	
Pyrrol-2-carboxylic acid	78	64	73	58	87
2-Oxyglutaric acid	70	58	70	18	
L-Glutamine	64	45	61	15	17
Oxoproline	58	34	50	10	18
L-Proline	55	39	54	46	20

Layer	L1	=	Silica gel
Solvent	S1	=	Methanol-water (7:3)
	S2	=	Ethanol-water (7:3)
	S3	=	Ethanol-water (55:45)
	S4	=	n-propanol-33% ammonia (67:33)
	S5	=	n-butanol-acetic acid-water (4:1:1)
Detection	D1	=	Ninhydrin

REFERENCE

1. Drawert, F. and Barton, H., *Hoppe Seyler's Z. Physiol. Chem.*, 355, 902, 1974.

Reproduced by permission of Verlag Walter de Gruyter & Co.

Table TLC 2
CYSTINE AND GLYCINE DERIVATIVES

Layer	L1	L1	
Solvent	S1	S2	
Detection	D1	D1	
Literature	1	1	
Compound	$R_F \times 100$		Color with ninhydrin (collidine)
S-Carboxymethylcysteine	46	11	Blue
N-Gly-S-carboxymethyl-cysteine	40	11	Yellow
Glycine	36	18	Purple
S-CarboxymethylCysGly	44	13	Olive-gray
Cystine	19	15	Blue-purple
Monoglycylcystine	22	21	Gray
Cystinylglycine	24	18	Purple
N,N'-Diglycylcystine	24	25	Yellow
Cystinyl-bis-diglycine	23	23	Faint gray

Layer	L1	=	Polygram Cel 300 (Brinkmann, Westbury, N.Y.)
Solvent	S1	=	1-butanol-acetic acid-water (4:1:2)
	S2	=	Pyridine-acetone-3 N NH$_4$OH (50:30:25)
Detection	D1	=	Ninhydrin (collidine)

REFERENCE

1. Armstrong, M. D., *J. Chromatogr.*, 175, 216, 1979.

Reproduced by permission of the Elsevier Scientific Publishing Company.

Table TLC 3
AMINO ACIDS AND THEIR METHYLATED DERIVATIVES

Layer	L1	L1	L1
Solvent	S1	S2	S3
Detection	D1	D1	D1
Literature	1	1	1
Compound		$R_F \times 100$	
Asp	80	79	82
Thr	79	78	80
Ser	80	80	81
Glu	80	79	80
Gly	78	64	79
Ala	72	60	72
Pro	51	49	51
Val	65	61	65
Met	58	50	57
Ile	52	49	52
Leu	53	50	53
Trp	51	41	50
Phe	54	50	55
Try	10	—	11
Asn	69	—	70
Gln	61	—	60
Cys	77	—	79
Cys$_2$	60	—	62
HyPro	75	—	78
His	47	16	48
1-MeHis	37	15	36
3-MeHis	26	12	24
Lys	59	25	58
MML[a]	38	19	39
DML	23	16	21
TML	14	12	13
Arg	29	8	28
MMA	21	8	20
DMA	14	9	13
DMA	15	8	16

[a] MML = N$^\epsilon$-monomethyl-D,L-lysine hydrochloride; DML = N$^\epsilon$, N$^\epsilon$-D,L-lysine hydrochloride; TML = N$^\epsilon$,N$^\epsilon$,N$^\epsilon$-trimethyl-D,L-lysine dihydrochloride. The guanidino-methylated arginines are: MMA = N^G-monomethyl-L-arginine; DMA = N^G,N^G-dimethyl-L-arginine; DMA = N^G,N$'^G$-dimethyl-L-arginine.

| Layer | L1 | = | Fixion 50-X8 plates. |

Solvent	Component	Buffer solution		
		S1	S2	S3
	Hydrated citric acid (g)	100.0	24.6	105.0
	Hydrochloric acid (mℓ)	14.0	6.5	—
	Sodium hydroxide (g)	60.0	14.0	60.0
	Sodium chloride (g)	—	—	58.5
	Sodium ions (N)	1.5	0.35	2.5

| Detection | D1 | = | Ninhydrin. |

REFERENCE

1. Tyihak, E., Ferenczi, S., Jazai, I., Zoltán, S., and Patthy, A., *J. Chromatogr.*, 102, 257, 1974.

Reproduced by permission of the Elsevier Scientific Publishing Company.

Table TLC 4
DIAMINODICARBOXYLIC ACIDS IN FREE FORM AND AS DANSYL DERIVATIVES

Layer	L1	L2	L2
Solvent	S1	S2	S3
Detection	D1	D2	D2
Literature	1	1	1
Compound	$R_F \times 100$		
Meso derivatives:			
Diaminosuccinic	5	0	8
Diaminoadipic	64	4	13
Diaminopimelic	25	6	26
D,D + L,L-derivatives			
Diaminosuccinic	16	1	12
Diaminoadipic	23 (28)	22	38
Diaminopimelic	25 (37)	21	79
Meso derivatives			
Diaminosuberic	—	8	37
Diaminoazelaic	—	11	46
Diaminosebacic	—	13	54
D,D + L,L-derivatives			
Diaminosuberic	—	28	71
Diaminoazelaic	—	32	75
Diaminosebacic	—	34	81

Layer L1 = Cellulose, compounds separated as didansyl derivatives
       L2 = Silica gel G thin layers (Merck), compounds separated in free form
Solvent S1 = Methanol-water-acetic acid (40:20:2)
        S2 = Benzene-pyridine-acetic acid (40:10:1)
        S3 = Benzyl-alcohol-chloroform-ethyl acetate-acetic acid (8:12:10:1)
Detection D1 = Ninhydrin
          D2 = Spots visualized by UV fluorescence of the derivative

REFERENCE

1. Chimiak, A. and Polonski, T., *J. Chromatogr.*, 115, 635, 1975.

Table TLC 5
TRYPTOPHAN DERIVATIVES, METABOLITES, AND RELATED PEPTIDES

Layer	L1	L1	L1	L1	L1	L1
Solvent	S1	S2	S3	S4	S5	S6
Detection	D1	D1	D1	D1	D1	D1
Literature	1	1	1	1	1	1
Compound	$R_F \times 100$					
3-Methylindole	96	96	90	97	93	91
Indole-3-acetic acid	91	94	89	92		28
5-Hydroxyindole-3-acetic acid DCA salt	89					
3-Indoleacetone	95	95	90	92	69	90
5-Methoxy-3-indole-acetic acid	90	67	87	94	88	28
DL-3-Indoleacetic acid	47	67	44	92(26)	2	24
3-Indolepyruvic acid	36	94	88	92	0	21
3-Indoleacetonitrile	95	95	89	95	67	90

Table TLC 5 (continued)
TRYPTOPHAN DERIVATIVES, METABOLITES, AND RELATED PEPTIDES

Indomethacin	91	88	88	93	79	83
Indoxyl-β-D-glucoside	88	71	85	45	50	20
Indoxyl sulfate	79	65	84	20	18	40
L-Tryptophan	45	61	34	5	1	22
Tryptamine-HCl	13	69	8	15	2	68
5-Hydroxy-DL-tryptophan	32					
5-Methoxy-DL-tryptophan	21	58	27	2	0	21
5-Hydroxytryptamine oxalate salt	38					
5-Methoxytryptamine-HCl	14(38)	70	26(7)	17	4,8	70(65)
N-Acetyl-L-tryptophan	56	74	52	57	7	27
N-Methyltryptamine	6	59,68	4	15	2	67
N,N-Dimethyltryptamine	7	57	4	12	8(86)	90
N-Acetyl-5-methoxy-tryptamine	93	86	86	95	74	88
N-Acetyl-5-hydroxy-tryptamine	92					
Tryptophol	95	93	90	95	85	88
5-Methoxytryptophol	95	89	89	95		88
6-Hydroxymelatonin	73					
5-Hydroxytryptophol	95	90	88	81		
N-Carbobenzoxy-L-tryptophan	79		80	92	44	40
Gly-L-Trp	18	59	9	1	0	17
L-Pro-L-Trp	10	59	5	2	1	17
L-Leu-L-Trp	49	73	33	13	2	33
L-Phe-L-Trp-acetate	52	72	41	15	4	31
L-Val-L-Trp	40	71	23	11(8)	0	29
Gly-L-Trp-Gly	11	53	4	1	0	14
L-Lys-L-Trp-L-Lys	0	23	0	0	0	2
L-Leu-L-Trp-L-Leu	63	85	59(46)	27(33)	5	55,62
Gly-Gly-L-Trp	10	53	3	1	0	12
Kynurenic acid	56	62	51	11	3	37
DL-Kynurenine sulfate	41	57	23	3	2	0,21
3-Hydroxyanthranilic acid	10	41	4	0(2)	0	0

Layer	L1	=	Silica gel 60.
Solvent	S1	=	Chloroform-isopropanol-water (2:7:1).
	S2	=	n-butanol-acetic acid-water (5:2:3).
	S3	=	Ethyl acetate-isopropanol-water (4:5:1).
	S4	=	Chloroform-ethanol-acetic acid (13:6:1).
	S5	=	Benzene-dioxane-methanol (5:3:2).
	S6	=	n-hexane-n-propanol-ammonia (2:7:1).
Detection	D1	=	Spray with 70% perchloric acid, 5 sec.; plates developed with solvent S6 were sprayed with the reagent for 10 sec.

REFERENCE

1. Nakamura, H. and Pisano, J. J., J. Chromatogr., 152, 167, 1978.

Reproduced by permission of the Elsevier Scientific Publishing Company.

Table TLC 6
TRYPTOPHAN DERIVATIVES AND RELATED PEPTIDES

Layer	L1	L1	L1	L1
Solvent	S1	S2	S3	S4
Detection	D1	D1	D1	D1
Literature	1	1	1	1
Compound		$R_F \times 100$		
L-Tryptophan	56	38	46	33
DL-Tryptophanamide·HCl	60	42	30	24
L-Tryptophan hydroxamate	55	33	5	31
DL-Tryptophan methyl ester·HCl	57	38	46	32
DL-Tryptophan ethyl ester·HCl	57	38	46	33
DL-Tryptophan butyl ester·HCl	57	38	45	33
DL-Tryptophan octyl ester·HCl	57	38	45	31
L-Tryptophan benzyl ester·HCl	57	38	46	33
4-Methyl-DL-tryptophan	60	42	49	35
5-Methyl-DL-tryptophan	60	43	49	36
6-Methyl-DL-tryptophan	60	42	48	35
7-Methyl-DL-tryptophan	59	41	47	37
DL-5-Benzyloxytryptophan	65	48	53	44
DL-6-Benzyloxytryptophan	64,70[a]	48,46,40[a]	52	44,49,0[a]
DL-α-Methyltryptophan	61	44	48	40
DL-4-Fluorotryptophan	58	39	45	35
DL-5-Fluorotryptophan	59	41	46	38
DL-6-Fluorotryptophan	60	42	45	36
L-Trp-L-Ala	65	44	43	23
L-Trp-β-Ala	62	30	31	13
L-Trp-Gly	54	32	35,41[a]	17
L-Trp-L-Glu	51,55		26,22[a]	7,0
L-Trp-L-Ile	76	69	56,60[a]	45,52[a]
L-Trp-L-Leu	76	66	55,60[a]	40,47[a]
L-Trp-α-L-Lys	46,47,51	12,15,7	2	0
L-Trp-L-Phe	72	59	55	42
L-Trp-L-Trp	74	62	54	42
L-Trp-L-Tyr	69	56	52	43
L-Trp-L-Val	73,69	62	49	36,42
L-Trp-Gly-Gly	49	22	25	8
L-Trp-L-Met-L-Phe·NH$_2$·HCl	67,64[a],54[a]	47	50,30,34[a]	40,33[a],25[a],16[a],8[a]
Tryptamine·HCl	67	52	9	8
5-Methyltryptamine·HCl	69	55	10	8
7-Methyltryptamine	68	54	10	9
5-Benzyloxytryptamine·HCl	74	60	12	10
5-Fluorotryptamine·HCl	69	55	9	9
6-Fluorotryptamine·HCl	70	56	20	12
6-Hydroxytryptamine creatinine sulfate complex			53	0

[a] Trace.

Layer   L1 = Silica gel 60 plates.
Solvent   S1 = n-butanol-acetic acid-water (5:2:3).
          S2 = n-butanol-acetic acid-water (4:1:5) upper phase.
          S3 = Chloroform-isopropanol-water (2:7:1).
          S4 = Ethyl acetate-n-propanol-water w (5:4:1).
Detection D1 = Fluorescamine dip reagent after separation of free compounds.

REFERENCE

1. Nakamura, H. and Pisano, J. J., *J. Chromatogr.*, 152, 153, 1978.

Reproduced by permission of the Elsevier Scientific Publishing Company.

Table TLC 7
TRYPTOPHAN DERIVATIVES AND RELATED PEPTIDES (AS FLUORESCAMINE DERIVATIVES)

Layer	L1	L1	L2	L1
Solvent	S1	S2	S3	S4
Detection	D1	D1	D1	D1
Literature	1	1	1	1
Compound		$R_F \times 100^a$		
L-Tryptophan	33(83)	39(78)	66	56(68)
DL-Tryptophanamide·HCl	(87)	76(94)	83	68(88)
L-Tryptophan hydroxamate	(85)	40	65	(71)
DL-Tryptophan methyl ester·HCl	32(84)	40(77)	66	53(68)
DL-Tryptophan ethyl ester.HCl	31(83)	40(77)	66	54(68)
DL-Tryptophan butyl ester.HCl	33(84)	40(77)	66	53(68)
DL-Tryptophan octyl ester.HCl	32	40(78)	65	53(68)
L-Tryptophan benzyl ester.HCl	32(84)	42(79)	63	52(70)
4-Methyl-DL-tryptophan	34(85)	40(78,82)	63	56(70)
5-Methyl-DL-tryptophan	34(85)	40(79)	64	55(70)
6-Methyl-DL-tryptophan	35(85)	40(78)	64	57(70)
7-Methyl-DL-tryptophan	34(85)	40(78)	63	56(70)
DL-5-Benzyloxy-tryptophan	51(86)	40(80)	67	66(71)
DL-6-Benzyloxytryptophan	51(87)	39(79)	68	65(71)
DL-7-Benzyloxytryptophan			66	66(71)
DL-α-Methyl-tryptophan	30(76)	39(82)	64	52
DL-4-Fluorotryptophan	30(85)	39(79)	62	53(70)
DL-5-Fluorotryptophan	34(84)	39(77)	63	55(70)
DL-6-Fluorotryptophan	34(84)	39(77)	63	57(70)
L-Trp-L-Ala	35(81)	36(74)	70	37(68)
L-Trp-β-Ala	29(85)	38(83)	70	43(70)
L-Trp-Gly	31(85)	26(60)	69	31(66)
L-Trp-L-Glu	6(66)	20(38)	56	26
L-Trp-L-Ile	59(88)	73(84)	75	65(70)
L-Trp-L-Leu	(88)	75(86)	75	62(70)
L-Trp-α-L-Lys	13,30(80)	12,38	64,67	30,48(65)
L-Trp-L-Phe	(85)	39(80)	73	61(69)
L-Trp-L-Tyr	(86)	39(82)	70	57(68)
L-Trp-L-Val	(86)	40(82)	72	55(69)
L-Trp-Gly-Gly	32(72)	22(40)	65	31
L-Trp-L-Met-L-Asp-L-Phe.NH$_2$.HCl	18(81)	39,43(72)	66,77	31,53(68)
Tryptamine.HCl	36(72,79,83)	86(94)	87	73(94)
5-Methyltryptamine.HCl	87	87(95)	88	74(94)
7-Methyltryptamine	88	86(94)	87	73(94)
5-Benzyloxytryptamine.HCl	87	87(94)	89	75(94)
5-Fluorotryptamine.HCl	89	85(94)	85	74(93)
6-Fluorotryptamine.HCl	88	86(94)	86	74(93)
6-Hydroxytryptamine creatinine sulfate complex	19	39(83)	62	30

^{a} The $R_F \times 100$ values of spots which could not be detected under longwave UV light and were detected by the successive 40% PCA spray are shown in parentheses.

Layer	L1	= Silica gel 60.
	L2	= Silica gel Q5 (Whatman).
Solvent	S1	= Dioxane-triethanolamine-methanol (6:1:1).
	S2	= Chloroform-isopropanol-water (2:8:1).
	S3	= Ethanol-chloroform-28% ammonia-water (5:2:1:1).
	S4	= Ethylacetate-n-hexane-methanol-water (60:20:25:10).
Detection	D1	= Fluorescence (365 nm).

REFERENCE

1. Nakamura, H. and Pisano, J. J., *J. Chromatogr.*, 152, 153, 1978.

Reproduced by permission of the Elsevier Scientific Publishing Company.

Table TLC 8
AMINO ACID DERIVATIVES (INCLUDING SOME PEPTIDES)

Layer   L1
Solvent   S1
Detection   D1
Technique   T1
Literature   1

Compound	$R_F \times 100$	Color	Limit of detection (nmole)
N-DNP-L-Alanine	82	Green	0.2
N,N-di-DNP-L-Lysine	84	Green	0.1
N-DNP-L-Phenylalanine	85	Green	0.2
ONPS-L-Alanine.DCA salt	86	Reddish brown	0.2
ONPS-Glycine.DCA salt	83	Reddish brown	0.2
ONPS-L-Proline.DCA salt	89	Reddish brown	0.2
N-Pht-Glycine	76	Grayish green	0.7
N-Pht-L-Leucine	85	Grayish green	0.7
N-Pht-L-Phenylalanine	84	Grayish green	0.7
Tosyl-L-arginine	81	Pink	0.8
Tosyl-DL-arginine methyl ester.HCl	74	Pink	0.8
PTC-L-Alanine	92	Red	0.08
PTC-L-Ala-L-His	67	Red	0.08
PTC-Gly-L-His-Gly	64	Red	0.08
α-N-Benzoyl-DL-alanine	83	Red	0.6
N-Benzoyl-DL-phenylalanine	89	Red	0.6
Dns-Glycine	79	Yellowish green	0.2
Dns-L-β-Phenylalanine	86	Yellowish green	0.2
Dns-L-Valine	88	Yellowish green	0.2
FLA-L-Alanine	81	Bluish green	0.2
FLA-L-Ala-L-His	67	Bluish green	0.2
FLA-Gly-L-His-Gly	66	Bluish green	0.2
BOC-L-Valine	—	(Grayish red)	>10
BOC-L-Glutamine	—	(Grayish red)	>10
BOC-L-Glutamine-γ-benzyl ester.DCS salt	—	(Grayish red)	>10
CBZ-L-Valine	—	(Faint red)	>10
CBZ-Glycine	—	(Faint red)	>10
CBZ-L-Phenylalanine	—	(Faint red)	>10
CBZ-L-Arginine.HCl	—	(Faint red)	>10
CBZ-Gly-Gly-L-Leu	—	(Faint red)	>10
CBZ-Nitro-L-arginine	77	Reddish brown	0.3
Glycine-p-nitroanilide	78	Bluish green	0.2
L-Leucine-p-nitroanilide.HCL	85	Bluish green	0.2
L-Leucine-β-naphthylamide.HCL	87	Red	0.3
L-Valine-β-naphthylamide	85	Red	0.3
L-Glutamic acid dibenzyl ester.HCL	82	Red	8
CBZ-L-Glutamic-α-benzyl ester	95	Red	8
CBZ-L-Alanine-p-nitrophenyl ester	90	Green	0.2
CBZ-L-Proline-p-nitrophenyl ester	91	Green	0.2
BOC-L-Glutamine-p-nitrophenyl ester	90	Green	0.2
BOC-L-Valine TCP ester	95	Light purple	8
BOC-Glycine TCP ester	95	Light purple	8
L-Pyroglumatic acid	—	(Faint red)	>10
Histidine thiohydantoin	68	Red	0.7
Tryptophan thiohydantoin	91	Red	0.7
Benzoyl-DL-arginine-p-nitroanilide.HCl	83	Bluish green	0.05
α-CBZ-L-lysine benzyl ester-p-tosylate	69	Red	7
PTH-DL-Alanine	88	Red	0.08
PTH-DL-Phenylalanine	90	Red	0.08
PTH-L-Arginine	77	Red	0.08
MTH-Glycine	83	Red	0.1

Table TLC 8 (continued)
AMINO ACID DERIVATIVES (INCLUDING SOME PEPTIDES)

MTH-DL-Asparagine	77	Red	0.1
MTH-DL-Histidine	68	Red	0.1
N-Acetyl-L-(+)-tryptophan	79	Red	0.5
N-Acetyl-L-(+)-phenylalanine	—	None	>10
N-Acetyl-L-(−)-leucine	—	None	>10
L-Glutamic-γ-hydrazide	—	None	>10
N-Methyl-DL-alanine	—	None	>10
N-Methyl-DL-leucine	—	None	>10
N-Methyl-DL-valine	—	None	>10
N,N-Dimethylglycine.HCl	—	None	>10
L-Tryptophan	72	Red	0.3
L-Tryptophyl-L-phenylalanine	79	Red	0.3
L-Valyl-L-tryptophan	74	Red	0.3
L-Tyrosine	63	Red	1
L-Phenylalanine	69	Red	2
L-Cystine	43	Red	2

Note: Colors in parentheses were observed only before development. Abbreviations: DNP = 2,4-dinitrophenyl; ONPS = o-nitrophenylsulfenyl; Pht = phthalyl; tosyl = p-toluenesulfonyl; PTC = phenylthiocarbamyl; PTH = phenylthiohydantoin; Dns = 1-dimethylamino-naphthalene-5-sulfonyl; FLA = fluorescamine; BOC = tert.-butyloxycarbonyl; CBZ = N-benzyloxycarbonyl; DCA = dicyclohexylamine; TCP = 2,4,5-trichlorophenyl; MTH = methylthiohydantoin.

Layer   L1 = Wakogel FM plates (silica gel)
Solvent   S1 = n-butanol-acetic acid-water (5:2:3)
Detection   D1 = UV light
Technique   T1 = 10 cm distance of development

REFERENCE

1. Nakamura, H., Pisano, J. J., and Tamura, Z., *J. Chromatogr.,* 175, 153, 1979.

Reproduced by permission of the Elsevier Scientific Publishing Company.

Table TLC 9
AMINO ACID DERIVATIVES: 2-p-ISOTHIOCYANOPHENYL-3-PHENYLINDENONE (DIITC) AND MONO- AND BIS(DIPHENYL)INDENONYLTHIOUREAS (MITU AND BITU)

Layer	L1	L1	L1
Solvent	S1	S2	S3
Detection	D1	D1	D1
Literature	1	1	1
Compound		$R_F \times 100$	
Alanine	61 ± 2	37 ± 2	—
Arginine	3 ± 1	88 ± 2	—
Asparagine	18 ± 2	48 ± 2	—
Aspartic acid	12 ± 1	32 ± 2	—
Cysteic acid	0	6 ± 2	—
Glutamine	29 ± 3	50 ± 2	—
Glutamic acid	25 ± 3	35 ± 2	—
Glycine	52 ± 2	41 ± 2	—
Histidine	12 ± 2	90 ± 2	—
Hydroxyproline	35 ± 3	39 ± 2	—
Isoleucine	75 ± 2	26 ± 2	55 ± 1
Leucine	75 ± 2	24 ± 2	52 ± 1
Lysine	31 ± 1	1 ± 0	—
Methionine	65 ± 2	27 ± 3	63 ± 1
Methionine sulphone	23 ± 2	50 ± 2	—
Phenylalanine	67 ± 3	27 ± 2	71 ± 1
Proline	84 ± 3	32 ± 3	—
Serine	18 ± 3	44 ± 2	—
Threonine	28 ± 2	44 ± 2	—
Tryptophan	29 ± 3	17 ± 2	—
Tyrosine	11 ± 2	23 ± 2	—
Valine	72 ± 2	30 ± 2	—
DIITC	93 ± 3	14 ± 1	—
MITU	44 ± 3	14 ± 1	—
BITU	39 ± 1	0	—

Layer	L1	=	Polyamide sheets, 5 × 5 cm.
Solvent	S1	=	Toluene-n-pentane-glacial acetic acid (60:30:15).
	S2	=	60% aqueous acetic acid.
	S3	=	n-butanol-glacial acetic acid (9:1).
Detection	D1	=	The spots are naturally colored.

REFERENCES

1. Mancheva, I. N. and Vladovska Yulchovska, Y. B., *J. Chromatogr.*, 151, 207, 1978.

Reproduced by permission of the Elsevier Scientific Publishing Company.

Table TLC 10
AMINO ACIDS, 4-*N,N*-DIMETHYLAMINOAZOBENZENE-4'-THIOHYDANTOINS

Layer	L1	L1	L1
Solvent	S1	S2	S3
Detection	D1	D1	D1
Literature	1	1	1
Compound		$R_F \times 100$	
Proline	93	86	100
Isoleucine	74	59	98
Leucine	78	64	98
Valine	63	51	97
Phenylalanine	58	46	96
Methionine	53	42	96
Alanine	37	35	88
Glycine	25	18	79
Tryptophan	30	32	84
Tyrosine	13	21	68
Hydroxyproline	18	22	73
Asparagine	1	6	34
Glutamine	1	6	38
Serine	5	9	40
Threonine	10	16	60
Glutamic acid	0	1	13
Histidine	0	1	37
Aspartic acid	0	0	3
Arginine	0	0	0
DABTH-Ser	—	44	—
Carboxymethylcysteine	0	0	4

Layer	L1	=	Silica gel Merck G 60.
Solvent	S1	=	Chloroform-methanol (100:2).
	S2	=	Chloroform-ethanol (100:3).
	S3	=	Chloroform-methanol (9:1).
Detection	D1	=	Amino acid derviatives appear as red spots after exposure to HCl vapor.

REFERENCE

1. Chang, J. Y., Creaser, E. H. and Hughes, G. J., *J. Chromatogr.*, 140, 125, 1977.

Table TLC 11
2-THIOHYDANTOINS OF AMINO ACIDS

Layer	L1	L1
Solvent	S1	S2
Detection	D1	D1
Literature	1	1
Compound	$R_F \times 100$	
α-Aminobutyric acid	68	95
Ala	53	94
Arg	12	21
Asn	22	52
Asp	34	50
CM-cys	37	57
PE-cys	11	74

Table TLC 11 (continued)
2-THIOHYDANTOINS OF AMINO ACIDS

CH_3-cys	61	96
Gln	22	63
Glu	35	80
pGlu	35	96
Gly	37	89
His	10	23
Ile	73	97
Leu	78	96
n-Leu	79	97
Lys	32	84
Met	69	93
Phe	73	97
Thr	78	95
Trp	77	97
Tyr	67	96
Val	74	97

Layer   L1 = Silica gel plates.
Solvent S1 = Heptane-n-butanol-anhydrous formic acid (10:7:3).
             S2 = Chloroform-95% ethanol-acetic acid (100:50:15).
Detection D1 = Ninhydrin.

REFERENCE

1. Dwulet, F. E. and Gurd, F. R. N., *Anal. Biochem.*, 82, 385, 1977.

Reproduced by permission of Academic Press Inc.

Table TLC 12
AMINO ACID DERIVATIVES. IODOTYROSINES AND IODOHISTIDINES

Layer	L1
Solvent	S1
Detection	D1
Literature	1

Compound	$R_F \times 100$
Diiodohistidine	22
3,5-diiodotyrosine	28
3-monoiodotyrosine	46
Monoiodohistidine	59
Thyroxine	66
3,3',5-triiodothyronine	71
Iodide	95

Layer   L1 = Eastman chromagram silica gel sheets No. 13179 without fluorescent indicator.
Solvent S1 = Benzyl alcohol-acetone-1 NNH_4OH (1:4:1).
Detection D1 = Amino acids were visualized with ninhydrin, Pauly's or Folin's phenol reagent. Iodide was visualized with starch-iodate or ceric sulfate-arsenious acid-methylene blue reagent.

REFERENCE

1. Goldberg, W. M., *J. Chromatogr.*, 134, 246, 1977.

Table TLC 13
AMINO ACID DERIVATIVES. NITROSAMINO ACIDS

Layer	L1	L1	L1
Solvent	S1	S2	S3
Detection	D1	D1	D1
Literature	1	1	1

Compound	$R_F \times 100$		
N-Nitrososarcosine	64, 70	71	75
Sarcosine	23	10	2
N-Nitroso-L-proline	63, 69	72	21
L-Proline	33	12	3
N-Nitroso-4-hydroxy-L-proline	60, 66	67	13
4-Hydroxy-L-proline	28	9	1

Layer   L1 = Silica gel Eastman Chromagram sheets No. 13179.
Solvent   S1 = 95% ethanol-benzene-water (4:1:1).
        S2 = Methanol-chloroform (4:1).
        S3 = Acetonitrile-chloroform-95% ethanol-acetic acid (100:100:97:3).
Detection   D1 = Fluorescamine spray.

REFERENCE

1. Young, J. C., *J. Chromatogr.*, 151, 215, 1978.

Reproduced by permission of the Elsevier Scientific Publishing Company.

Table TLC 14
AMINO ACIDS, DANSYL DERIVATIVES

Layer	L1	L1	L1	L1	L1	L1	L1	L1	L1	L1
Solvent	S1	S2	S3	S4	S5	S6	S7	S8	S9	S10
Detection	D1	D1	D1	D1	D1	D1	D1	D1	D1	D1
Literature	1	1	1	1	1	1	1	1	1	1

Compound	$R_F \times 100$									
Ala	53	48	49	69	69	57	81	68	43	79
Arg	5	3	3	91	39	9	76	22	1	6
Asp	8	7	10	69	88	10	88	37	12	19
Cys	3	3	4	19	43	22	78	9	3	6
Glu	15	10	15	66	88	2	88	34	5	30
Gly	32	21	32	69	63	48	80	48	28	69
His	7	5	13	96	76	32	84	36	6	18
Ile	77	54	65	40	57	71	78	76	60	84
Leu	70	49	59	34	57	71	78	75	54	80
Lys(mono)	35	31	38	22	9	63	72	58	9	79
Lys(di)	53	37	48	78	69	35	82	40	39	76
Met	52	36	51	43	59	68	80	62	55	81
Phe	57	38	53	31	43	68	77	62	51	81
Pro	85	66	71	55	74	46	84	75	69	90
Ser	12	7	16	81	71	49	82	42	10	44
Thr	15	10	26	81	74	57	82	56	16	56
Tyr	63	47	61	0	0	84	73	65	58	91
Val	72	56	61	47	67	71	81	80	61	88
Dns-OH	0	1	0	51	54	16	74	0	4	4
Dns-NH$_2$	51	38	47	71	17	96	49	60	40	91

Table TLC 14 (continued)
AMINO ACIDS, DANSYL DERIVATIVES

Layer	L1	=	Polyamide.
Solvent	S1	=	Benzene-acetic acid (9:1).
	S2	=	Toluene-acetic acid (9:1).
	S3	=	Toluene-ethanol-acetic acid (17:1:2).
	S4	=	Water-formic acid (200:3).
	S5	=	Water-ethanol-ammonium hydroxide (17:2:1).
	S6	=	Ethyl acetate-ethanol-ammonium hydroxide (20:5:1).
	S7	=	Water-ethanol-ammonium hydroxide (14:15:1).
	S8	=	n-Heptane-n-butanol-acetic acid (3:3:1).
	S9	=	Chlorobenzene-acetic acid (9:1).
	S10	=	Ethyl acetate-methanol-acetic acid (20:1:1).
Detection	D1	=	Fluorescence.

REFERENCE

1. Metrione, R. M., *J. Chromatogr.*, 154, 247, 1978.

Reproduced by permission of Elsevier Scientific Publishing Company.

Table TLC 15
ALIPHATIC AMINES

	L1	L1
Layer	L1	L1
Solvent	S1	S2
Detection	D1	D1
Amine	$R_F \times 100$	
n-Butylamine	21	52
t-Butylamine	95	92
Dibutylamine	82	90
Methylamine	55	
Trimethylamine		68
Diethylene triamine	18	32
Ethylene diamine	25	40
Triethylene tetramine	12	21

Layer	L1	=	TLC plates, 0.6 mm thick, were prepared from 2 parts of silica gel and one part of calcium oxalate (mesh size 200) in 2 parts of water. The plates were activated at 60°C for 24 hr.
Solvent	S1	=	Butanol-ethanol-ammonia (80:20:20).
	S2	=	Ethanol-ammonia (80:20).
Detection	D1	=	After development the chromatoplates were sprayed with a solution of 50% CS_2 in methanol heated at 40°C for 5 min and then sprayed with 1% $AgNO_3$. Some of the amines gave black spots immediately and some after heating at 40°C for 15 min.

REFERENCE

1. Srivastava, S. P. and Dua, V. K., *Z. Anal. Chem.*, 279, 367, 1976.

Table TLC 16
ALIPHATIC AMINES

Layer		L1	L1	L1	L1	L1	L2	L2	L3					
Solvent		S1	S2	S3	S4	S5	S6	S7	S7					
Detection							D1	D1	D1	D1	D2	D3	D4	D5
Literature			1	1	1	1	1							
Compound					$R_F \times 100$					\multicolumn{5}{l}{Detection reaction (positive +, negative −)}				

Formula	Compound													
CH_5N	Methylamine	3.5	6							+	+ −	−	+ −	+
$C_2H_5NO_2$	Glycine	0	2	36	47	80				+	+ −	+	+	+
C_2H_7N	Ethylamine	7	16							+	+ −	−	+ −	+
C_2H_7N	Dimethylamine	4	7	11	12					+ −	+	+	+	−
C_2H_7NO	Ethanolamine	4	10	37	44	58				+	+	+	+	+
$C_2H_8N_2$	Ethylenediamine	2	4	15	20	40				+	+	+	+	+
C_3H_9N	Propylamine	16	35							+	+ −	+	+	+
C_3H_9N	Isopropylamine	17.5	36	73						+ −	+	+	+ −	+
C_3H_9N	Trimethylamine		43							−	+ −			−
C_3H_9OH	Propanolamine	4	8	28	31	50				+	+	+	−	+
$C_3H_{10}N_2$	Propylenediamine	3	10	35	40	55				+	+	+	+	+
C_4H_9NO	Morpholine	4.3	71	86						+ −	+	+	+	−
$C_4H_{10}N_2$	Piperazine	3	5	23	25	40				+	+	+	+	−
$C_4H_{11}N$	Butylamine	22	48							+	+	+	+	+
$C_4H_{11}N$	Isobutylamine	11	58	84						+	+	+	+	+
$C_4H_{11}N$	Diethylamine	16	32	57	41					+ −	+ −	+	−	−
$C_4H_{11}NO$	3-Methoxypropyl-amine	18	43	63						+	+	+	+	+
$C_4H_{11}NO$	Ethylethanolamine	11	23	54						+	+	+	+	−
$C_4H_{11}NO_2$	Diethanolamine	5	16	50						+ −	+	−	+	+ −
$C_5H_{13}N$	Diethylenetriamine	0	0	7	8.5	30				+	+	+	−	+
$C_5H_{13}N$	Pentylamine	29	55							+	+	+	+	+
$C_5H_{13}N$	Isoamylamine	30	56	84						+	+	+	+	+
$C_5H_{13}N$	2-Methylbutylamine	36	68	85						+	+	+	+	+
$C_6H_{13}N$	Cyclohexylamine	33	63	85						+ −	+	+	+	+
$C_6H_{13}NO$	N-Ethylmorpholine	95	100							−	+	+	−	−
$C_6H_{15}N$	Hexylamine	34	65				70	86	44.5	+	+	+	+	+
$C_6H_{15}N$	3-Amino-2,2'-dimethylbutane	51	90							−	+	+	+	+ −
$C_6H_{15}N$	2-Amino-3-methylpentane	47	78	88						−	+	+	+	+ −
$C_6H_{15}N$	2-Amino-4-methylpentane	42	73	82						−	+	+	+	+
$C_6H_{15}N$	Di-n-propylamine	51	80	91						+ −	+	+	+	−
$C_6H_{15}N$	Diisopropylamine	33	66	90						−	−	+	−	−
$C_6H_{15}N$	Triethylamine		75							+ −	+			−
$C_6H_{15}NO_2$	Ethyldiethanolamine	30	52	84						+ −	+	+	−	−
$C_6H_{15}NO_3$	Triethanolamine	18	36	75						−	+	+	−	−
$C_7H_{17}N$	Heptylamine	36	70				56	82	40	+	+	+	+	+
$C_7H_{17}NO_2$	Propyldiethanolamine	52	69	92						+ −	+	+	−	−
$C_8H_{19}N$	Octylamine	37.5	74				49	78	37	+	+	+	+	+
$C_8H_{19}N$	2-Ethylhexylamine	54	88							+	+	+	+ −	+
$C_8H_{19}N$	Di-n-butylamine	63	95							+	+	+	+	+ −
$C_8H_{19}N$	Diisobutylamine	85	99							+	+	+	+	+ −
$C_8H_{19}N$	tert.-Octylamine	52	87							−	+ −	+	−	−
$C_8H_{23}N_5$	Tetraethylene-pentamine			0	3.0	20				+	+	+	−	+

Table TLC 16 (continued)
ALIPHATIC AMINES

Formula	Name											
$C_9H_{20}N_2$	N-(3-Aminopropyl) cyclohexylamine	5	18	61				+	+	+	+	+
$C_9H_{21}N$	Nonylamine	39	77		36	74	33.3	+	+	+	+	+
$C_9H_{21}NO_3$	Triisopropanolamine	52	85					−	+	+	−	−
$C_{10}H_{23}N$	Decylamine	40.5	78		27	70	31	+	+	+	+	+
$C_{10}H_{23}NO$	2-Ethylhexylethanolamine	65	93					+	+	+	+	+−
$C_{11}H_{25}N$	Undecylamine	42	79		19	65	28	+	+	+	+	+
$C_{12}H_{27}N$	Dodecylamine	44	79		10	58	26	+	+	+	+	+
$C_{13}H_{29}N$	Tridecylamine	47	80		6.5	50	23	+	+	+	+	+
$C_{14}H_{31}N$	Tetradecylamine	50	82		4.5	43	21.5	+	+	+	+	+
$C_{15}H_{33}N$	Pentadecylamine	52	83		3.2	38	19.5	+	+	+	+	+
$C_{16}H_{35}N$	Hexadecylamine	55	85		2.5	30	17	+	+	+	+	+
$C_{16}H_{35}N$	Di-2-ethylhexylamine	100						+	+	+	+	−
$C_{17}H_{37}N$	Heptadecylamine	58	85		2.0	24	14.8	+	+	+	+	+
$C_{18}H_{39}N$	Stearylamine	60	85		1.5	18	13.3	+	+	+	+	+

Layer   L1 = Silica gel G
           L2 = Paraffin oil saturated kieselguhr
           L3 = Silianized silica gel
Solvent S1 = Chloroform-methanol-17% ammonia (82.5:15:5.2)
           S2 = Chloroform-methanol-17% ammonia (70:26:4)
           S3 = Chloroform-methanol-17% ammonia (40:40:20)
           S4 = Chloroform-methanol-17% ammonia (25:50:25)
           S5 = Methanol-17% ammonia (35:65)
           S6 = Acetone-17% ammonia (55:45)
           S7 = Acetone-17% ammonia (70:30)
Detection D1 = 1% ninhydrin in ethanol acetic acid (95:5)
           D2 = 1% potassium permanganate - 1% potassium persulphate (1:1)
           D3 = Iodine: 25% methanolic solution
           D4 = 5% sodium nitroprussiate in acetaldehyde, mixed with an equal volume of 2% sodium carbonate solution
           D5 = 1% 2,5 dimethoxytetrahydrofuran buffered solution of pH 6.6; after spraying the plate must be heated in an oven at 110° for 5 min. and then sprayed again with a 1% p-dimethylaminobenzaldehyde solution in 3% hydrochloric acid

REFERENCE

1. Prandi C., *J. Chromatogr.*, 155, 149, 1978.

Reproduced by permission of Elsevier Scientific Publishing Company.

Table TLC 17
PRIMARY AMINES (SOAP CHROMATOGRAPHY)

Layer	L1	L2	L3	L4	L5	L5	L5	L5	L5	L6	L7
Solvent	S1	S1	S1	S1	S1	S2	S3	S4	S5	S1	S1
Detection	D1	D1	D1	D1	D1	D1	D1	D1	D1	D1	D1
Literature	1	1	1	1	1	1	1	1	1	1	1

Compound	$R_F \times 100$										
Octopamine	92	81	80	63	41	50	66	69	76	37	38
Amphetamine	67	53	42	30	16	25	30	32	42	11	9
Histamine	90	72	60	36	15	34	46	54	73	10	7
Tryptamine	67	49	38	22	15	20	25	26	39	10	9
3-Hydroxytyramine	84	79	72	60	43	50	61	61	68	34	36
Tyramine	80	71	62	48	35	41	50	52	56	26	26
1-Phenylethylamine	70	53	42	30	18	26	31	33	41	12	12
2-Phenylethylamine	71	55	42	27	16	25	30	32	42	10	9
Noradrenaline	83	83	78	69	54	57	65	67	68	46	48

Layer L1 = Silanized silica gel 60 HF.
       L2 = Silanized silica gel 60 HF, impregnated with 0.25% anionic detergent (DBS).
       L3 = Silanized silica gel 60 HF, impregnated with 0.5% DBS.
       L4 = Silanized silica gel 60 HF, impregnated with 1% DBS.
       L5 = Silanized silica gel 60 HF, impreganted with 2% DBS.
       L6 = Silanized silica gel 60 HF, impregnated with 3% DBS.
       L7 = Silanized silica gel 60 HF, impregnated with 4% DBS.
Solvent S1 = Water-methanol-acetic acid (54.3:40:5.7).
         S2 = 0.05% Hydrochloric acid in water-methanol (40%)-acetic acid (5.7%).
         S3 = 0.1% Hydrochloric acid in water-methanol (40%)-acetic acid (5.7%).
         S4 = 0.25% Hydrochloric acid in water-methanol (40%)-acetic acid (5.7%).
         S5 = 0.5% Hydrochloric acid in water-methanol (40%)-acetic acid (5.7%).
Detection D1 = Ninhydrin.

REFERENCE

1. Lepri, L., Desideri, P. G., and Heimler, D., *J. Chromatogr.*, 153, 77, 1978.

Table TLC 18
PRIMARY MONO AND DIAMINES

Layer	L1	L1	L2	L2	L3	L3	L4	L4
Solvent	S1	S2	S1	S2	S1	S2	S1	S2
Detection	D1	D1	D1	D1	D1	D1	D1	D1
Literature	1	1	1	1	1	1	1	1

Compound	$R_F \times 100$							
Methylamine	94	93	77	82	65	78	54	77
Ethylamine	93	92	73	80	61	76	52	74
Ethanolamine	94	94	83	85	71	83	62	82
n-Propylamine	89	92	66	79	54	74	49	63
n-Butylamine	84	88	51	66	32	58	20	43
Isobutylamine	88	92	54	69	38	64	25	50
n-Amylamine	72	70	33	44	19	42	10	20
Isoamylamine	75	73	34	45	20	43	11	22
n-Hexylamine	59	56	18	37	10	25	4	15
n-Heptylamine	47	44	10	21	4	13	2	7
n-Octylamine	33	30	4	12	2	7	1	3
n-Decylamine	13	10	1	4	0	3	0	2
n-Dodecylamine	4	3	0	0	0	0	0	0
n-Tetradecylamine	2	1	0	0	0	0	0	0
Benzylamine	79	73	37	52	23	49	14	32
1,2-Diaminoethane	94	92	52	78	25	74	15	57

Table TLC 18 (continued)
PRIMARY MONO AND DIAMINES

1,2-Diaminopropane	94	92	50	77	24	73	15	55
1,3-Diaminopropane	94	92	53	86	28	83	17	63
1,4-Diaminobutane	94	92	50	86	27	83	16	63
1,5-Diaminopentane	94	92	50	84	24	80	11	60
1,6-Diaminohexane	94	90	49	79	22	73	9	52
1,7-Diaminoheptane	94	90	35	77	18	64	7	37
1,8-Diaminooctane	94	90	22	63	8	48	4	22
Spermine	94	92	3	67	1	35	0	6
Spermidine	94	93	15	78	8	64	2	19

Layer   L1   =   Silanized silica gel 60 HF (C_2) Merck.
        L2   =   As L1, impregnated with 1% triethanolamine dodecylbenzene sulfonate.
        L3   =   As L1, impregnated with 2% triethanolamine dodecylbenzene sulfonate.
        L4   =   As L1, impregnated with 4% triethanolamine dodecylbenzene sulfonate.
Solvent   S1   =   Water-methanol-acetic acid (64:3.30:5.7).
        S2   =   0.25 mol/l Hydrochloric acid in water-methanol (30%)-acetic acid (5.7%).
Detection   D1   =   1% Ninhydrin in pyridine-acetic acid (5:1).

REFERENCE

1. Lepri, L., Desideri, P. G., and Heimler, D., *J. Chromatogr.*, 173, 119, 1979.

Reproduced by permission of Elsevier Scientific Publishing Company.

Table TLC 19
POLYAMINES

Layer	L1
Solvent	S1
Detection	D1
Literature	1

Compound	$R_F \times 100$
Ethylenediamine	19
Diethylenetriamine	12
Triethylenetetramine	7
Propylenediamine	33
Ethylenepropylenetriamine	27
Dipropylenetriamine	24
Tripropylenetetramine	10
Hexamethylenediamine	56
Trimethylhexamethylenediamine	
2,2,4-isomer	78
2,4,4-isomer	73
Isophorondiamine	78
m-Xylylenediamine	90
p-Xylylenediamine	74
Monoethanolamine	46
Diethanolamine	60
Triethanolamine	74

Layer   L1   =   Silufol.
Solvent   S1   =   Pyridine-*tert.* butyl alcohol-conc. ammonia (1:1:1).
Detection   D1   =   Spray with 0.08-0.1% phenol red in 60% alcohol.

REFERENCE

1. Wiesmer, I. and Wiesnerova, L., *J. Chromatogr.*, 114, 411, 1975.

Reproduced by permission of the Elsevier Scientific Publishing Company.

Table TLC 20
POLYAMINES BY ION-EXCHANGE THIN-LAYER CHROMATOGRAPHY

Layer	L1
Solvent	S1
Detection	D1
Technique	T1

Compound	$R_F \times 100$
Agmatine	8
Arginine	38
Methylamine	75
Ornithine	66
Putrescine	26
Spermidine	11
Spermine	5

Layer   L1 = Fixion 50 × 8 ion-exchange thin-layer chromato-sheets (Na⁺).
Solvent S1 = 200 mmole/*l* potassium hydrogen phosphate + 2 mole/*l* NaCl, brought to pH 7.5 with NaOH.
Detection D1 = Ninhydrin reagent containing cadmium acetate.
Technique T1 = 4 hr run at room temperature.

REFERENCE

1. Bardocz, S. and Karsai, T., *J. Chromatogr.*, 223, 198, 1981.

Table TLC 21
N,N'-ALKYLATED DIAMINES

Layer	L1	L1	L1	L1
Solvent	S1	S2	S3	S4
Detection	D1	D1	D1	D1
Technique	T1	T1	T1	T1
Literature	1	1	1	1

Compound		$R_F \times 100$		
N-cyclohexyl propylenediamine-1,3	—ᵃ	15	30	46
N,N'-diisopropylpropylenediamine-1,3	14	26	44	66
N-cyclohexyl hexamethylenediamine-1,6	—ᵃ	21	45	64
N,N'-dicyclohexyl hexamethylenediamine-1,6	—ᵃ	27	52	77
N-ethyl trimethylhexamethylenediamine-1,6	—ᵃ	19	46	65
		26	58	74
N-hexyl trimethylhexamethylenediamine-1,6	—ᵃ	50	74	—ᵇ
		62	83	
N,N'-dimethyl trimethylhexamethylene diamine-1,6	—ᵃ	—ᵃ	34	51
			46	62
N,N'-diisopropyl trimethylhexamethylene diamine-1,6	17	70	87	—ᵇ
	28	81	91	
N,N'-dicyclohexyl trimethylhexamethylene diamine-1,6	57	—ᵇ	—ᵇ	—ᵇ
	71			
N,N'-dimethyl isophorondiamine	—ᵃ	—ᵃ	39	58
			51	67
N,N'-diisopropyl isophorondiamine	26	74	—ᵇ	—ᵇ
	37	84		

Table TLC 21 (continued)
N,N'-ALKYLATED DIAMINES

^a R_F values too low for practical determination.
^b Separation of isomers not achieved.

Layer	L1	=	Silufol (Kavalier, Czechoslovakia).
Solvent	S1	=	Ethanol-acetic acid isobutyl ester-25% ammonia (10:6:1).
	S2	=	Ethanol-acetic acid isobutyl ester-25% ammonia (10:6:2).
	S3	=	Ethanol-acetic acid isobutyl ester-25% ammonia (10:6:4).
	S4	=	Ethanol-acetic acid isobutyl ester-25% ammonia (10:6:6).
Detection	D1	=	Iodine in chloroform (spray).
Technique	T1	=	Development distance 10-12 cm.

REFERENCE

1. Klemm, D., Haase, L., Bellstedt, K., Schubert, K., and Hörhold, H.-H., *J. Chromatogr.*, 171, 512, 1979.

Table TLC 22
AROMATIC AMINES

Layer	L1	L1	L1	L1	L1	L1
Solvent	S1	S2	S3	S4	S5	S6
Detection	D1	D1	D1	D1	D1	D1
Literature	1	1	1	1	1	1

Compound	\multicolumn{6}{c}{$R_F \times 100$}
Aniline	16	20	74	81	77	76
Diphenylamine	79	82	88	—	96	95
Phenyl-2-naphthylamine	0	14	35	80	89	—
1,4-Phenylenediamine	0	1	4	5	7	9
2-Methyl-1,4-phenylenediamine	0	2	9	20	27	35
N,N-Dimethyl-1,4-phenylenediamine	8	8	17	21	21	25
2-Methylamino-5-aminotoluene	0	1	3	6	8	7
N,N'-Dimethyl-1,4-phenylenediamine	0	2	4	6	8	8
N,N,N'-Trimethyl-1,4-phenylenediamine	0	4	8	11	10	9
2,5-Bis(methylamino)toluene	0	3	6	7	8	8
N,N'-Diethyl-1,4-phenylenediamine	0	1	2	2	3	3
N,N'-Diisopropyl-1,4-phenylenediamine	0	2	4	6	8	8
N,N'-Di-sec.-butyl-1,4-phenylenediamine	0	1	2	2	3	3
N-Phenyl-N'-isopropyl-1,4-phenylenediamine	5	11	36	40	34	28
N-Phenyl-N'-sec.-butyl-1,4-phenylenediamine	16	28	68	83	81	72
N,N'-Diphenyl-1,4-phenylenediamine	25	29	68	97	97	96
N-Phenyl-N'-tosyl-1,4-phenylenediamine	2	13	60	95	96	95
N,N'-2-naphthyl-1,4-phenylenediamine	18	21	76	98	96	96

Layer	L1	=	Silufol plates (silica gel).
Solvent	S1	=	Ethanol - n-hexane (5:95).
	S2	=	Ethanol - n-hexane (10:90).
	S3	=	Ethanol - n-hexane (20:80).
	S4	=	Ethanol - n-hexane (30:70).
	S5	=	Ethanol - n-hexane (40:60).
	S6	=	Ethanol - n-hexane (50:50).
Detection	D1	=	Iodine vapor.

REFERENCE

1. Uchytil, B., *J. Chromatogr.*, 93, 447, 1974.

Table TLC 23
AROMATIC AMINES

Layer	L1	L1	L1	L1	L1	L1	L1	L1
Solvent	S1	S2	S3	S4	S5	S6	S7	S8
Detection	D1	D1	D1	D1	D1	D1	D1	D1
Literature	1	1	1	1	1	1	1	1
Compound				$R_F \times 100$				
Aniline	15	22	34	65	80	79	78	75
Diphenylamine	75	82	84	89	94	95	94	94
Phenyl-2-naphthylamine	40	49	68	68	85	94	—	—
1,4-Phenylenediamine	0	1	3	3	5	7	14	21
2-Methyl-1,4-phenylenediamine	0	2	6	7	14	19	28	32
N,N-Diethyl-1,4-phenylenediamine	6	11	25	34	38	45	52	54
2-Methylamino-5-aminotoluene	0	2	4	5	7	14	24	24
N,N'-Dimethyl-1,4-phenylenediamine	1	2	6	7	10	14	27	20
N,N,N'-Trimethyl-1,4-phenylenediamine	4	5	9	10	12	18	26	20
2,5-Bis(methylamino)toluene	2	3	6	8	11	17	24	23
N,N'-Diethyl-1,4-phenylenediamine	1	1	2	3	3	4	7	8
N,N'-Diisopropyl-1,4-phenylenediamine	0	0	1	2	3	5	10	13
N,N'-Di-sec.-butyl-1,4-phenylenediamine	3	12	13	21	39	39	31	30
N-Phenyl-N'-isopropyl-1,4-phenylenediamine	5	15	27	39	43	55	44	42
N-Phenyl-N'-sec.-butyl-1,4-phenylenediamine	19	32	45	69	84	84	82	78
N,N'-Diphenyl-1,4-phenylenediamine	20	32	37	67	94	96	96	96
N-Phenyl-N'-tosyl-1,4-phenylenediamine	2	12	29	61	92	94	93	93
N,N'-Di-2-naphthyl-1,4-phenylenediamine	13	22	34	68	94	98	97	96

Layer	L1	=	Silufol plates (Silica gel).
Solvent	S1	=	5% ethanol in n-hexane containing 0.1% triethylamine.
	S2	=	10% ethanol in n-hexane containing 0.1% triethylamine.
	S3	=	15% ethanol in n-hexane containing 0.1% triethylamine.
	S4	=	20% ethanol in n-hexane containing 0.1% triethylamine.
	S5	=	25% ethanol in n-hexane containing 0.1% triethylamine.
	S6	=	30% ethanol in n-hexane containing 0.1% triethylamine.
	S7	=	40% ethanol in n-hexane containing 0.1% triethylamine.
	S8	=	50% ethanol in n-hexane containing 0.1% triethylamine.
Detection	D1	=	Iodine vapors.

REFERENCE

1. Uchytil, B., *J. Chromatogr.*, 93, 447, 1974.

Table TLC 24
AROMATIC AMINES ON AMMONIUM TUNGSTOPHOSPHATE

Layer	L1	L2	L3	L4
Solvent	S1	S1	S1	S1
Detection	D1	D1	D1	D1
Literature	1	1	1	1
Compound	\multicolumn{4}{c}{$R_F \times 100$}			

Compound	L1	L2	L3	L4
Aniline	50	21	15	12
o-Toluidine	45	20	14	11
m-Toluidine	43	19	12	9
p-Toluidine	37	19	10	8
o-Bromoaniline	n.d.	18	11	9
m-Bromoaniline	29	11	7	7
p-Bromoaniline	33	13	8	7
o-Chloroaniline	n.d.	18	11	9
m-Chloroaniline	40	17	10	8
p-Chloroaniline	45	20	12	9
2,4-Dichloroaniline	n.d.	18	11	8
o-Nitroaniline	82	62	52	41
m-Nitroaniline	50	20	14	8
p-Nitroaniline	78	55	45	35
o-Aminobenzoic acid	68	45	35	20
m-Aminobenzoic acid	46	23	15	10
p-Aminobenzoic acid	72	48	36	23
4-Amino-3,5-dimethylbenzoic acid	68	48	37	24
3,4-Diaminobenzoic acid	37	22	15	11
3,5-Diaminobenzoic acid	14	7	4	2
o-Aminophenylarsonic acid	80	70	65	56
p-Aminophenylarsonic acid	75	65	59	50
o-Aminophenylsulfonic acid	90	77	73	60
m-Aminophenylsulfonic acid	90	78	74	61
p-Aminophenylsulfonic acid	91	85	84	75
1,3,6-Xylidene-4-sulfonic acid	90	77	75	62
2-Aminotoluene-5-sulfonic acid	90	82	80	69
4-Aminotoluene-3-sulfonic acid	90	73	68	55
6-Amino-4-chloro-3-toluene-sulfonic acid	90	74	70	58
o-Phenylenediamine	2	1	1	0
m-Phenylenediamine	2	1	1	0
p-Phenylenediamine	1	1	1	0
a-Naphthylamine	e.s.	4	3	3
1-Naphthylamino-7-sulfonic acid	85	69	60	45
1-Naphthylamino-4-sulfonic acid	88	78	70	55

Note: n.d. = Not determined; e.s. = elongated spot.

Layer	L1	=	Ammonium tungstophosphate - $CaSO_4 \cdot 1/2\ H_2O$ (0.5:2).
	L2	=	Ammonium tungstophosphate - $CaSO_4 \cdot 1/2\ H_2O$ (2:2).
	L3	=	Ammonium tungstophosphate - $CaSO_4 \cdot 1/2\ H_2O$ (4:2).
	L4	=	Ammonium tungstophosphate - $CaSO_4 \cdot 1/2\ H_2O$ (8:2).
Solvent	S1	=	Water.
Detection	D1	=	5% N,N-dimethyl-p-aminobenzaldehyde.

REFERENCE

1. Lepri, L., Desideri, P. G., and Heimler, D., *J. Chromatogr.*, 207, 29, 1981.

Reproduced by permission of the Elsevier Scientific Publishing Company.

Table TLC 25
AROMATIC AMINES ON TUNGSTOPHOSPHATE LAYERS

Layer	L1	L1	L1	L1	L1	L1
Solvent	S1	S2	S3	S4	S5	S6
Detection	D1	D1	D1	D1	D1	D1
Literature	1	1	1	1	1	1
Compound	\multicolumn{6}{c}{$R_F \times 100$}					

Compound						
Aniline	35	47	51	61	67	63
p-Toluidine	33	44	49	59	66	61
o-Toluidine	29	41	47	57	64	59
m-Toluidine	25	39	45	55	63	57
p-Bromoaniline	20	31	38	44	54	46
o-Bromoaniline	18	30	38	46	56	48
m-Bromoaniline	15	27	35	43	53	45
p-Chloroaniline	32	43	49	56	63	58
o-Chloroaniline	22	37	44	51	62	54
m-Chloroaniline	21	34	41	49	59	52
2,4-Dichloroaniline	20	30	38	47	56	n.d.
p-Nitroaniline	30	37	42	47	55	53
o-Nitroaniline	30	36	39	43	51	52
m-Nitroaniline	19	26	31	36	48	39
p-Aminobenzoic acid	48	50	56	63	70	66
o-Aminobenzoic acid	32	38	46	54	64	60
m-Aminobenzoic acid	30	36	44	51	62	55
4-Amino-3,5-dimethylbenzoic acid	31	34	42	50	58	55
3,4-Diaminobenzoic acid	35	38	46	55	63	55
3,5-Diaminobenzoic acid	5	13	20	30	44	33
p-Aminophenylarsonic acid	44	52	56	66	73	75
o-Aminophenylarsonic acid	36	40	45	56	64	75
p-Aminophenylsulphonic acid	81	81	82	85	86	82
o-Aminophenylsulphonic acid	68	66	67	72	73	72
m-Aminophenylsulphonic acid	72	70	71	77	77	75
1,3,6-Xylidene-4-sulphonic acid	71	70	71	75	76	75
2-Aminotoluene-5-sulphonic acid	76	75	78	81	83	79
4-Aminotoluene-3-sulphonic acid	63	60	64	68	70	68
6-Amino-4-chloro-3-toluene-sulphonic acid	63	60	64	64	64	66
p-Phenylenediamine	2	10	20	34	51	35
o-Phenylenediamine	2	13	23	36	52	37
m-Phenylenediamine	2	12	22	36	52	37
a-Naphthylamine	8	13	19	26	36	28
1-Naphthylamino-4-sulphonic acid	70	65	67	70	72	68
1-Naphthylamino-7-sulphonic acid	62	59	60	62	63	61

Note: n.d. = Not determined. R_F value of the first solvent front in S6 = 0.75.

Layer	L1 =	Ammonium tungstophosphate - $CaSO_4 \cdot 1/2 H_2O$ 4:2.
Solvent	S1 =	1 mol/l HNO_3.
	S2 =	1 mol/l HNO_3 + 0.25 mol/l NH_4NO_3.
	S3 =	1 mol/l HNO_3 + 0.5 mol/l NH_4NO_3.
	S4 =	1 mol/l HNO_3 + 1 mol/l NH_4NO_3.
	S5 =	1 mol/l HNO_3 + 2 mol/l NH_4NO_3.
	S6 =	1 mol/l NH_4NO_3.
Detection	D1 =	5% N,N-dimethyl-p-aminobenzaldehyde.

REFERENCE

1. Lepri, L., Desideri, P. G., and Heimler, D., *J. Chromatogr.*, 207, 29, 1981.

Reproduced by permission of Elsevier Scientific Publishing Company.

Table TLC 26
PRIMARY AROMATIC AMINES (SOAP CHROMATOGRAPHY)

Layer	L1	L2	L4	L4	L4	L4	L4	L3	L4	L5	L6
Solvent	S1	S1	S2	S3	S4	S5	S6	S1	S1	S1	S1
Detection	D1	D2	D1	D1	D1	D1	D1	D1	D1	D1	D1
Literature	1	1	1	1	1	1	1	1	1	1	1
Compound						$R_F \times 100$					
Aniline	70	52	34	42	50	60	70	42	30	21	17
m-Toluidine	64	38	23	29	38	49	55	31	29	14	11
o-Toluidine	65	43	27	35	45	56	61	36	27	19	15
p-Toluidine	64	38	23	30	40	51	57	32	22	14	11
2,4-Dimethylaniline	56	29	16	21	25	34	45	23	13	9	6
2,6-Dimethylaniline	44	31	21	27	33	40	54	27	18	13	10
m-Nitroaniline	41	36	35	36	47	60	65	36	35	29	28
o-Nitroaniline	32	29	26	28	29	29	29	27	26	23	22
p-Nitroaniline	39	36	35	36	39	49	57	36	35	30	30
2,4-Dinitroaniline	33	31	24	25	24	24	25	29	24	21	16
m-Bromoaniline	37	27	17	21	29	40	46	23	17	13	10
o-Bromoaniline	70	51	24	28	43	57	65	43	24	21	19
p-Bromoaniline	42	26	15	19	25	35	43	20	14	9	7
p-Chloroaniline	44	31	18	22	30	40	49	25	16	11	7
2,4-Dichloroaniline	19	16	13	16	23	33	45	15	13	9	7
m-Anisidine	62	40	28	34	42	52	59	35	24	18	14
o-Anisidine	64	43	33	41	48	58	65	37	28	21	16
p-Anisidine	70	45	30	38	44	55	62	37	25	18	14
m-Phenylenediamine	86	68	53	60	69	76	80	62	46	38	30
o-Phenylenediamine	79	58	40	48	55	66	76	47	35	28	22
p-Phenylenediamine	87	75	58	67	75	80	82	68	49	40	32
2,6-Diaminotoluene	85	61	42	58	62	69	77	58	41	31	22
2,4-Diaminotoluene	79	57	38	49	56	66	76	48	34	26	18
3,4-Diaminotoluene	69	43	25	33	39	53	72	50	44	39	33
α-Naphthylamine	38	24	13	15	21	28	37	17	12	7	5
4-Aminodiphenylamine	44	17	7	9	13	18	28	11	5	3	2
2-Aminodiphenylamine	22	11	4	6	8	12	17	8	4	2	1
3-Methoxy-4-amino-diphenyl-amine	38	15	6	8	11	15	24	8	4	2	1
4-Methoxy-4'-amino-diphenyl-amine	42	17	7	9	13	20	36	11	5	3	2
4-4'-Diaminodiphenyl-amine	84	48	21	37	48	72	79	32	13	6	4
2,4-Dinitro-4'-amino-diphenyl-amine	24	14	8	9	12	17	26	10	7	3	2
Benzidine	68	29	15	27	45	62	80	24	12	6	4
o-Tolidine	49	16	8	16	31	51	76	11	7	3	2
o-Dianisidine	29	14	10	21	37	61	80	10	8	5	4

Layer
- L1 = Silanized silica gel (Merck 60 HF) without impregnation.
- L2 = Silanized silica gel Merck 60 HF impregnated with 0.5% dodecylbenzenesulfonate.
- L3 = Silanized silica gel Merck 60 HF impregnated with 1% dodecylbenzenesulfonate.
- L4 = Silanized silica gel Merck 60 HF impregnated with 2% dodecylbenzenesulfonate.
- L5 = Silanized silica gel Merck 60 HF impregnated with 3% dodecylbenzenesulfonate.
- L6 = Silanized silica gel Merck 60 HF impregnated with 4% dodecylbenzenesulfonate.

Solvent
- S1 = 1 M acetic acid in 30% methanol.
- S2 = HCl in water-methanol (30%)-acetic acid (5.7%), pH 1,95.
- S3 = HCl in water-methanol (30%)-acetic acid (5.7%), pH 1,60.
- S4 = HCl in water-methanol (30%)-acetic acid (5.7%), pH 1,25.
- S5 = HCl in water-methanol (30%)-acetic acid (5.7%), pH 1,05.
- S6 = HCl in water-methanol (30%)-acetic acid (5.7%), pH 7,5.

Detection
- D1 = 5% N,N-dimethyl-p-aminobenzaldehyde in a 5:1 v/v mixture of ethanol and glacial acetic acid.

REFERENCE

1. Lepri, L., Desideri, P. G., and Heimler, D., *J. Chromatogr.*, 155, 119, 1978.

Reproduced by permission of Elsevier Scientific Publishing Company.

Table TLC 27
AROMATIC AMINES: CATECHOLAMINES AND THEIR 3-O-METHYL DERIVATIVES AND RELATED PHENYLETHYLAMINES

Layer	L1	L1	L1	L1	L1
Solvent	S1	S2	S3	S4	S5
Detection	D1	D1	D1	D1	D1
Literature	1	1	1	1	1
Compound			$R_F \times 100$		
L-DOPA	50	0	21	1	5
Dopamine.HCl	58	2	18	3	4
5-Hydroxydopamine.HCl	51	4	9	1	2
6-Hydroxydopamine.HBr	57	1	ND	ND	3
L-Norepinephrine.HCl	54	2	18	5	3
DL-Normetanephrine.HCl	60	4	18	26	19
3-Methoxytyramine.HCl	62	6	17	13	22
DL-α-Methyl-DOPA	52	0	31	1	50
L-3-O-Methyl-DOPA	55	1	19	2	38
4-Methyl-α-ethyl-m-tyramine.HCl	76	21	ND	ND	53
Tyramine.HCl	66	6	30	16	31
DL-Octopamine.HCl	62	5	31	31	21
O-Methyl-L-tyrosine.HCl	61	2	24	2	45
DL-p-Chlorophenylalanine	67	2	27	3	50
L-Phenylalanine	61	2	25	3	46
L-Tyrosine	56	1	25	2	43

Layer	L1	=	Silica gel.
Solvent	S1	=	n-Butanol-acetic acid-water (5:2:3).
	S2	=	Benzene-dioxane-acetic acid (4:1:5).
	S3	=	Ethyl acetate-ethanol-acetic acid (3:2:1).
	S4	=	Benzene-acetone-methanol (1:2:1).
	S5	=	n-Butanol-ethanol-water (2:1:1).
Detection	D1	=	Fluorescamine.

REFERENCE

1. Nakamura, H. and Pisano, J. J., *J. Chromatogr.*, 154, 51, 1978.

Reproduced by permission of the Elsevier Scientific Publishing Company.

Table TLC 28
AROMATIC AMINES. DETECTION WITH PERSULFATE

Layer	L1
Solvent	S1
Detection	D1
Literature	1

Compound	$R_F \times 100$	Color of the spots
p-Aminophenol	70	Reddish brown
m-Aminophenol	78	Brown
p-Phenylenediamine	51	Bluish black
o-Phenyleneamine	69	Yellow
m-Phenylenediamine	60	Black
o-Tolidine	77	Dark yellow

Table TLC 28 (continued)
AROMATIC AMINES. DETECTION WITH PERSULFATE

Compound	$R_F \times 100$	Color
3,5-Diaminobenzoic acid	68	Yellowish brown
3,4-Diaminobenzoic acid	72	Yellow
Diphenylamine	92	Black
Dimethoxybenzidine	73	Violet
N,N'-Diphenylbenzidine	96	Black (tailing)
Diaminobenzidine	56	Black
Benzidine	74	Yellowish black
4,4'-Diaminodiphenyl 3,3'-Diglycolic acid	76	Pink
Sulfapyridine	81	Light brown
Sulfanilic acid	32	Brown (tailing)

Layer	L1 =	Silica gel GF.
Solvent	S1 =	n-butanol-acetic acid-water (50:10:40).
Detection	D1 =	Spray with 1% solution of $K_2S_2O_8$ which is 0.001 M with respect to $AgNO_3$.

REFERENCE

1. Khulbe, K. C. and Mann, R. S., J. Chromatogr., 150, 554, 1978.

Reproduced by permission of the Elsevier Scientific Publishing Company.

Table TLC 29
AMINES, DANSYL DERIVATIVES

Layer	L1
Solvent	S1
Detection	D1
Literature	1

Compound	$R_F \times 100$
p-Tyramine(bis-dansyl)	51
3-Methoxytyramine(bis-dansyl)	42
p-Octopamine(bis-dansyl)	17
Tryptamine	19
Histamine(bis-dansyl)	33
β-Phenylethylamine	62
Dibenzylamine	65
N-Methylbenzylamine	77
n-Butylamine	56
Phenylethanolamine	26
Normetanephrine(bis-dansyl)	11
Dopamine(tris-dansyl)	47
Aniline	25
Phenylalanine	3
γ-Aminobutyric acid	60

Layer	L1 =	Silica gel 60.
Solvent	S1 =	Benzene-triethylamine (8:1).
Detection	D1 =	Fluorescence.

REFERENCE

1. Davis, B. A., J. Chromatogr., 151, 252, 1978.

Table TLC 30
HISTIDINE, HISTAMINE, HISTIDYL PEPTIDES, AND RELATED COMPOUNDS AS FLUORESCAMINE DERIVATIVES

Layer	L1	L1	L1	L1	L1	L1
Solvent	S1	S2	S3	S4	S5	S6
Detection	D1	D1	D1	D1	D1	D1
Literature	1	1	1	1	1	1

Compound			$R_F \times 100$			
L-Histidine	65	49	76	84	62	81
Histamine dihydrochloride	74	70	86	97	78 (69)	80
DL-2-Methylhistidine dihydrochloride hydrate	64 (75)	40 (46)	75,62	76 (86)	24	79
L-3-Methylhistidine	54 (66)	23	14 (11)	32	6 (3)	49
DL-4-Methylhistidine	63 (52)	44 (18,24,31)	78	87 (6,10)	57	81
L-Histidine methyl ester dihydrochloride	64,81	47,78	76,87	86,99	63,87	82,88
L-Histidine ethyl ester hydrochloride	65,85(70,78)	48,86	76,90	85,99	62,89	83,88
L-Histidine hydroxamate	65	47	76	86	63	82
L-Histidinol dihydrochloride	76 (53)	76 (21)	86 (2)	97	79 (70)	87
L-Histidinol phosphate	57	26	45	12	11	71
1,4-Methylhistamine dihydrochloride	55 (65)	22 (40)	2 (14,18)	30	1	27
L-His-L-Leu	80	75	78	95	85	88
L-His-Gly	66	52	52	86	72	81
L-His-L-Lys hydrobromide	56 (72)	26 (58)	18	20	5	75 (88)
L-His-L-Tyr	75	70	79	92	19	87
L-His-L-Ala	68	57	53	90	75	84
DL-His-DL-His	56,58	27,29	44	22,35	6	71,74
L-His-L-Ser	62 (58)	43 (34)	50	76	59 (49)	82
L-His-L-Phe	76	70	80	96	83	88
Glucagon	60	18	7(0,44,46)	8 (1)	2	92

Layer	L1	=	Silica gel 60 Merck.
Solvent	S1	=	n-Butanol-acetic acid-water (5:2:3).
	S2	=	n-Butanol-acetic acid-water (4:1:5) (upper phase).
	S3	=	Ethyl-acetate-methanol-water (60:25:10).
	S4	=	Chloroform-methanol-acetic acid (6:2:2).
	S5	=	Benzene-dioxan-acetic acid (2:5:3).
	S6	=	Isopropanol-acetic acid-water (6:2:2).
Detection	D1	=	Derivatives viewed in UV light.

REFERENCE

1. Nakamura, H., *J. Chromatogr.*, 131, 215, 1977.

Table TLC 31
FLUORESCAMINE DERIVATIVES OF PRIMARY CATECHOLAMINES, THEIR 3-O-METHYL DERIVATIVES AND RELATED PHENYLETHYLAMINES

Layer	L1	L1	L1	L1
Solvent	S1	S2	S3	S4
Detection	D1	D1	D1	D1
Compound		$R_F \times 100$		
L-DOPA	26	45	84	80
Dopamine HCl	70	84	91	89
5-Hydroxydopamine HCl	1	75	ND[b]	0
6-Hydroxydopamine HBr	69	83	ND	ND
L-Norepinephrine HCl	59	81	90	87
DL-Normetanephrine HCl	67	83	92	90
3-Methoxytyramine HCl	69	83	93	92
DL-α-Methylnorepinephrine HCl	66	83	93	92
DL-α-Methyl-DOPA	19	ND	87	ND
L-3-O-Methyl-DOPA	27	45	84	75
4-Methyl-α-ethyl-m-tyramine HCl[a]	69	88	95	96
4-O-Methyldopamine	69	NA[c]	NA	NA
Tyramine HCl[a]	71	87	94	92
DL-Octopamine HCl[a]	68	87	94	90
O-Methyl-L-tyrosine-HCl[a]	39	45	83	81
DL-p-Chlorophenylalanine[a]	45	46	84	77
L-Phenylalanine[a]	44	46	84	82
L-Tyrosine[a]	35	46	85	77

[a] Compounds whose bluish green fluorescence of fluorescamine derivatives disappears on spraying with 70% HClO$_4$.
[b] ND = not detected.
[c] NA = not analyzed.

Layer	L1	=	Silica gel 60.
Solvent	S1	=	Ethyl acetate-n-hexane-methanol-water (60:20:25:10).
	S2	=	Chloroform-isopropanol-water (2:8:1).
	S3	=	n-Butanol-acetic acid-water (5:2:3).
	S4	=	Benzene-dioxane-acetic acid (2:5:1).
Detection	D1	=	The compounds spotted on the plate were buffered by overspotting 1 μl of 0.5 M phosphate buffer pH 8.0, were air dried and derivatized on the origin with fluorescamine by the pre-dipping method.

REFERENCE

1. Nakamura, H. and Pisano, J. J., *J. Chromatogr.*, 154, 51, 1978.

Reproduced by permission of the Elsevier Scientific Publishing Company.

Table TLC 32
ANILS

Layer	L1	L1	L1	L1	L1	L1	L1	L1
Solvent	S1	S2	S3	S4	S5	S6	S7	S8
Detection	D1	D1	D1	D1	D1	D1	D1	D1
Literature	1	1	1	1	1	1	1	1
Compound				$R_F \times 100$				
R'=N–C$_6$H$_5$	87	89	72	46	93	87	—	—
R'=N–C$_6$H$_4$–N(CH$_3$)$_2$	85	87	64	5	87	94	—	—
R'=N–C$_6$H$_4$–C$_6$H$_4$–N=R'	97	93	74	3	95	93	—	—
R'=N–C$_6$H$_4$–N=R'	96	48	55	0	88	94	—	—
R'=N–(naphthyl, OH, SO$_3$H)	88	77	76	56	96	95	—	—
R'=N–(1-naphthyl)	92	95	87	12	95	92	—	—
R'=N–(2-naphthyl)	72	92	69	17	97	94	—	—
R'=N–C$_6$H$_4$(o-OCH$_3$)	91	56	55	10	92	74	—	—
R'=N–C$_6$H$_4$(p-OCH$_3$)	93	90	73	19	95	94	—	—
R'=N–C$_6$H$_4$(o-Cl)	97	88	69	77	92	90	—	—
R'=N–C$_6$H$_4$(m-Cl)	83	93	71	86	95	91	—	—
R'=N–C$_6$H$_4$(o-OH)	90	90	77	2	95	92	—	—
R'=N–C$_6$H$_4$(m-OH)	97	95	83	0	98	87	—	—
R'=N–C$_6$H$_4$(o-CH$_3$)	92	88	67	47	92	96	45	6

Table TLC 32 (continued)
ANILS

R'=N—⌬—CH₃ (with CH₃ on ring)	89	87	63	48	94	82	65	20
R'=N—⌬—CH₃	95	92	70	62	95	67	92	40

Layer	L1	=	Silica gel plates.
Solvent	S1	=	Diethyl ether.
	S2	=	Acetone.
	S3	=	Methanol.
	S4	=	Benzene.
	S5	=	Benzene-methanol (1:1).
	S6	=	Acetic acid.
	S7	=	Chloroform.
	S8	=	Benzene-chloroform (4:1).
Detection	D1	=	Natural color (yellow or brown).

REFERENCE

1. Upadhyay, R. K. Singh, V. P., and Vajpai, U., *J. Chromatogr.*, 93, 494, 1974.

Note: R' = (decalin with CO·CH substituent)

Reproduced by permission of the Elsevier Scientific Publishing Company.

Section II
Techniques

II.I.	Procedures for the Determination of Amino Acids by Liquid Chromatography
II.II.	Ion Exchange Resins for Amino Acid Determinants
II.III.	Gas Chromatographic Determination of Amino Acids
II.IV.	Determination of Amino Acid Enantiomers by Gas Chromatraphy
II.V.	Detection Systems for Amino Acids in Automatic Analyzers

Section II.I

PROCEDURES FOR THE DETERMINATION OF AMINO ACIDS BY LIQUID CHROMATOGRAPHY*

The following tables and figures summarize typical liquid chromatographic procedures for amino acid determination.

---

* Tables 1-4 were prepared by A. J. Smith.

Table 1
TYPICAL PROCEDURES FOR ION-EXCHANGE CHROMATOGRAPHY OF HYDROLYSATE AMINO ACIDS

System	Time (hr)	Column (cm)	Resin[a]	pH	Buffer Composition Na$^+$(N)	Buffer Composition Cit.(M)	Misc.[b]	Temp.(°C)	Ref.
Two-column									
Standard									
Acidic, neutral	2.0	0.9 × 56	HP-AN90	3.49, 4.30	0.2	0.067	—	53.5	1
Basic		0.9 × 5	HP-B80	5.36	0.35	0.105	—	53.5	
Accelerated									
Acidic, neutral	1.5	0.2 × 25	DC-4a	3.25, 4.25	0.2	—	—	60.0	2
Basic		0.2 × 7.5	DC-4a	5.28	0.35	—	—	60.0	
Single-column									
Stepwise elution									
Standard	4.5	0.28 × 60	A-7 (or DC-6a, or M-82)	3.25	0.2	0.067	NaCl	60.0	3
				4.25	0.4	0.067	NaCl		
				6.40	1.05	0.105			
Accelerated	1.5	0.4 × 27	Ultropac8	3.20	0.2	—	PrOH	52, 56, 80	4
				4.25	0.2	—	—		
				6.45	1.2	—	—		
Constant molarity	1.0	0.32 × 15	DC-5a	3.25	0.2	0.067	—	45, 65	5
				4.25	0.2	0.067	—		
				5.25	0.2	0.067	—		
				10.0	0.2	0.049(borate)	NaCl		
	0.5	0.172 × 48	DC-X10	3.15	0.2	—	EtOH		5
				3.25	0.2	—	—		
				4.25	0.2	—	—		
				9.45	0.2	—	—		
Gradient elution									
Standard	1.0	0.3 × 25	DC-4a	2.6	0.2	—	—	53.0	6
				6.3	1.2	—	—		
Constant molarity	1.0	0.26 × 35	Rank Hilger 7μ	2.2	0.2	—	—	60.0	7
				11.5	0.2	—	—		

[a] The resins are products of: Hamilton - HP-AN90, HP-B80; Durrum - DC-4a, DC-5a, DC-6a, DC-X10; Rank Hilger - Rank Hilger 7μ; Bio-Rad® -A-7; LKB - Ultropac 8; Beckman - M82.
[b] Abbreviations: PrOH, propanol; EtOH, ethanol.

REFERENCES

1. Benson, J. V., *Anal. Biochem.*, 50, 477, 1972.
2. Hare, P. E., *Methods Enzymol.*, 47, 3, 1977.
3. Liao, T. H., Robinson, G. W., and Salnikow, J., *Anal. Chem.*, 45, 2286, 1973.
4. LKB Applications Report No. PCN10, LKB Biochrom Ltd., Cambridge, England
5. Benson, J. V., *Methods Enzymol.*, 47, 19, 1977.
6. Voelter, W. and Zech, K., *J. Chromatogr.*, 112, 643, 1975.
7. Murren, C., Stelling, D., and Felstead, G., *J. Chromatogr.*, 115, 236, 1975.

Table 2
TYPICAL PROCEDURES FOR ION-EXCHANGE CHROMATOGRAPHY OF COLLAGEN HYDROLYSATE AMINO ACIDS

System	Time (hr)	Column (cm)	Resin[a]	Buffer Composition pH	Na⁺(N)	Cit.(M)	Misc.[b]	Temp.(°C)	Ref.
Two-column									
Acidic, neutral	3.5	0.9 × 56	UR-30	3.12,4.35	0.2	—	—	50.0	1
Basic		0.9 × 5.3	PA-35	5.25	0.35	—	—	50.0	
Acidic, neutral	3.5	0.9 × 59	UR-30	3.25,3.80	0.2	0.2	—	24—67	2
Basic		0.9 × 9	PA-35	5.85	0.35	0.35	—		
Single-column									
Standard	4.75	0.9 × 54	AA-15	2.77	0.2	—	MeCell	55,67	3
				3.29,3.61	0.2	—	—		
				4.70	0.35	—	PrOH		
Microbore (low back-pressure — 500 psi)	4.0	0.28 × 69	AA-20	2.77,3.34	0.2	0.2	—	55.0	4
				4.14	0.14	0.2	—		
				6.40	1.0	0.2	—		
Microbore (medium back-pressure — 750 psi)	1.5	0.4 × 27	Ultropac8	3.00	0.2	—	iPrOH	45,62,83	5
				3.42,4.25	0.2	—	—		
				6.45	1.2	—	—		
Microbore (high back-pressure — 1600 psi)	2.0	0.18 × 48	DC-4a	2.90	0.18	0.067	MeOH	48,54	6
				3.32,4.45	0.18	0.067	—		
				6.25	1.0	0.133	—		

[a] The resins are products of: Beckman - UR-30, PA-35, AA-15, AA-20; LKB - Ultropac 8; Durrum - DC-4a.
[b] Abbreviations: MeCell, methyl cellosolve; PrOH, propanol; MeOH, methanol; iPrOH, isopropanol.

REFERENCES

1. Page, R. C., Jones, D., and Hansen, R., *Anal. Biochem.*, 37, 293, 1970.
2. Osborne, R. M., Longton, R. W., and Lamberts, B. L., *Anal. Biochem.*, 44, 317, 1971.
3. Guire, P., Riguetti, P., and Hudson, B. G., *J. Chromatogr.*, 90, 350, 1974.
4. Vergnes, J. P. and Freeman, I. L., *Anal. Biochem.*, 99, 427, 1979.
5. LKB Applications Report No. PCN18, LKB Biochrom Ltd., Cambridge, England.
6. Georgiadis, A. G., Coffey, J. W., Hamilton, J. G., and Miller, O. N., *Anal. Biochem.*, 67, 453, 1975.

Table 3
TYPICAL PROCEDURES FOR ION-EXCHANGE CHROMATOGRAPHY OF PHYSIOLOGICAL FLUID AMINO ACIDS

System	Time (hr)	Column (cm)	Resin[a]	pH	Na+(N)	Cit.(M)	Misc.[b]	Temp. (°C)	Ref.
Two-column									
Sodium									
Acidic, neutral	6.0	0.9 × 56	HP-AN90	3.20, 4.30	0.2	0.066	—		1
Basic			HP-B80	4.26	0.38	0.127	—	32.5, 62.0	
				6.32	0.50	0.166	—		
Lithium									
Acidic, neutral	3.5	0.9 × 56	HP-AN90	2.88	0.3	0.053	—	40.0	1
				4.15	0.3	0.050	—		
Basic	2.2	0.7 × 15	A-5	4.52	0.6	0.2	—	55.2	2
Lithium/sodium									
Acidic, neutral	2.5	0.4 × 22	C3	2.70	0.3	0.05	MeCell	42.0	3
				4.50	0.3	0.05	—		
Basic		0.4 × 22	C3	4.30	0.5	0.05	—	57.0	
				6.00	1.0	0.05	—		
Single-column									
Stepwise elution									
Sodium	6.0	0.9 × 27	UR-40	3.17, 4.24	0.18	—	—	—	4
				4.15	0.40	—	—		
				5.36	1.00	—	—		
Lithium	7.5	0.9 × 55	M-82 or M-72	2.80	0.3	0.05	—	39, 60	5
				4.10	1.2	0.21	—		
	4.5	0.46 × 27	Ultropac 8	2.80	0.2	—	iPrOH	35, 60	6
				3.00	0.3	—	iPrOH		
				3.02	0.6	—	—		
				3.45	1.0	—	—		
				3.30	1.6	—	—		

Table 3 (continued)
TYPICAL PROCEDURES FOR ION-EXCHANGE CHROMATOGRAPHY OF PHYSIOLOGICAL FLUID AMINO ACIDS

System	Time (hr)	Column (cm)	Resin[a]	pH	Buffer Composition Na⁺(N)	Buffer Composition Cit.(M)	Misc.[b]	Temp. (°C)	Ref.
Gradient	8.0	0.63 × 75	C2	2.75	0.275	0.07	MeOH	39,60	7
				2.875, 3.80	0.275	0.07	—		
				6.50	1.5	0.07	LiCl		
				3.10[c]	0.275	0.07	—		
	4.0	0.3 × 35	Rank Hilger 7μ	1.90	0.15	0.10	—	42,63	8
				11.5	0.3	0.05	EDTA		

[a] The resins are products of: Hamilton — HP-AN90, HP-B80; Bio-rad® — A-5; Technicon — C2, C3; Sondell — UR-40; Beckman — M-72, M-82; LKB — Ultropac 8; Rank Hilger — Rank Hilger 7μ.

[b] Abbreviations: MeCell, methyl cellosolve; iPrOH, isopropanol; MeOH, methanol; EDTA, ethylenediaminetetra-acetic acid.

[c] This buffer is used for reequilibration of the column.

REFERENCES

1. Benson, J. V., *Anal. Biochem.*, 50, 477, 1972.
2. Mondino, A., Bongiovanni, G., Noe, V., and Raffaele, I., *J. Chromatogr.*, 63, 411, 1971.
3. Ersser, R. S., *Med. Lab. Sci.*, 33, 257, 1976.
4. Efron, K. and Wolf, P. L., *Clin. Chem.*, 18, 621, 1972.
5. Kedenburg, C. P., *Anal. Biochem.*, 40, 35, 1971.
6. LKB Applications Report No. PCN 12, LKB Biochrom Ltd., Cambridge, England.
7. Houpert, Y., Tarallo, P., and Siest, G., *J. Chromatogr.*, 115, 33, 1975.
8. Rank Hilger Application Report No. CP 17, Rank Hilger, Margate, England.

Table 4
TYPICAL PROCEDURES FOR HIGH PERFORMANCE LIQUID CHROMATOGRAPHY OF AMINO ACID DERIVATIVES

Derivative	Column (cm)	Packing[a]	Eluant	Temp.(°C)	Ref.
o-Phthaldialdehyde-ethanethiol	30 × 0.39	µ-Bondapak-C_{18} (10µm)	A = 0.0125 M sodium phosphate, pH 7.2 B = Acetonitrile	—	1
Phenylthiohydantoin Isocratic elution	25 × 0.46	Zorbax-ODS (5—6µm)	Acetonitrile-0.01 N sodium acetate, pH 4.5 Polar derivatives (24:76 v/v) Non-polar derivatives (42:58 v/v)	62	2
Gradient elution	25 × 0.46	Zorbax-ODS (5—6µm)	A = 0.01 N Sodium acetate, pH 4.5 B = Acetonitrile	62	2
	25 × 0.46	Zorbax-CN	A = 0.015—0.04 M Sodium acetate, pH 5.4 B = Methanol-acetonitrile (17:3 v/v)	31	3
Dinitrophenyl	25 × 0.46	Zorbax-ODS (5—6µm)	A = 1% Acetic acid B = Acetonitrile	62	4

[a] Zorbax-ODS & -CN are products of DuPont and µ-Bondapak is a product of Waters Associates.

REFERENCES

1. Hill, D. W., Walters, F. H., Wilson, T. D., and Stuart, J. D., *Anal. Chem.*, 51, 1338, 1979.
2. Zimmerman, C. L., Appella, C., and Pisano, J. J., *Anal. Biochem.*, 77, 569, 1977.
3. Johnson, N. D., Hunkapiller, M. W., and Hood, L. E., *Anal. Biochem.*, 100, 335, 1979.
4. Zimmerman, C. L., and Pisano, J. J., *Methods, Enzymol.*, 47, 45, 1977.

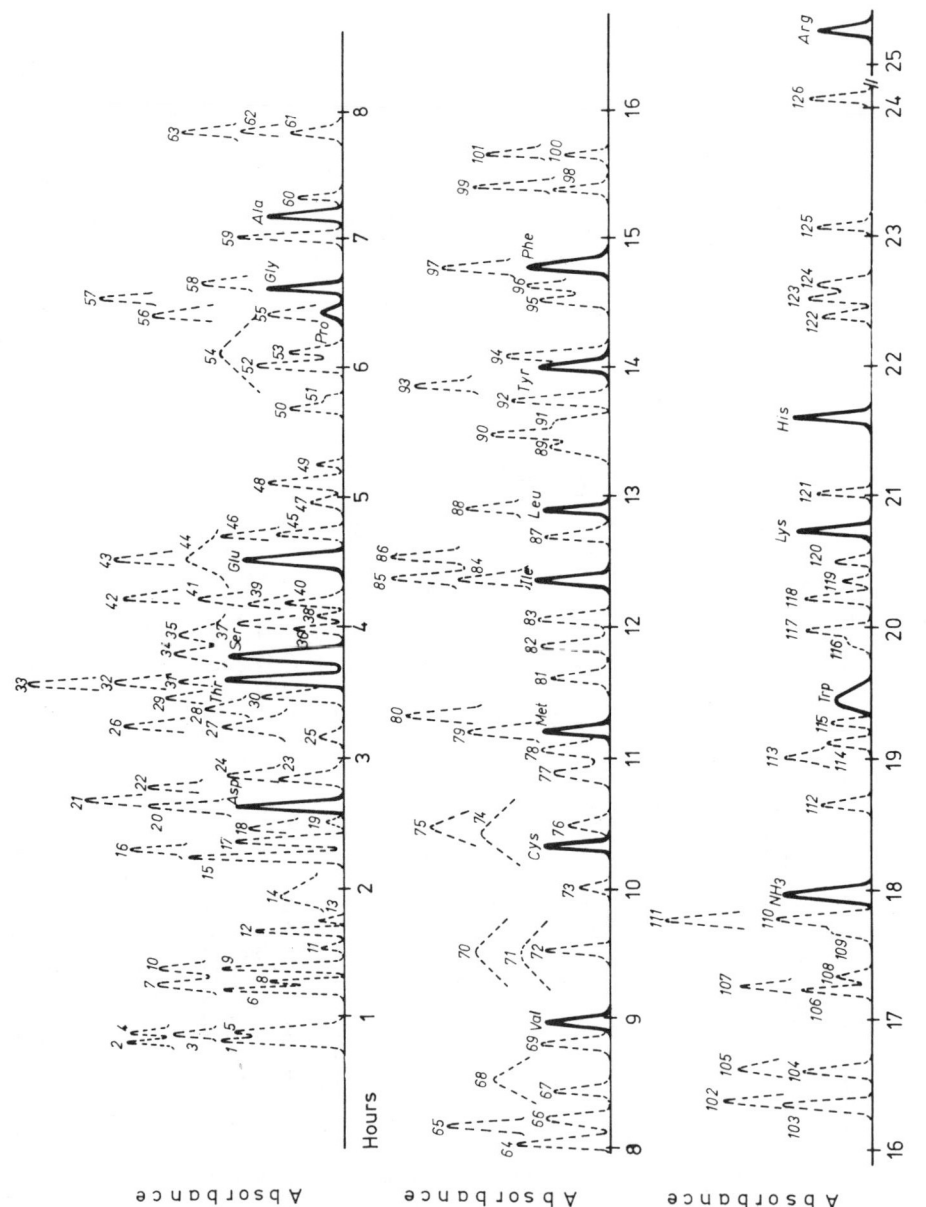

FIGURE 1. Elution positions of 145 ninhydrin-positive compounds on a 140 × 0.6 cm column of Chromobeads Type B (a strongly acidic cation exchange resin) in the lithium form. A nine-chambered autograd was used to obtain the gradient, solutions used being as follows:

Chamber No.	Buffer 1. (pH 2.75), ml	Buffer 2. (pH 3.01), ml	Buffer 3. (pH 6.50), ml
1	98*	—	—
2	50	50	—
3,4,5,6	—	100	—
7,8,9	—	—	100

* Plus 2 ml of isopropyl alcohol

The 18 physiological amino acids and ammonia are indicated by solid lines. Numbered peaks are identified as follows: 1. cysteic acid, 2. homocysteic acid, 3. cysteinesulfinic acid, 4. O-phosphothreonine, 5. O-phosphoserine, 6. taurine, 7. penicillaminic acid, 8. threo-β-hydroxyaspartic acid, 9. phosphoethanolamine, 10. levulinic acid, 11. dithiothreitol, 12. erythro-β-hydroxyaspartic acid, 13. urea, 14. S-carboxymethylglutathione, 15. allo-γ-hydroxyglutamic acid, 16. S-methylcysteine sulfoxides, 17. allo-β-hydroxyglutamic acid, 18. cephalosporin C, 19. 3-hydroxypipecolic acid, 20. S-carboxymethylcysteine, 21. S-carboxymethylpenicillamine, 22. diaminosuccinic acid (peak 1), 23. glutathione (reduced), 24. S-methylglutathione, 25. 4-hydroxyproline, 26. diaminosuccinic acid (peak 2), 27. penicilloic acid of isopenicillin N, 28. γ-(L-glutamyl)-L-cysteine, 29. methionine sulfoxide (peak 1), 30. methionine sulfone, 31. penicillamine sulfoxide (peak 2), 32. allo-threonine, 33. β-hydroxyvaline, 34. δ-(L-α-aminoadipyl)-L-cysteine, 35. δ-(L-α-aminoadipyl)-L-cysteinyl-L-valine, 36. O-methylthreonine, 37. O-methylserine, 38. Allo-4-hydroxyproline, 39. muramic acid, 40. asparagine, 41. allo-4-hydroxypipecolic acid, 42. S-carbamylcysteine, 43. β-methoxyvaline, 44. δ-(L-α-aminoadipyl)-L-cysteinyl-D-valine, 45. glutamine, 46. homoserine, 47. 4-oxopipecolic acid, 48. sarcosine, 49. 5-hydroxypipecolic acid, 50. cysteine, 51. threo-thiolbutyrine, 52. S-methylcysteine, 53. α-aminoadipic acid, 54. glutathione (oxidized), 55. erythro-thiolbutyrine, 56. S-carboxymethyl-homocysteine, 57. β-hydroxyleucine, 58. penicillamine (reduced), 59. isoserine, 60. lanthionine (peak 1), 61. citrulline, 62. lanthionine (peak 2), 63. α-aminiosobutyric acid, 64. glucosamine, 65. S-ethylcysteine, 66. α-aminobutyric acid, 67. mannosamine, 68. Bis-γ-(L-glutamyl)-L-cysteine, 69. galactosamine, 70. bis-δ-(L-α-aminoadipyl)-L-cystine, 71. Bis-δ-(L-α-aminoadipyl)-L-cystinyl-bis-L-valine, 72. α-aminopimelic acid, 73. pipecolic acid, 74. Bis-α-(L-α-aminoadipyl)-L-cystinyl-bis-D-valine, 75. 6-aminopenicillanic acid, 76. homocysteine, 77. phenylglycine, 78. homocitrulline, 79. norvaline, 80. mixed disulfide of L-cysteine and D-penicillamine, 81. Allo-isoleucine, 82. ethionine, 83. djenkolic acid, 84. penicillamine (oxidized), 85. cystathionine, 86. allo-cystathionine, 87. α-amino-β-hydroxybutyric acid, 88. 3,4-dihydroxyphenylalanine, 89. isoglutamine, 90. α,ε-diaminopimelic acid, 91. norleucine, 92. cycloserine, 93. α-amino-β-ethylvaleric acid, 94. mixed disulfide of L-cysteine and DL-homocysteine, 95. β-alanine, 96. Mixed disulfide of DL-homocysteine and D-penicillamine, 97. O-benzylserine, 98. β-aminoisobutyric acid, 99. δ-aminolevulinic acid, 100. L-cysteinyl-L-valine, 101. L-cysteinyl-D-valine, 102. argininosuccinic acid, 103. homocystine, 104. γ-aminobutyric acid, 105. S-benzylcysteine, 106. 5-hydroxytryptophan, 107. α-aminocaprylic acid, 108. ethanolamine, 109. kynurenine, 110. L-cystinyl-bis-L-valine, 111. L-cystinyl-bis-D-valine, 112. L-cystinyl-bis-L-valinol, 114. 5-hydroxylysine, 115. allo-5-hydroxylysine, 116. creatinine, 117. α,γ-diaminobutyric acid, 118. ornithine, 119. valinamide, 120. ε-aminocaproic acid, 121. 1-methylhistidine, 122. 3-methylhistidine, 123. carnosine, 124. homocarnosine, 125. α-amino-β-guanidinopropionic acid, 126. homocysteine thiolactone.

(Reproduced from Adriaens, P., Meesschaert, B., Wuyts, W., Vanderhaeghe, H., and Eyssen, H., *J. Chromatogr.*, 140, 103, 1977. With permission.)

LIQUID COLUMN CHROMATOGRAPHIC ANALYSIS OF AMINO ACIDS*

Development of Amino Acid Analysis

Amino acid analysis has found wide use in biochemistry for determination of purity of proteins, their quantitation, their composition, and in protein sequencing. Chromatographic techniques form the basis of most modern methods and derive from the pioneering work of Martin and Synge and the later developments of Moore and Stein. Although gas-liquid chromatography has gained wide attention, ion-exchange chromatography still provides the prime method for separation. The development of high performance liquid chromatography has provided a fast sensitive method for separation of many amino acid derivatives, particularly amino acid-phenylthiohydantoins, but this technique is still lacking for the separation of underivatized samples. Automation of the amino acid analyzer has enabled fast results to be obtained with high precision and accuracy; while early separations of protein hydrolysates took several hours, modern developments with micro-bore columns now allow run times of about 1 hr to be achieved. Although ninhydrin still finds the widest use in the detection of amino acids, fluorimetric techniques have been developed that extend the limits of detection. The aim of this chapter is to present some of the factors involved in achieving good analysis of amino acids by liquid column chromatography and the rationale behind their use.

Sample Preparation

Analysis of a protein requires prior hydrolysis to its constituent amino acids and this necessitates a hydrolytic procedure of wide specificity to cleave all the peptide bonds present. This is probably the area of amino acid analysis where the greatest error can appear and careful selection of a procedure that achieves maximum hydrolysis of the peptide bonds present with minimum destruction of the amino acids is required. Different areas of the same molecule may be hydrolyzed to different extents by a particular reagent because of steric hindrance arising from bulky amino acid side chains and also from the degree of accessibility allowed by the secondary and tertiary structure of the protein. Destruction of amino acids, particularly during acid hydrolysis, results in losses which may be corrected for,[1] but unfortunately the rate of destruction of free amino acids will not necessarily correspond to that of the residues in a polypeptide or protein. A kinetic study to determine the yield with respect to time during hydrolysis is required for reliable results.[2-4] The need to correct for losses can be largely obviated by the addition of a suitable agent to the hydrolytic medium, which will minimize amino acid destruction during hydrolysis. Certain individual amino acids will require special hydrolytic procedures to determine their content separately from the main protein hydrolysis, such as the mild acid hydrolytic procedure for phosphoserine.[5] Details of sample preparation are given elsewhere in this Handbook.

Hydrochloric acid has probably found the widest use in protein hydrolysis, but the nonoxidizing sulfonic acids provide good recoveries of most of the amino acids.[6-8] Alkaline hydrolysis is not generally used because of the considerable destruction of amino acids that occurs. It is still, however, the favored method in the determination of tryptophan.[9] Ultimately, the hydrolytic procedure chosen will be a compromise between achieving hydrolysis of the peptide bonds and minimizing amino acid destruction.

The use of proteolytic enzymes[10-12] for protein hydrolysis overcomes many of the disadvantages of acid hydrolysis, but requires complex incubation procedures, a

---

* This article was written by A. J. Smith.

knowledge of the susceptibility of the protein to a particular enzyme and suffers from the lack of stability of these enzymes.

The greatly increased sensitivity of modern analytical methods has led to the problem of contamination in amino acid determination. While contamination from hands[13,14] and glassware can provide considerable sources of error, the variability of reagent purity can be a large contributing factor in any consideration of contamination. Fluorimetric reagents for detection, HCl for hydrolysis and solvents for HPLC require particular attention. Care in the preparation of distilled or deionized water and ultrafiltration of eluting buffers is recommended.

Ion-Exchange Chromatography

The separation of a mixture of amino acids on an ion-exchange column results from ionic and hydrophobic interactions between the amino acid molecules and the resin. The degree of resolution achieved is dependent on these interactions and on the flow of eluting buffers through the column. A large number of variables influence these interactions; changing some of these parameters can rearrange the order of elution to a limited extent, but generally the amino acids will be eluted in the group order acidic, neutral and basic amino acids. This elution order arises from the pKa values of the amino acids[15] and the hydrophobic interactions between nonpolar side chains and the resin;[16] branching and shortness of the nonpolar side chains decrease elution time.[17]

To achieve good separation of a mixture of amino acids, the two prime requirements are adequate resolution and good sensitivity. The resolution of two adjacent peaks is dependent on the separation of their band centers and the widths of the bands. Figure 2 shows how overlapping peaks may be resolved by separating band centers or by reducing the peak widths. Although the top and middle profiles show peaks with the same widths, resolution is achieved in the middle profile because the band centers are further apart. The bottom profile shows adequately resolved peaks with the same band centers as the top profile due to the narrower width of the former peaks. The resolution between two adjacent peaks may be defined by the resolution ratio

$$Rr = \frac{v_a - v_b}{\sigma_a + \sigma_b}$$

where v_a and v_b are the elution volumes at peak maxima and σ_a and σ_b are the variances of the elution peaks (σ = baseline width/ 4). Resolution is essentially complete when $Rr \geq 1.5$. Good sensitivity of a system requires a high concentration of amino acid at the band center, since this provides a more readily observed response than the same amount in a larger volume.

Some of the factors involved in achieving separation of amino acid mixtures will be discussed in the following sections.

Operating Parameters
Column Dimensions

The width of a peak is directly proportional to the column cross-sectional area and to the square root of its length. While earlier analyzers used a column of 0.9 cm diameter and 150 cm length, this may be converted to a diameter of 0.65 cm to increase resolution and sensitivity. Modern developments include micro-bore systems with columns of 0.29 cm[18] or 0.175 cm[19] diameter and lengths shortened as far as 25-30 cm.[20,21]

Resin

Sulfonated polystyrene resins with about 8% cross-linking are generally used in amino acid analysis, the properties varying from one supplier to another. Information on resins commercially available is presented elsewhere in this Handbook.

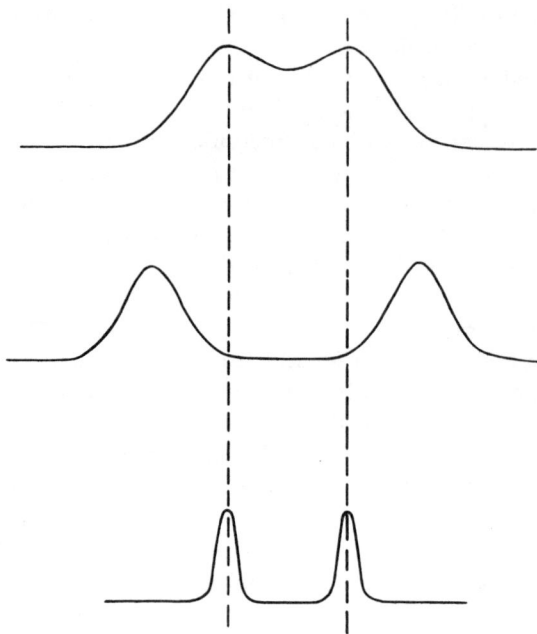

FIGURE 2. Separation of adjacent peaks. Overlapping peaks can be resolved by separating band centers or by reducing peak widths. (From Karger, B. L., Snyder, L. R., and Horvath, C., *Chromatography: An Introduction to Separation Science,* John Wiley & Sons, New York, 1973, 121. With permission.)

Resins vary in the divinylbenzene used in polymerization, extent of sulphonation, degree of cross-linking, and particle shape and size.

While early systems used crushed resins of irregular shape, the introduction of spherical resins has increased performance. The introduction of resins of small particle size (8 μm) has allowed sharper peaks to be obtained in micro-bore column systems, but increased back-pressure with diminished particle size limits their usage and standard-bore columns operate better with resins of 13-25 μm diameter. A mixture of various diameter particles can increase resolution.[22]

With careful treatment, a resin will last indefinitely. Contamination, particularly with microbial growth, metal ions, grease, or proteins should be avoided but techniques for cleaning resins are available should contamination occur.[17]

Sample Application

The volume of a sample should not exceed 6% of the bed volume of the column and a smaller sample is preferable. The pH of the loading buffer is rather a matter of individual choice. It has been reported that the pH of the sample can favorably affect the resolution of the analysis in the early parts of the chromatogram, but alteration of elution buffers could probably achieve the same effects even with samples applied in the pH 2.2 buffer of Moore and Stein.[23]

Flow Rate

The peak dimensions will be affected by the flow rate through the column. Peak width varies directly as a function of the square root of the flow rate and the height varies inversely to this same function.[24,25] In practice it is difficult to achieve stable pump delivery at low flow rates and since pressure is directly proportional to flow rate, column back-pressure becomes a limiting factor. Consequently, little alteration of flow rate to improve resolution is performed.

Buffers

The buffers used for elution provide one of the most powerful ways of altering resolution. Many different systems have been described and may be classified by the counter ion used (Na or Li) and by the method of buffer change (step-wise or gradient elution). Procedures for both single and two-column analysis of amino acids are available. Details of individual procedures are given elsewhere in this Handbook.

The choice between step-wise and gradient elution depends on the equipment available. The majority of commercially available analysers operate with step-wise elution, but programed gradient elution analyzers offer the advantage of not requiring valves and timers. Use of a single or two-column system again depends on the equipment available and individual choice of the operator, both systems having their own merits. A single column system requires half the amount of sample needed for a two-column procedure, simplifies calibration and automation, but suffers from elution of a large, broad ammonia peak unless a filter column is incorporated in the system; in the two-column system, ammonia elutes at more than twice the phenylalanine elution volume and hence provides no problems.

Citrate is the most commonly used buffer component. Citrate buffers (0.2 N) will elute all the acidic and neutral amino acids, although higher ionic strengths are required to elute the basic amino acids. In step-wise elution, three buffers are generally used with the pH ranges of 2.5 to 3.5 for acidic and some neutral amino acids, 3.8 to 4.5 for the remaining neutral amino acids, and 5.0 to 6.5 for the basic amino acids. Buffers of higher pH (up to 11) and lower ionic strength are sometimes used for resolution of basic amino acids when a buffering component such as phosphate[26] or borate[27] is required. Poor recoveries can be experienced with such conditions.[28]

The counter ion to the buffering citrate is usually either sodium or lithium. Sodium has probably found wider use for protein hydrolysates, but lithium is favored for physiological fluids and enzymatic digests because of its ability to distinguish the amides asparagine and glutamine from each other. Systems using a combination of Na and Li buffers have been reported.[29,30] Step-wise elution typically uses a sodium concentration of 0.2 N for elution of acidic and neutral amino acids and 0.35 to 0.4 N for elution of basic amino acids, while lithium concentrations are usually slightly higher at 0.28 to 0.3 N for acidic and neutral amino acids and 0.6 N for basic amino acids. In gradient elution, the final sodium concentration may be as high as 1.2 N, while lithium buffers may be higher at 1.5 N.[31]

The addition of small amounts of organic solvents to buffers has been observed to improve resolution in certain regions of the chromatogram. The addition of varying concentrations of alcohols has proved particularly useful in improving the resolution between threonine and serine. The organic solvent should only be added to one buffer in a system and should not be used in the equilibration of the column otherwise separation of some of the other amino acids may be adversely affected.

Antioxidants such as thiodiglycol[28] or ascorbic acid[7] are often added to buffers to prevent oxidation of methionine to the sulphoxide. While ultrafiltration of buffers is recommended, the addition of an inhibitor of microbial growth such as 0.01% (v/v)-octanoic acid or 0.05% (w/v)-pentachlorophenol is worthwhile.[32] Use of a nonionic detergent, such as Brij-35, alters solvent surface tension to permit a faster flow rate to be achieved without an undue increase in column back-pressure.[28]

Temperature

The degree of separation of the amino acids is a function of column temperature, which influences the interaction of the amino acids with the resin, the dissociation of citrate buffers and the column resistance to buffer flow.

The distribution of amino acids between resin and buffer phases at lower temperatures results in broad peaks, but as the column temperature is increased, the peaks become sharper with a corresponding improvement in resolution. Generally, a constant temperature of 50 to 60°C is used for protein hydrolysate analysis. The order of elution of some amino acids is influenced by temperature; aspartic acid elutes after threonine at lower temperatures but precedes it at temperatures above 45°C. Temperature shifts are utilized to achieve the resolution required in the analysis of physiological fluids.

As the temperature is raised, the pKa values for the acid dissociations of citric acid become higher; this may result in altered resolution because of the pH changes.

As the column temperature is increased, the resistance to flow is found to decrease.

Ligand-Exchange Chromatography

This method relies on the formation of metal-ion complexes of the amino acids prior to their separation on ion-exchange resins conventionally used in amino acid analysis.

$$R-\underset{\underset{NH_3^+}{|}}{\overset{\overset{COO^-}{|}}{C}}-H + M^{++} \rightarrow \left[R-\underset{\underset{NH_2}{|}}{\overset{\overset{COO^-}{|}}{C}}-H \cdots M^{++} \right]^+ + H^+$$

The metal ion utilized may be Zn^{++}, Cu^{++}, Ni^{++}, Cd^{++}, or Co^{++}, although Zn^{++} is favored[33] because of the good elution characteristics of its amino acid complexes and its ability to chelate with citrate prior to reaction of the column effluent with ninhydrin. There are greater differences in the stability constants for zinc-amino acid complexes than the ionization constants of amino acids, which theoretically should provide a powerful system for resolving the amino acids. The sensitivity of the system is in the 0.1 to 0.5 μmol. range utilizing the ninhydrin detection system, although this may be increased to the nmol. range by the use of a fluorometric detection procedure.[34] The method utilizes two columns, which are eluted with different concentrations of sodium acetate buffer containing zinc acetate.

Although this method has not found wide use, it has the advantage of only requiring a single buffer to elute each column hence obviating the requirement for valves and timers.

Pyridoxal Method

The formation of pyridoxyl derivatives of amino acids, and their subsequent detection by fluorometry or radioactive monitoring, increases the sensitivity over that of ninhydrin by several orders of magnitude.[35] Reaction of the amino acid with pyridoxal followed by reduction produces the corresponding pyridoxyl derivative, which fluoresces and can be tritiated by use of tritiated borohydride in the reduction step.

The chromatographic separation of the pyridoxal derivatives may be carried out on a single column system using conventional resins, but requires slightly different conditions to underivatized amino acids. A higher column temperature of 70°C is required and the buffer is 0.4 *M*-sodium citrate, pH 3.33 to 7.5.

High Performance Liquid Chromatography (HPLC)

HPLC has stimulated much interest in amino acid analysis because of the speed, economy, and the ease of use of this technique compared with conventional ion-exchange techniques. Nonpolar stationary phase columns have gained widest use and care is required in the choice of both this stationary phase and the mobile phase. Although the separation of underivatized amino acid mixtures has been reported,[36] the resolution is not as good as that obtained in conventional ion-exchange chromatography. The developments in column packing materials and the small particle diameters that are being introduced will almost certainly improve the degree of resolution, although HPLC is unlikely to replace ion-exchange techniques in routine amino acid analysis at the present time.

The use of HPLC in the analysis of phenylthiohydantoin amino acids has already found wide use. Protein sequencing studies using the Edman degradation procedure requires rapid quantitative determination at high sensitivity of all the PTH-amino acids obtained at each cycle. In the past, gas-liquid chromatography or thin-layer chromatography have been commonly used for this purpose, but the introduction of HPLC has overcome many of the disadvantages of these other techniques. Octadecyl-silane columns eluted with acetonitrile-sodium acetate gradients have produced good results,[37] but use of a cyanopropylsilane column improves resolution, sensitivity, column lifetime, and reproducibility.[38] Again, the introduction of smaller diameter particles for the column packing should improve resolution.

REFERENCES

1. Reese, M. W., The estimation of threonine and serine in proteins, *Biochem. J.*, 40, 632, 1946.
2. Robel, E. J. and Crane, A. B., An accurate method for correcting unknown amino acid losses from protein hydrolysates, *Anal. Biochem.*, 48, 233, 1972.
3. Robel, E. J., Method of extrapolation for yield-decay-type data, *J. Agr. Food Chem.*, 21, 906, 1973.
4. Hapner, K. D. and Hamilton, K. R., Basic computer program for amino acid analysis data, *J. Chromatogr.*, 93, 99, 1974.
5. Cohen-Solal, L., Lian, J. B., Kossiva, D., and Glimcher, M. J., Identification of organic phosphorus covalently bound to collagen and non-collagenous proteins of chicken-bone matrix, *Biochem. J.*, 177, 81, 1979.
6. Lui, T.-Y. and Chang, Y. H., Hydrolysis of proteins with p-toluene sulfonic acid: determination of tryptophan, *J. Biol. Chem.*, 246, 2842, 1971.
7. Niece, R. L., Automated single-column analysis of amino acids using ascorbic acid as reductant for air-stable ninhydrin, *J. Chromatogr.*, 103, 25, 1975.
8. Penke, B., Ferenczi, R., and Kovacs, K., A new acid hydrolysis method for determining tryptophan in peptides and proteins, *Anal. Biochem.*, 60, 45, 1975.
9. Hugli, T. E. and Moore, S., Determination of the tryptophan content of proteins by ion exchange chromatography of alkaline hydrolysates, *J. Biol. Chem.*, 247, 2828, 1972.
10. Sletten, K., Dus, K., De Klerk, H., and Kamen, M. D., Cytochrome c_2 of Rhodospirillum rubrum. I. Molecular properties of the protein and amino acid sequences of its peptides derived by the action of trypsin and thermolysin, *J. Biol. Chem.*, 243, 5492, 1968.
11. Salnikow, J., Liao, T.-H., Moore, S., and Stein, W. H., Bovine pancreatic deoxyribonuclease A. Isolation, composition and amino acid sequences of the tryptic and chymotryptic peptides, *J. Biol. Chem.*, 248, 1480, 1973.
12. Garner, M. H., Garner, W. H., and Gurd, F. R. N., Recognition of primary sequence variations among sperm whale myoglobin components with successive proteolysis procedures, *J. Biol. Chem.*, 249, 1513, 1974.
13. Hamilton, P. B., Amino acids on hands, *Nature (London)*, 205, 284, 1965.
14. Oro, J. and Skewes, H. B., Free amino acids on human fingers: The question of contamination in microanalysis, *Nature (London)*, 207, 1042, 1965.
15. Eichhorn, M. M., Data on the naturally occurring amino acids, *Handbook of Biochemistry*, Sober, H. A., Ed., CRC Press, Boca Raton, Fla., 1970, B3.
16. Stein, W. H. and Moore, S., Chromatographic determination of the amino acid composition of proteins, *Cold Spring Harbor Symp. Quant. Biol.*, 14, 179, 1950.
17. Robinson, G. W., Factors affecting separation of amino acids on ion-exchange columns, *Amino acid determination*, Blackburn, S., Ed., Marcel Dekker Inc., New York, 1978, 47.

18. Liao, T. H., Robinson, G. W., and Salnikow, J., Use of narrow-bore columns in amino acid analysis, *Anal. Chem.*, 45, 2286, 1973.
19. Durrum Chemical Corp., Palo Alto, California.
20. Atkin, G. E. and Ferdinand, W., Accelerated amino acid analysis: Studies on the use of lithium citrate buffers and the effect of n-propanol in the analysis of physiological fluids and protein hydrolysates, *Anal. Biochem.*, 38, 313, 1970.
21. Voelter, W. and Zech, K., High performance liquid chromatographic analysis of amino acids and peptide-hormone hydrolysates in the picomole range, *J. Chromatogr.*, 112, 643, 1975.
22. Benson, J. V., Jr., Multipurpose resins for analysis of amino acids and ninhydrin-positive compounds in hydrolysates and physiological fluids, *Anal. Biochem.*, 50, 477, 1972.
23. Moore, S. and Stein, W. H., Procedures for the chromatographic determination of amino acids on four percent cross-linked sulfonated polystyrene resins, *J. Biol. Chem.*, 211, 893, 1954.
24. Hamilton, P. B., Bogue, D. C., and Anderson, R. A., Ion exchange chromatography of amino acids — analysis of diffusion (Mass transfer) mechanisms, *Anal. Chem.*, 32, 1782, 1960.
25. Hamilton, P. B., Effect of variables on column pressure, elution volume, peak width, height and separation (resolution) in the ion exchange chromatography of amino acids on sulfonated styrene-divinylbenzene copolymer resins, *Handbook of Biochemistry*, Sober, H. A., Ed., CRC Press, Boca Raton, Fla., 1970.
26. Murren, C., Stelling, D., and Felstead, G., An improved buffer system for use in single-column gradient-elution ion-exchange chromatography of amino acids, *J. Chromatogr.*, 115, 236, 1975.
27. Hare, P. E., Ion-exchange chromatography in lunar organic analysis *Space Life Sciences*, 3, 354, 1972.
28. Moore, S. and Stein, W. H., Chromatography of amino acids on sulfonated polystyrene resins, *J. Biol. Chem.*, 192, 663, 1951.
29. Ellis, J. P., Jr. and Garcia, J. B., Jr., An accelerated operation of a three sample amino acid analyser, *J. Chromatogr.*, 87, 419, 1973.
30. Oulevey, J. and Heitefuss, R., Ein verbessertes einsäulenprogramm zur analyse der freien aminosäuren und säureamide in pflanzenextrakten, *J. Chromatogr.*, 94, 283, 1974.
31. Thomson, A. R. and Miles, B. J., Ion-exchange chromatography of amino acids: improvements in the single column system, *Nature (London)*, 203, 483, 1964.
32. Atkin, G. E. and Ferdinand, W., Accelerated amino acid analysis with lithium citrate buffers. The effect of resin cross-linkage on elution patterns and the avoidance of microbial contamination of columns, *J. Chromatogr.*, 62, 373, 1971.
33. Wagner, F. W. and Shepherd, S. L., Ligand exchange amino acid analysis: resolution of some amino sugars and cysteine derivatives, *Anal. Biochem.*, 41, 314, 1971.
34. Maeda, M., Tsuji, A., Ganno, S., and Onishi, Y., Fluorophotometric assay of amino acids by using automated ligand-exchange chromatography and pyridoxal-zinc(II) reagent, *J. Chromatogr.*, 77, 434, 1973.
35. Lustenberger, N., Lange, H.-W., and Hempel, K., Amino acid analysis on the picomole scale, *Angew. Chem. (Int. Ed.)*, 11, 227, 1972.
36. Hancock, W. S., Bishop, C. A., and Hearn, M. T., The analysis of nanogram levels of free amino acids by reverse-phase high pressure liquid chromatography, *Anal. Biochem.*, 92, 170, 1979.
37. Harris, J. U., Robinson, D., and Johnson, A. J., Analysis of (3-phenyl, 2-thio)hydantoin amino acids by high pressure liquid chromatography: Comparison of 3 programs with particular reference to the glutamic and aspartic derivatives, *Anal. Biochem.*, 105, 239, 1980.
38. Johnson, N. D., Hunkapiller, M. W., and Hood, L. E., Analysis of phenylthiohydantoin amino acids by high performance liquid chromatography of DuPont Zorbax cyanopropylsilane columns, *Anal. Biochem.*, 100, 335, 1979.
39. Karger, B. L., Snyder, L. R., and Horvath, C., *Chromatography An Introduction to Separation Science*, John Wiley & Sons, New York, 1973, 121.

Section II.II.

ION-EXCHANGE RESINS FOR AMINO ACID DETERMINANTS

The resin generally used in amino acid analysis is sulfonated polystyrene with about 8% cross-linking. The first automated ion-exchange systems used crushed resins of irregular shape, but these have now been replaced by spherical resins because of their superior performance. Nevertheless, appreciable differences in performance may still be encountered between resins from different commercial suppliers and even between different batches of nominally the same resin. Atkin and Ferdinand[1] studied the elution pattern of acidic and neutral amino acids using eight batches of Aminex AS resin. Some resins gave good resolution of the threonine/serine and glycine/alanine pairs, others gave sharp peaks too close together, and others gave peaks that were too broad for satisfactory resolution of these pairs. With one exception the performance of all the resins was satisfactory in the analysis of basic amino acids. These authors considered the use of per cent cross-linkage to be of no practical value in classifying resins for amino acid analysis.

Some of the factors accounting for variations in repetitive batches of resin have been listed by Hamilton.[2] These comprise (1) irregular shape, (2) wide range of diameter, (3) variations in cross-linking from bead to bead, (4) nonhomogeneity, and (5) incomplete sulfonation. Rahm et al.[3] studied the relation between the reactions conditions of sulfonation of a styrene-divinylbenzene copolymer and the resolution efficiency of the resulting resin in separating neutral amino acids. The following parameters were varied in the sulfonation process: the swelling agent for the copolymer, the concentration of the sulfonating agent, the duration of sulfonation, and the temperature of sulfonation. A higher temperature, longer sulfonation time, and higher concentration of sulfonating agent allowed the preparation of resins which showed a better resolution of threonine-serine, but poorer resolution of glycine-alanine and particularly tyrosine-phenylalanine. The use of a swelling agent for the copolymer prior to sulfonation had the reverse effect.

The matrix of the cation-exchange resins used for separating amino acids is formed from a styrene-divinylbenzene (DVB) copolymer. The cross-linking comonomer (technical grade DVB) is not a single compound but contains in addition to *m*- and *p*-divinylbenzene, other monomers, e.g., ethylvinylbenzenes and also components, mainly diethylbenzenes, that do not polymerize but do affect the properties of the copolymer. *m*-Divinylbenzene is considered to be a more suitable cross-linking agent because, owing to its lower reactivity, it creates conditions for the formation of a copolymer with a more regular arrangement of cross-linkages.

Rahm[4] compared the performance of two resins, cross-linked by the same amounts of technical grade DVB and *m*-DVB in the separation of acidic and neutral amino acids. He drew the following conclusions: (1) Bands of amino acids eluted from the column packed with the *m*-DVB resins were narrower than those of the resin cross-linked with technical grade DVB, (2) With the *m*-DVB resin, the difference in the peak elution volume of threonine and serine was less than that of glycine and alanine was equal to, and that of tyrosine and phenylalanine greater than the corresponding values with the resin made from technical grade DVB. (3) The narrower elution peaks obtained with the *m*-DVB resin had a positive effect on the resolution. (4) The analysis time of an amino acid mixture was the same with both resins. (5) The hydrodynamic resistance of the *m*-DVB resin was slightly higher than that of the standard resin.

By paying greater attention to process variables, Benson[5] was able to prepare reproducible resins. All his resins were made in small-batch processes. Certain variations are inherent in large-batch production, e.g., use of practical grade monomers, wide particle size distribution, and consequent wide variation in cross-linking. By employing purified monomers and small-scale operation Benson was able to prepare a polymer relatively free of linear material and to narrow the size range of the batch. This author described a single-resin system capable of analyzing amino acids with improved resolution, shorter analysis times, and reduced operating back-pressures. The resin used, Hamilton Spherical Chromatographic Resins, type HP-AN90, was a 7.0% cross-linked resin with spherical particles having a diameter of 7 to 20 μ. Resins were made specifically in an effort to reproduce HP-AN90 when it was apparent that this was the optimal resin. The result was several batches, all giving the same resolving capabilities as the model resin. Mixing of batches produced no loss in resolution.

The resin may play a key role in the differences in performance between different analytical systems. Because of the high expense of resins, most laboratories manipulate other parameters to achieve desired separations. It is possible that many of the recommended variations in buffer recipes that are reported in the literature reflect variability of resins. Any attempt to standardize resin types and ensure their reproducibility between different batches is to be welcomed.

REFERENCES

1. Atkin, G. E. and Ferdinand, W., *J. Chromatogr.*, 62, 373, 1971.
2. Hamilton, P. B., *Advances in Chromatography*, Vol. 2, Giddings, J. C. and Keller, R. A., Eds., Marcel Dekker, New York, 1966.
3. Rahm, J., Weinova, H., and Prochazka, Z., *J. Chromatogr.*, 60, 256, 1971.
4. Rahm, J., *J. Chromatogr.*, 115, 455, 1975.
5. Benson, J. V., *Anal. Biochem.*, 50, 477, 1972.

Section II.III.

THE GAS CHROMATOGRAPHIC DETERMINATION OF AMINO ACIDS

Although the majority of amino acid determinations is at present carried out using the automatic analyzer, gas chromatographic methods are the subject of an increasing number of studies. The present article describes some of the factors relevant to choosing an appropriate gas chromatographic technique for determining amino acids.

COLUMNS

The advantages of using capillary columns for separating amino acid derivatives have been emphasized; a capillary column is much more sensitive than a packed column[1-3] (Figure 1). This feature is of importance, for example, in the analysis of plasma amino acids, since the plasma sample size can be scaled down to 20 to 50 μl in the case of premature or newborn infants. Picomole amounts of amino acids are readily quantitated. Open tubular or capillary columns are possibly the most powerful separating technique yet devised. The unrestricted flow path for the carrier gas through the column makes it possible to construct very long open tubular columns without prohibitive pressure drops.

Columns in which a thin film of the liquid phase is coated on the inside surface of the column are called wall coated open tubular (or WCOT) columns; those in which a porous layer is formed on the inside wall of the open tube and coated with the liquid phase film are termed support coated open tubular (or SCOT) columns. Generally open tubular columns give much better separations in equal or shorter time than packed columns. The advantages of SCOT columns over WCOT columns is that their sample capacity is higher, they have less stringent requirements concerning instrumentation, and in many cases they can provide an even shorter analysis time than WCOT columns.

MacKenzie and Tenaschuk point out that the resolving power of capillary columns is undoubtedly advantageous for the analysis of samples which cannot be adequately resolved using packed columns.[4] However, when applied to N-acyl alkyl esters of the protein amino acids, the quoted analysis times using capillary columns are readily achieved on packed columns with moderate program rates and excellent resolution. The advantages of capillary columns for ultramicro analysis are unquestioned, but in the opinion of MacKenzie and Tenaschuk the additional expense and delicacy of technique required with capillary columns do not outweigh the simplicity, cheapness, and ruggedness of the packed column for the routine analysis of protein amino acid derivatives.

Glass columns have long been considered the most inert for the chromatography of labile molecules. Fenimore et al. claimed that nickel tubing was comparable in "chemical inertness" to glass and in some cases better than glass.[5] Gehrke et al., however, chromatographed N-TFA amino acid n-butyl esters on a EGA bilayer packing in a 1 m × ¼ inch nickel tube and found that histidine was completely destroyed, while the response of arginine and cystine was greatly reduced.[6] In this instance glass columns were much superior to nickel columns. Smith and Wonnacott used 150 ft stainless steel capillary columns in their studies on the resolution of enantiomeric isopropyl esters of N-TFA amino acids, and thus claimed to avoid difficulties and costs encountered in loading and maintaining glass capillary columns.[7]

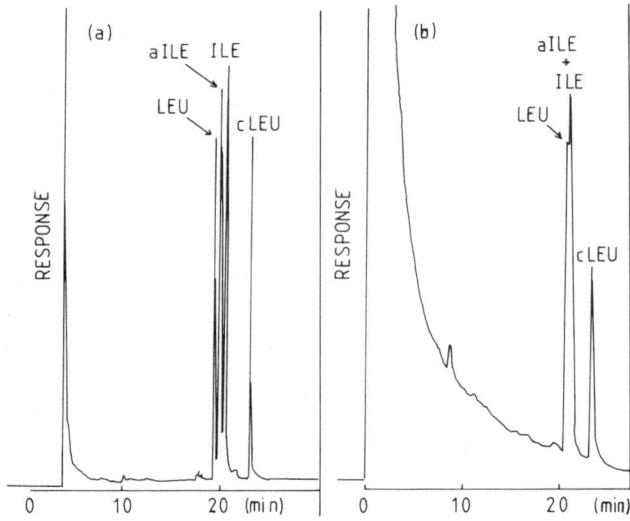

FIGURE 1. Gas chromatograms of a mixture of leucine, isoleucine, alloisoleucine, and cycloleucine as isobutyl ester, N(O)-heptafluorobutyrate derivatives (a) on an OV-101-coated glass capillary column, (b) on an OV-1-packed column. (Reproduced from Desgres, J., Boisson, D., and Padieu, P., J. Chromatogr., 162, 133, 1979. With permission.)

LIQUID PHASES

With one exception, good single-column separations of N-acyl amino acid esters have been obtained on nonpolar stationary phases such as SE-30. It is normal practice with such stationary phases operated over a temperature program of say 100 to 250°C to use a compensating column to counteract column bleed. In the experience of MacKenzie and Tenaschuk a virtually flat baseline can be obtained provided the columns are properly conditioned and the carrier gas flow is carefully controlled.[4] Under such conditions a reference column is not strictly necessary. Gehrke and Takeda have studied the separation characteristics of several siloxane liquid phases of different polarities.[8] No single phase or combination of these phases yielded a complete separation of the 20 protein amino acid N-TFA n-butyl esters on a single column. On an Apiezon M column no peak for histidine was observed and three distinct clusters of peaks that could not be further resolved were obtained.[9] A combination of a Tabsorb and a Tabsorb HAC column was essential for the separation of all 20 protein amino acids.

One essential for the successful determinations of amino acids by gas chromatography is stability of the derivatives on the column. The interactions of the N-TFA n-butyl ester derivatives of histidine, arginine, and cystine with the liquid phase ethylene glycol adipate (EGA) and the solid support have been studied, changes in the RWR (relative weight response) being used as an index of substrate or solid support interaction.[6] The essentially nonpolar Gas-Chrom® Q and the more reactive Chromosorb® W AW were coated with a nonpolar OV-101 polymer to determine whether the solid support contributed to the breakdown of the N-TFA n-butyl esters derivatives of these three amino acids. A bilayer column was prepared by coating OV-101/Gas-Chrom® Q with polar EGA to determine the interaction of the EGA polymer with these derivatives. The polar EGA was primarily responsible for the breakdown of histidine. The decomposition of arginine and cystine was caused mainly by the solid support, Chromosorb® W AW, and the more active support increased the extent of breakdown. As the loading of the EGA polymer on the Gas-Chrom® Q of the polymer bilayer was decreased, an increased histidine response was seen, while the arginine and cystine response was unaffected.

DETECTORS

The minimum detectable limit of the flame ionization detector has been reported as about 1 ng of an individual amino acid HFB derivative or 5 ng of amino acids. As the HFB derivatives have high electron affinity they are very suitable for electron capture detection. The sensitivity for HFB amines is increased more than 100-fold using electron capture detection,[10] while a ^{63}Ni source can detect 1 pg of methionine and 2 pg of cysteine.[11] Bengtsson and Odham found that sensitivity to HFB amino acid derivatives was increased 100-fold using electron capture detection with optimization of the technique.[2] The principal drawback of this type of detector in amino acid analysis is the difficulty of avoiding a heavy baseline drift due to the temperature programing. The effect is generally attributed to irregularities in carrier gas flow.

A nitrogen-selective flame ionization detector (NPD) showed a sensitivity improvement over the conventional flame ionization detector of approximately 80 times; enhancement values for individual N-acetyl amino acid n-propyl esters are shown in Table 1.[3] The detection limit of the system is about 5 picomoles. Mass fragmentography is a sophisticated detection system for individual amino acids at levels less than a picogram.[2] It is particularly useful in searching for specific amino acids present in small amounts in unknown complex biological samples. The reliability of the analysis at these levels where the amounts determined approach the normal background levels is of importance.

SYNTHESIS OF DERIVATIVES

Several derivatization procedures for making amino acids volatile enough for gas chromatographic separation have been devised; they mostly require two synthetic steps, esterification of the carboxyl group and acylation of the α-amino and other functional groups. Relatively few of these procedures have gained general acceptance. For example, the use of trimethylsilyl (TMS) amino acids, advocated by Gehrke and co-workers[12] while enabling Pocklington[13] to quantitatively determine 18 natural protein amino acids and other nonprotein amino acids after extraction from lyophilized seawater, failed to give reproducible analyses in the hands of Sarkar and Malhotra.[9]

The N-trifluoroacetyl n-butyl esters of amino acids have been the subject of intensive studies;[14,15] the closely related N-heptafluorobutyryl (HFB) derivatives have many desirable characteristics and are often the derivative of choice.[3] The HFB amino acid esters are relatively stable, are volatile enough for practical handling and permit a reasonably short analysis time. Additionally the fluoro derivatives give a very high response with electron capture detectors. HFB amino acid esters which have been used in gas chromatographic studies include the isopropyl, isoamyl, isobutyl and n-propyl derivatives.

Esterification

Careful control of the conditions during the preparation of the volatile derivatives is necessary if quantitative determination of amino acids is to be attained. MacKenzie and Tenaschuk investigated ab initio the process of amino acid isobutyl ester preparation.[4] 25 µℓ amounts of the amino acid standard, containing 2.5 µmoles/mℓ of each amino acid, were esterified using 100 and 200 µℓ of 3 M HCl-isobutanol and acylated. No significant differences in the RMR (relative molar response) values of the amino acid derivatives was observed.

Table 1
SENSITIVITY ENHANCEMENT OF THE NITROGEN-SELECTIVE DETECTOR (NPD) OVER THE FLAME IONIZATION DETECTOR (FID) FOR N-ACETYL AMINO ACID n-PROPYL ESTERS

Amino acid	Sensitivity enhancement
Alanine	133
Valine	85
Glycine	149
Isoleucine	79
Leucine	82
Proline	96
Threonine	112
Serine	105
Aspartic acid	81
Methionine	85
Cysteine	117
Phenylalanine	84
Hydroxyproline	80
Glutamic acid	80
Tyrosine	56
Ornithine	176
Histidine	140
Lysine	172
Arginine	180
Tryptophan	96

REFERENCE

Adams, R. F., Vandemark, F. L., and Schmidt, G. J., *J. Chromatogr. Sci.*, 15, 63, 1977.

Reproduced from the Journal of Chromatographic Science by permission of Preston Publications Inc.

Compared with evaporation of excess isobutanol at room temperature, no significant losses were observed following evaporation at 50°C, but at 75°C there was significant loss of the more volatile amino acid derivatives. The finding that with a short reaction time the yields of the HFB amino acid isobutyl esters are better than those of the corresponding n-butyl esters is probably dependent on a faster nucleophilic displacement reaction, assumed to be a mixed SN_1 and SN_2, at the carboxyl group of the amino acid.[3] Esterification of amino acids occurred at least as effectively in 3.5 M HCl-n-propanol as in 8 M HCl-n-propanol.[16]

Some authors have found it convenient to add acetyl chloride to the alcohol rather than to add HCl gas, considering the procedure to be an excellent alternative way of preparing the esterifying agent. A decrease of 8 to 10% in the RMR values of serine, lysine, and cystine, however, was found when adding redistilled acetyl chloride to isobutanol in place of using 3 M HCl-isobutanol.[4] Furthermore, the introduction of another reagent, acetyl chloride, increases the potential for the introduction of impurities with deleterious effects on the esterification.

Acylation

The acylation reaction is performed at an elevated temperature. MacKenzie and Tenaschuk have acylated amino acid isobutyl esters at different temperatures.[17] Acylation of most of the esters was essentially complete after heating at 75°C for 10 min, but heating for 5 min at 75°C was inadequate for complete acylation of valine and isoleucine. In general the time and temperature of acylation are not critical, although a relatively high temperature is necessary for the N-acylation of arginine alkyl esters.

The need for pure reagents for acylation has been emphasized.[18] HFBA (heptafluorobutyric anhydride) of purity greater than 99% should be used; considerably reduced responses for threonine, tyrosine, and arginine have been observed using unsatisfactory batches of reagent. It is also important that the ethyl acetate should be anhydrous. Accumulation of peroxides has a negative effect on the acylation of methionine and arginine; however, peroxides were not detected in reagent grade ethyl acetate after storage at 0°C for several months.

The imidazole nitrogen of histidine cannot be readily acylated using the perfluorocarboxylic and anhydrides. Several solutions to the problem have been proposed. On-column acylation by co-injecting the sample with an anhydride converts the monoacyl derivative of histidine to the diacyl derivative. Trifluoroacetic and heptafluorobutyric anhydrides have been used for this purpose.[19,20] MacKenzie and Tenaschuk found that on-column acylation using 0.2 $\mu\ell$ of acetic anhydride per microliter of sample was sufficient to obtain the maximal response for histidine. Similar results were obtained by adding acetic anhydride directly to the sample. Butyric and propionic anhydrides gave comparable results. The precision of the determination for each derivative was within ± 1.5%. N^α-TFA histidine n-butyl ester has been reacted with ethoxyformic anhydride (EFA) to form N^α-TFA-N^π-carbethoxy histidine n-butyl ester, which was considered to be a stable derivative suitable for gas chromatography.[21,22] MacKenzie and Tenaschuk, however, found that results of comparable precision to those using EFA could be obtained by on-column acylation and saw no benefit in using the former technique.[17]

It may be concluded that although the gas chromatographic procedure for the determination of amino acids is capable of yielding accurate and precise results, its principal disadvantage is inherent in the method itself, i.e., the necessity to convert the amino acid to a suitable derivative, with the possibility of introducing additional sources of error and variability which are not encountered using ion-exchange resin techniques.

REFERENCES

1. Desgres, J., Boisson, D., and Padieu, P., *J. Chromatogr.*, 162, 133, 1979.
2. Bengtsson, G. and Odham, G., *Anal. Biochem.*, 92, 46, 1979.
3. Adams, R. F., Vandemark, F. L., and Schmidt, G. J., *J. Chromatogr. Sci.*, 15, 63, 1977.
4. MacKenzie, S. L. and Tenaschuk, D., *J. Chromatogr.*, 171, 195, 1979.
5. Fenimore, D. C., Whitford, J. H., Davis, C. M., and Zlatkis, A., *J. Chromatogr.*, 140, 9, 1977.
6. Gehrke, C. W., Younger, D. R., Gerhardt, K. O., and Kuo, K. C., *J. Chromatogr. Sci.*, 17, 301, 1979.
7. Smith, G. G. and Wonnacott, D. M., *Anal. Biochem.*, 109, 414, 1980.
8. Gehrke, C. W. and Takeda, H., *J. Chromatogr.*, 76, 63, 1973.
9. Sarkar, S. K. and Malhotra, S. S., *J. Chromatogr.*, 170, 371, 1979.
10. Clarke, D. D., Wilk, S., and Gitlow, S. E., *J. Gas Chromatogr.*, 4, 310, 1966.
11. Zumwalt, R. W., Kuo, K., and Gehrke, C. W., *J. Chromatogr.*, 57, 193, 1971.
12. Gehrke, C. W. and Leimer, K., *J. Chromatogr.*, 57, 219, 1971.

13. Pocklington, R., *Anal. Biochem.*, 45, 409, 1972.
14. Gehrke, C. W., Kuo, K., and Zumwalt, R. W., *J. Chromatogr.*, 57, 209, 1971.
15. Kaiser, F. E., Gehrke, C. W., Zumwalt, R. W., and Kuo, K. C., *J. Chromatogr.*, 94, 113, 1974.
16. Kirkman, M. A., *J. Chromatogr.*, 97, 175, 1974.
17. MacKenzie, S. L. and Tenaschuk, D., *J. Chromatogr.*, 173, 53, 1979.
18. Felker, P., *J. Chromatogr.*, 153, 259, 1978.
19. Hall, N. T. and Nagy, S., *J. Chromatogr.*, 171, 392, 1979.
20. Jonsson, J., Eyem, J. and Sjoquist, J., *Anal. Biochem.*, 51, 204, 1973.
21. Moodie, I. M., *J. Chromatogr.*, 99, 495, 1974.
22. Pearce, R. J., *J. Chromatogr.*, 136, 113, 1977.

Section II.IV.

THE DETERMINATION OF AMINO ACID ENANTIOMERS BY GAS CHROMATOGRAPHY

Geochemical studies of D and L enantiomers of amino acids have greatly aided our understanding of the occurrence and fate of amino acids in terrestrial and extraterrestrial samples. Of various methods available for these kinds of studies, gas chromatography has probably been the most useful. The gas chromatographic determination of the enantiomers of amino acids has been approached in two different ways; (1) the enantiomers are converted to derivatives using an optically active reagent and the diastereomers thus formed are separated using a nonchiral chromatographic system,[1-3] and (2) the enantiomers are separated directly on a chiral stationary phase after derivative formation with a nonchiral reagent.[4-8]

The first method has four principal disadvantages; (1) the choice of derivatization reagents is limited by the requirement of optical activity, (2) systematic errors are introduced due to differences in the reaction kinetics of derivative formation between two enantiomers, (3) the derivatization reagent must be optically pure, and (4) errors may be encountered due to the increased probability of racemization at one of the asymmetric carbon atoms during derivative formation. Direct separation of enantiomers on a chiral stationary phase has the advantage that reagents generally used for derivatization in gas chromatography may be employed.

The separation of diastereomers may have some advantages. The method is applicable to a larger variety of compounds than the alternative procedure and separations can be achieved with conventional stationary phases on glass capillaries which are commercially available. Among diastereomeric derivatives that have been used for the gas chromatographic resolution of amino acids are the N-TFA amino acid (+)-2-butyl esters,[1] the N-PFP amino acid (+)-3-methyl-2-butyl esters[2] and the N-TFA-L-prolyl-peptide methyl esters.[3] The quantitative esterification of amino acids with higher alcohols is difficult, however, because of the poor solubility of the amino acids in the alcohols. High acid concentrations (7 N) may be necessary to obtain good yields of the appropriate esters.[2]

The use of chiral phases for the resolution of enantiomers of amino acids was pioneered by Gil-Av and his colleagues and developed by many later workers.[4-8] The majority of suitable phases are N-TFA dipeptide cyclohexyl esters. The diamide phases of Feibush are even more efficient; they show high resolution factors and generally lower polarity than the dipeptide phases, which reduces retention times.[9] Most protein amino acids can be separated in short analysis times, but some amino acid enantiomers, such as histidine, arginine, and tryptophan, have often given poor results. The separation of derivatives of proline and aspartic acid may also be difficult.[2]

In choosing an optically active stationary phase several criteria have to be considered. The phase should possess good thermal stability, high-coating efficiency, low-melting characteristics, good selectivity, and resolution efficiency for enantiomeric amino acids. Smith and Wonnacott have described columns which they consider meet these desirable requirements.[10] N-docosanoyl-L-valyl-tertbutylamide and N-octadecanoyl-L-valyl-L-valine cyclohexyl ester were used as mixed chiral phases on 150 ft stainless steel capillary columns to separate enantiomeric isopropyl esters of N-TFA amino acids. When the capillary columns were used below 150°C, they gave baseline resolution and lasted for a year with almost day and night operation, resolving thousands of enantiomeric mixtures.

The low thermal stability of the many chiral stationary phases led to excessively long retention times of the more complex amino acids. Chiral stationary phases with greatly improved thermal stability have been synthesized by coupling the chiral L-valine-*tert*-butylamide moiety to a copolymer of dimethyl- and carboxyalkylmethyl silicone.[11,12] This type of stationary phase, later named Chirasil-Val, is now available commercially (Applied Science Division, Milton Roy Company Laboratory Group). The thermal stability of these silicones is such that columns may be operated at temperatures up to 200°C without appreciable bleeding, with short-term operation up to 240°C. Racemization of the chiral moiety becomes significant above about 210°C. Glass capillaries coated with Chirasil-Val are capable in principle of resolving all protein amino acid enantiomers in a single run and with a short analysis time.

The elution characteristics of the individual amino acids are dependent on the chemical nature of the capillary wall surface, and a surface pretreatment is necessary if all the protein amino acids are to be analyzed. Etching of borosilicate glass with HCl followed by deposition of colloidal silicic acid was considered to be the most suitable pretreatment. Chirasil-Val is relatively nonpolar, but shows considerable hydrogen-bonding capacity. With this stationary phase the PFP amino acid isopropyl esters gave the best chromatographic separation with the minimum peak overlap.

Saeed et al. have also described the preparation of a thermostable chiral stationary phase, based on the incorporation of L-valine-*tert*-butylamide in a modified form of polycyanopropylmethyl phenylmethyl silicone (OV-225).[13] Some parts in this phase (diamide and the silicone matrix) are the same as in the phase described by Frank et al.[11] There are some structural differences, however, e.g., a propyl link instead of an ethyl link to the chiral center and the presence of phenyl groups on the silicone matrix. The two phases are therefore chromatographically not identical. Comparison of the elution order of the N-PFP amino acid isopropyl esters reflected this difference between the two phases; on the modified OV-225 phase, glycine elutes after the isoleucines, proline after serine, and cysteine before aspartic acid. The sequence of amino acids eluting after methionine is the same on both phases. The stationary phase proved to have good thermal stability. Good separation factors for all the DL-amino acid pairs studied were obtained except for DL-proline. The synthesis of chiral phases of this type affords the possibility of preparing "tailor-made" stationary phases for specific problems; such investigations are stated to be in progress.[13]

Bonner and Blair have used enantiomeric stationary phases for the gas chromatographic analysis of D- and L-leucine mixtures.[14] The enantiomeric phases, N-docosanoyl-L-valine *tert*-butylamide and N-docosanoyl-D-valine *tert*-butylamide, were each capable of resolving mixtures of N-TFA-isopropyl esters of D- and L-leucine with essentially the same accuracy and precision and, furthermore, the order of elution of the D- and L-leucine esters was reversed when the chirality of the phase was reversed. This inversion of the elution pattern of enantiomeric eluates may be used to advantage in peak identification, improving analytical precision, and eliminating systematic errors.

Van Dort and Bonner have combined the two methods of separating amino acid enantiomers using an optically active phase for the separation of N-TFA amino acid diastereomers.[15] When a mixture of the four diastereomeric N-TFA-esters prepared from a mixture of R- and S-leucine and (+)- and (−)-2-butanol was subjected to gas chromatography on a capillary column loaded with N-lauroyl-S-valyl-*tert*-butylamide, the optically active phase only gave a partial separation of all four stereoisomers. This contrasted with the behavior of an optically inactive phase, Ucon H 90,000, which resolved only the two enantiomeric pairs of diastereomers, R(+) and S(−), and R(−) and S(+), where R and S refer to the leucine enantiomer moieties and (+) and (−) to the 2-butanol enantiomer moieties of each diastereomeric ester. Analysis of known mixtures of the four stereoisomers on the optically active phase established that the order of elution of the stereoisomers with increasing time was R(+), R(−), S(−) and S(+). A complete resolution and quantitative analysis of the four stereoisomers was brought about using the optically active and the optically inactive phases in tandem. Konig has achieved complete resolution of the four N-PFP-RS-leucine (±)-3-methyl-2-butyl stereoisomers on the optically active N-TFA-L-phenylalanyl-L-aspartic acid-bis-cyclohexyl ester,[16] while it will be recalled that in their pioneering studies Gil-Av et al.[17] separated (±)-2-butyl esters of N-TFA-RS-alanine-valine and -leucine into four stereoisomeric components using N-TFA-L-isoleucine lauryl ester as the stationary phase.

It may be concluded that amino acid enantiomers can now be accurately determined by gas chromatography. The method of using amino acid racemization as a geochronometer or geothermometer will most probably make increasing use of these techniques.[18]

REFERENCES

1. Pollock, G. E., Oyama, V. I., and Johnson, R. D., *J. Gas Chromatogr.*, 3, 174, 1965.
2. König, W. A., Rahn, W., and Eyem, J., *J. Chromatogr.*, 133, 141, 1977.
3. Hoopes, E. A., Peltzer, E. T., and Bada, J. L., *J. Chromatogr. Sci.*, 16, 556, 1978.
4. Gil-Av., E., Feibush, B., and Charles-Sigler, R., *Tet. Lett.*, 1009, 1966.
5. Feibush, B. and Gil-Av., E., *J. Gas Chromatogr.*, 5, 257, 1967.
6. Nakaparksin, S., Birrell, P., Gil-Av., E., and Oro, J., *J. Chromatogr. Sci.*, 8, 177, 1970.
7. König, W. A. and Nicholson, G. J., *Anal. Chem.*, 47, 951, 1975.
8. Parr, W., Pleterski, J., Yang, C., and Bayer, E., *J. Chromatogr. Sci.*, 9, 141, 1971.
9. Feibush, B., *Chem. Commun.*, 544, 1971.
10. Smith, G. G. and Wonnacott, D. M., *Anal. Biochem.*, 109, 414, 1980.
11. Frank, H., Nicholson, G. J., and Bayer, E., *J. Chromatogr. Sci.*, 15, 174, 1977.
12. Nicholson, G. J., Frank, H., and Bayer, E., *J. High Resolution Chromat. Chromat. Commun.*, 2, 411, 1979.
13. Saeed, T., Sandra, P., and Verzele, M., *J. Chromatogr.*, 186, 611, 1979.
14. Bonner, W. A. and Blair, N. E., *J. Chromatogr.*, 169, 153, 1979.
15. van Dort, M. A. and Bonner, W. A., *J. Chromatogr.*, 133, 210, 1977.
16. König, W. A., *Chromatographia*, 9, 72, 1976.
17. Gil-Av., E., Feibush, B., and Charles-Sigler, R., *Tet. Lett.*, 1009, 1966.
18. Pollock, G. E., Cheng C-N., and Cronin, S. E., *Anal. Chem.*, 49, 2, 1977.

Section II.V.

DETECTION SYSTEMS FOR AMINO ACIDS IN AUTOMATIC ANALYZERS

NINHYDRIN REAGENTS

Over a period of many years the most widely used reagent for determining amino acids following their separation on columns of ion-exchange resin has been ninhydrin. Reduced ninhydrin (hydrindantin) can be added to the reagent mixture,[1] used with amino acid analyzers for color development, or it can be formed in the reagent mixture by adding a reducing agent such as stannous chloride.[2] When hydrindantin is added directly to the reagent mixture, care must be exercised to prevent air oxidation taking place during transfer to the analyzer reservoir, and a delay of 12 hr is recommended before the reagent is ready for use. When the hydrindantin is formed by the addition of stannous chloride, precipitation of tin salts eventually occurs in the reagent reservoir and flow lines of the analyzer. For this reason stannous chloride has been often replaced by other reducing agents.

Reducing agents which have been used include hydrazine, cyanide, sodium borohydride, ascorbic acid, and titanous chloride.

When using hydrazine, ninhydrin, in buffer-methyl cellosolve solution, is pumped into a flowing stream of hydrazine sulfate, which in a mixing coil reduces sufficient ninhydrin to hydrindantin.[3] The reagent does not need a nitrogen atmosphere for storage.

Cyanide[4,5] is introduced into the column buffers rather than into the reagent itself. In this way the cyanide and ninhydrin are brought together only minutes before the reaction begins. Sodium acetate in the ninhydrin buffer may be replaced by sodium propionate to prevent clogging of the instrument lines.[5] The reagent, which is stable indefinitely, is prepared from sodium propionate, 201.8 g; propionic acid, 93 mℓ; methyl cellosolve, 500 mℓ; ninhydrin, 20 g; and water up to 1 ℓ. It is a canary yellow color but gives a water clear blank when mixed and heated at 95°C with an equal quantity of buffer. 0.01 N NaCN is added to the column buffer. Sodium cyanide gives a significantly lower baseline than hydrazine and is preferable when maximum sensitivity is required, but is more noxious and less stable.

The following procedure has been recommended when using sodium borohydride as reducing agent.[6] 80 g of ninhydrin in 3 ℓ of dimethylsulfoxide is placed in the ninhydrin reservoir of the analyzer and nitrogen is bubbled through for 30 min. 0.3 g of sodium borohydride is weighed in a small piece of Teflon® film and added together with the film to the solution. The solution is stirred magnetically for 1 hr while nitrogen is bubbled through. One liter of lithium acetate, 4 M, pH 5.2, which has been freed from oxygen by bubbling nitrogen through, is added. The color of the solution changes from pale reddish-yellow to red on addition of the sodium borohydride, and then to dark reddish purple on addition of the buffer. Nitrogen is bubbled through the final solution for a further 30 min. Sodium borohydride can be added after solution of ninhydrin in a mixture of dimethyl sulfoxide and lithium acetate buffer. In this case a 4 to 5% decrease of the ninhydrin color yield is seen.

Ascorbic acid used as the reducing agent[7] is incorporated into the column buffers rather than added directly to the ninhydrin solution. Ascorbic acid is stored frozen as a 1 M solution and added to the final citrate buffer to give concentrations of 1.3 to 2.0 mM.

When using titanous chloride in place of stannous chloride, 60 g of ninhydrin is dissolved in 3 ℓ of methyl cellosolve and 1 ℓ of 4 M acetate buffer is added under a nitrogen atmosphere.[8,9] 10 mℓ of 15% titanous chloride is added when the color of the preparation changes from yellow-green to deep red. This red color persists for some time after transfer of the reagent to the analyzer reservoir, but does not interfere with the amino acid analysis.

FLUOROGENIC REAGENTS

Since the description of the first amino acid analyzer by Spackman et al.,[10] improvements in instrumental design, sample preparation, and buffer composition, together with the introduction of the microbore column and microbead ion exchange resins, have led to better resolution, shorter analysis time, and higher sensitivity. The sensitivity of the instrumentation, however, generally remained in the nanomole range until fluorogenic reagents for amino acid detection were introduced. Two fluorogenic reagents, fluorescamine and o-phthalaldehyde have been principally used, and fluorometric detection systems have been added to a number of amino acid analyzers.

FLUORESCAMINE

Udenfriend et al.[11] first showed that fluorescamine could be used for the assay of primary amines, including amino acids. At pH 9 reaction with primary amines proceeds at room temperature, with a half-time of a fraction of a second, while excess reagent is destroyed with a half-time of several seconds. The reaction proceeds to near completion even when fluorescamine is not present in large excess. For assay, primary amines and amino acids are buffered to an appropriate pH, and fluorescamine, dissolved in a water miscible, nonhydroxylic solvent such as acetone, is added. The resulting fluorescence is proportional to the amine concentration and the fluorophors are stable over several hours. These properties lend themselves to automation, and apparatus has been developed for automated amino acid assay based on the use of the reagent.[12,13]

A number of problems have been associated with the use of the fluorescamine analyzer.[14] Fluorescamine is unstable in aqueous buffer solution and decomposes when exposed to moisture. The reagent must therefore be dissolved in acetone and delivery made using a separate pump with seals resistant to organic solvents. On account of the low viscosity of acetone, heterogeneous delivery may be observed and evaporation may result in inconsistencies in the reagent concentration.

The method suffers from the major drawback that detection of the imino acids proline and hydroxyproline is not possible. Proline can be detected with fluorescamine after it has been treated with N-chlorosuccinimide, which converts proline to 4-aminobutyraldehyde.[15] The procedure was adapted to enable proline[16] and hydroxyproline[17] to be determined using the automatic analyzer with fluorescamine.

o-PHTHALALDEHYDE

An alternative fluorometric method for the estimation of amino acids in column effluents following ion-exchange separation was introduced by Roth and Hampai.[18] The reaction uses o-phthalaldehyde and 2-mercaptoethanol, which at pH 9 give an intense blue fluorescence with all α-amino acids. The sensitivity is greater than that of the ninhydrin reaction, and the fluorescent product is formed at room temperature within 2 min, which makes the long heating coil used with ninhydrin detection unnecessary.

With o-phthalaldehyde and 2-mercaptoethanol as the detection reagent, N-chlorosuccinimide cannot be used for the oxidation of imino acids because of interference with the detection reaction. Claims have been made that other oxidants such as sodium hypochlorite and chloramine-T may be used in conjunction with o-phthalaldehyde for the determination of imino acids, but in many cases detailed information on the relevant procedures has not been published. Bohlen and Mellet,[20] however, reported experimental details of a method based on the reaction of imino acids with dilute sodium hypochlorite at alkaline pH which uses o-phthalaldehyde for detection.

Svedas et al.[21] have made a detailed study of the kinetics of interaction between amino acids and o-phthalaldehyde in the presence of mercaptoethanol. The optimal conditions for spectrophotometric determination of amino acids in this way were formulated as follows: the pH of the solution under analysis must be 9.7-10,0; the mercaptoethanol concentration must be higher than $(0.6-1.0) \times 10^{-3}$ M; and the o-phthalaldehyde concentration must be two- to threefold higher than that of the amino acid. The optimal time of the optical density measurement is 3 to 5 min after mixing. It is especially important to choose the proper time of measurement for amino acids such as glycine and histidine, whose derivatives degrade faster than those of other amino acids. The reproducibility of the determination was ± 2%. o-Phthalaldehyde and 2-mercaptoethanol react with primary amines to give intensely fluorescent products. The fluorescent adducts are 1-alkylthio-2-alkyl substituted isoindoles.[22]

Substitution of ethanethiol for 2-mercaptoethanol in its combination with o-phthalaldehyde yields an isoindole with physical properties that greatly extends the utility of the fluorescent adduct. Ethanethiol increases the stability of the adduct and its spectral responses to changes in solvent polarity, while leaving practically unchanged the observed fluorescence intensity.[23] Variations in thiol and amine substituents and solvent polarity have a large effect on the fluorescence spectra of the isoindoles.[24]

Cronin and Hare showed that the o-phthalaldehyde and 2-mercaptoethanol reagent could be applied to the simultaneous chromatographic analysis of amines and nonprotein amino acids, the majority of the compounds being determined quantitatively with high sensitivity. The technique was especially suited for the analysis of β-amino acids. Amino compounds that lack an α-hydrogen atom are detected with lower sensitivity, while secondary amino compounds cannot be detected.

Cronin et al.[26] substantially improved the fluorescent yield and thus the analytical sensitivity for amino acids and amines lacking an α-hydrogen atom by the insertion of a reaction coil between the reagent mixing tee and the fluorometer flow cell, and the substitution of methanethiol or ethanethiol for 2-mercaptoethanol in the o-phthalaldehyde reagent. Amino acids lacking an α-hydrogen atom could be recognized in complex mixtures by the substantial increase in fluorescence yield at a reaction temperature of 100°C rather than 25°C.

The two fluorogens, fluorescamine and o-phthalaldehyde, have different properties and specific advantages. Compared with ninhydrin, fluorescamine shows no greater sensitivity in amino acid analyzers with microbore columns. It would not be economical at present to use fluorescamine in analyzers with columns of large bore. Fluorescamine is usually dissolved in acetone and a buffer solution must therefore be added to the column effluent, requiring an additional pump.

o-Phthalaldehyde is stable in aqueous solvent and is at least 10 times more sensitive than fluorescamine in amino acid detection.[27] It has been claimed that high performance liquid chromatographic analysis of amino acids with o-phthaladehyde/2-mercaptoethanol is one of the most sensitive procedures available for amino acid analysis.[28] Fluorescamine does not yield a fluorescent adduct with ammonia and is thus suitable for the amino acid analysis of the hydrolysates of protein-containing polyacrylamide gel slices.[29]

The analysis of unaltered cysteine and cystine using o-phthalaldehyde and 2-mercaptoethanol presents a number of problems. Cysteine and cystine yield adducts with the reagent that are much less fluorescent than most of the other amino acid adducts, thus rendering their determination difficult. The problem is aggravated by the generally low abundance of cysteine and cystine in proteins and by the fact that the peak resolution between cystine and alanine is very dependent on pH. Bohlen and Mellet conclude that the reason for the poor response of cystine to o-phthalaldehyde is the lack of sufficient reaction time in the amino acid analyzer.

Lee et al. have described the construction and operation of a very sensitive microbore automatic analyzer which can use either fluorescamine or o-phthalaldehyde for detection.[30] A constant molarity buffer which is used has the advantages of giving a lower column pressure, achieving faster chromatographic speeds and requiring shorter regeneration and equilibration times. An important consideration is that this buffer produces a stable baseline. Detection of the amino acids is achieved by continuously pumping an oxidant, such as dilute sodium hypochlorite solution, to the column effluent throughout the entire chromatographic cycle. The simple modification of a conventional amino acid analyzer to obtain pmole sensitivity using o-phthalaldehyde for detection has been described.[31]

OTHER DETECTION REAGENTS

Detection reagents other than those already discussed have been used for the quantitative estimation of amino acids in the eluate from automatic analyzers but have not gained general acceptance. Harmeyer et al.[32] have recommended the use of 2,4,6-trinitrobenzene sulfonic acid. If light absorption can be measured at 350 nm, the reagent can replace ninhydrin without further technical modifications to the equipment. 2,4,6-Trinitrobenzenesulfonic acid is added to the elution buffers in a concentration of 60 mg/100 ml. The ninhydrin reagent is replaced by 6 N NaOH per 100 ml of solution which is continuously added to the elution buffer in a 1:2 ratio. The reaction between amino acids and 2,4,6-trinitrobenzene-sulfonic acid occurs at pH 9.2-9.3 within 10 to 20 min in a Teflon® coil at room temperature. The sensitivity of the method is about half that of the ninhydrin procedure, but the reproducibility of measurements is stated to be significantly better. The reagents used are stable for 2 to 4 weeks.

Maeda et al.[33] have described the application of a fluorescent reaction with pyridoxal and Zn(II) to an automatic amino acid analyzer. The fluorescent reaction is based on the formation of chelates between n-pyridoxylidene amino acids and Zn(II). Citrate buffers cannot be used for the chromatographic separation because citrate is a strong chelating agent and produces a fluorophor. Ligand-exchange chromatography based on the ability of amino acids to form complexes with Zn(II) is therefore used for the separation. The reagent solution is freshly prepared before use by dissolving 0.100 g of pyridoxal hydrochloride and 1.000 g of zinc acetate in 1000 ml of 2.0% pyridine-methanol. The effluent from the column is mixed with the reagent (4 parts of reagent to 1 part of effluent) and allowed to react at 65 to 75°C for 10 min in a reaction coil of PTFE capillary tubing. After reaction the mixture is passed through the flow cell of the fluorophotometer. The tentative conclusion may be drawn that fluorogenic reagents will be increasingly used in the future for amino acid estimation in automatic analyzers.

REFERENCES

1. Moore, S. and Stein, W. H., *J. Biol. Chem.*, 211, 907, 1954.
2. Moore, S. and Stein, W. H., *J. Biol. Chem.*, 176, 367, 1948.
3. Technicon Technical Publication No. TAO-0155-00. Technicon Instrument Corp., Tarrytown, New York, 1970.
4. Schwerdtfeger, E., *J. Chromatogr.*, 7, 418, 1962.
5. Rosen, H., Bernard, C. W., and Levenson, S. M., *Anal. Biochem.*, 4, 213, 1962.
6. Takahashi, S., *J. Biochem. Japan*, 83, 57, 1978.
7. Niece, R. L., *J. Chromatogr.*, 103, 25, 1975.
8. James, L. B., *J. Chromatogr.*, 59, 178, 1971.
9. James, L. B., *J. Chromatogr.*, 152, 298, 1978.
10. Spackman, D. H., Stein, W. H., and Moore, S., *Anal. Chem.*, 30, 1190, 1958.
11. Udenfriend, S., Stein, S., Böhlen, P., Dairman, W., Leimgruber, W., and Weigele, M., *Science*, 178, 871, 1972.
12. Stein, S., Böhlen, P., Stone, J., Dairman, W., and Udenfriend, S., *Arch. Biochem. Biophys.*, 155, 202, 1973.
13. Voelter, W. and Zech, K., *Anal. Biochem.*, 112, 643, 1975.
14. Lee, H-M, Forde, M. D., Lee, M. C., and Bucher, D. J., *Anal. Biochem.*, 96, 298, 1979.
15. Weigele, M., De Bernardo, S. L., and Leimgruber, W., *Biochem. Biophys. Res. Commun.*, 50, 352, 1973.
16. Felix, A. M. and Terkelsen, G., *Arch. Biochem. Biophys.*, 157, 177, 1973.
17. Felix, A. M. and Terkelsen, G., *Anal. Biochem.*, 56, 610, 1973.
18. Roth, M. and Hampai, A., *J. Chromatogr.*, 83, 353, 1973.
19. Drescher, D. G. and Lee, K. S., *Anal. Biochem.*, 84, 559, 1978.
20. Bohlen, P. and Mellet, M., *Anal. Biochem.*, 94, 313, 1979.
21. Svedas, V-J. K., Galaev, I. J., Borisov, I. L., and Berezin, I. V., *Anal. Biochem.*, 101, 188, 1980.
22. Simons, S. S., Jr. and Johnson, D. F., *J. Am. Chem. Soc.*, 98, 7098, 1976.
23. Simons, S. S., Jr. and Johnson, D. F., *Anal. Biochem.*, 82, 250, 1977.
24. Simons, S. S., Jr. and Johnson, D. F., *Anal. Biochem.*, 90, 705, 1978.
25. Cronin, J. R. and Hare, P. E., *Anal. Biochem.*, 81, 151, 1977.
26. Cronin, J. R., Pizzarello, S., and Gandy, W. E., *Anal. Biochem.*, 93, 174, 1979.
27. Benson, J. R. and Hare, P. E., *Proc. Nat. Acad. Sci. U.S.*, 72, 619, 1975.
28. Lee, K. S. and Drescher, D. G., *J. Biol. Chem.*, 254, 6248, 1979.
29. Stein, S., Chang, C. H., Böhlen, P., Imai, K., and Udenfriend, S., *Anal., Biochem.*, 60, 272, 1974.
30. Lee, H-M, Bucher, D. J., and Seid, R. C., Jr., *Ind. Eng. Chem. Product, Res. Dev.*, 18, 122, 1979.
31. Beecher, G. R., Dohm, G. L., and Peacock, I., *J. Chromatogr.*, 175, 183, 1979.
32. Harmeyer, J., Sallmann, H.-P., and Ayoub, L., *J. Chromatogr.*, 32, 258, 1968.
33. Maeda, M., Tsuji, A., Ganno, S., and Onishi, Y., *J. Chromatogr.*, 77, 434, 1973.

Section III
Detection Reagents

III.I. Detection Reagents for Paper Chromatography and Thin-Layer Chromatography
III.II. Summary Tables for Detection Reagents

Section III.I.

DETECTION REAGENTS FOR PAPER CHROMATOGRAPHY AND THIN-LAYER CHROMATOGRAPHY

The detection reagents, principally for paper and/or thin-layer chromatography, are arranged as far as possible according to chemical classes. Within each class tests are in alphabetical order. Some reagents may detect more than one class of compound.

ALPHABETICAL INDEX OF DETECTION REAGENTS

Ammonium hydrogen carbonate 177
N-Chloro-5-dimethylaminonaphthalene-1-sulfonamide (NCDA) 177
Dimethoxytetrahydrofuran-p-dimethylaminobenzaldehyde 178
Fluorescamine 178
Fluorescamine-perchloric acid 180
Iodine 181
Perchloric acid 181
o-Phthalaldehyde 182
Stannous chloride-dimethylaminobenzaldehyde 182
7,7,8,8-Tetracyanoquinodimethan (TCNQ) 183
4,4'-Tetramethyldiamino-diphenylmethane (TDM) 183
2,4,6-Trinitrobenzene sulfonate (TNBS) 183

AMMONIUM HYDROGEN CARBONATE

Procedure: The chromatoplates are placed inside a sealed tank of 4-gallon capacity containing 6 g ammonium hydrogen carbonate in the bottom. The tank is placed inside an oven and heated to temperatures between 110° and 150° for 2 to 12 hr according to the type of adsorbent and the nature of the compounds studied. The plates are removed and fluorescent products measured *in situ* with a spectrofluorimeter.

Results: The fluorescent derivatives formed by organic compounds are detected by examination under UV light (365 nm), when they show a greenish blue fluorescence.

Comment: Fluorescent derivatives are formed from most organic compounds separated on chromatoplates containing inorganic adsorbents (silica gel, alumina, Florisil, kieselguhr).

Reference: Segura, R. and Gotto, A. M., Jr., *J. Chromatogr.*, 99, 643, 1974.

N-CHLORO-5-DIMETHYLAMINONAPHTHALENE-1-SULFONAMIDE (NCDA)

Preparation: To prepare NCDA reagent, to a solution of 35 mg of N,N'-dimethylamino naphthalene-5-sulfonamide (DNS-amide) in 40 mℓ of MeOH is added hypochlorite reagent (commercial 6% sodium hypochlorite solution diluted 2-fold with redistilled water) dropwise under UV illumination until the fluorescence of DNS-amide has almost disappeared. Addition of excess hypochlorite reagent after disappearance of the fluorescence decomposes the NCDA. To this solution is then added 50 ml of 1.8% (v/v) C7A (cycloheptaamylose) solution, and the mixture is made up to 100 ml with MeOH. The ratio of chlorine to DNS-amide in this reagent, measured by titration with 0.1 N $Na_2S_2O_3$, is 1:1.

Procedure: After chromatography on a silica gel plate, the solvent is removed by blowing cold air from a hair dryer. The plate is sprayed with NCDA reagent (25 μl/cm^2) and the spots detected under UV irradiation.

Results: Detects PTH-amino acids. the limit of detection is 40 to 60 pmol.

Comment: To determine PTH-amino acids, the plate is dried over P_2O_5 under reduced pressure for 60 min, is sprayed with NCDA reagent and the fluorescence intensity is measured with a scanning fluorometer.

Reference: Murayama, K. and Kinoshita, T., *J. Chromatogr.*, 205, 349, 1981.

DIMETHOXYTETRAHYDROFURAN-p-DIMETHYLAMINOBENZALDEHYDE

Preparation: Solution a. 1% 2,5-dimethoxytetrahydrofuran solution buffered at pH 6.6
Solution b. 1% p-dimethylaminobenzaldehyde solution in 3 M HCl.

Procedure: Spray the TLC plate with solution a, heat in an oven at 110°C for 5 min and then spray with solution b.

Results: Detects primary amines.

Reference: Prandi, C., *J. Chromatogr.*, 155, 149, 1978.

FLUORESCAMINE

Technique I.

Preparation: Solution a. 0.05 M sodium borate buffer, pH 10.5
Solution b. 20 mg of fluorescamine in 100 ml of acetone.
Solution c. 0.2 M taurine in 0.2 M sodium phosphate buffer pH 7.5.

Procedure: Precoated silica gel 60 HPTLC plates on which chromatography has been conducted are heated at 110°C for 10 min in an electric oven. If the plate still smells of acetic acid or ammonia from the solvent it is dried further under a hot stream from a hair-dryer until free from smell. The plate is sprayed with solution a and heated at 110°C for 15 min. After cooling the plate is sprayed with solution b and allowed to stand at room temperature for 15 min in the dark. It is then sprayed with solution c and heated at 60°C for 5 min. The fluorescence is observed in the dark with a UV lamp which provides continuous light at 250-400 nm.

Results: Detects 20 to 30 pmole of some secondary amines. It is not suitable for the detection of volatile amines.

Comments: Fluorescamine gives no fluorescent products with secondary amines but produces aminoenone chromophores having an absorption maximum at 300-330 nm. These compounds are converted to fluorescent pyrrolinones by reaction with primary amines. The method is not applicable to the detection of amides and compounds with a peptide bond, such as hippuric acid. The fluorescence produced in the background reduces the sensitivity of the procedure. The background fluorescence probably originates from the hydrolysis product of fluorescamine in the reaction with taurine.

References: Nakamura, H., Tsuzuki, S., Tamura, Z., Yoda, R., and Yamamoto, Y., *J. Chromatogr.*, 200, 324, 1980.

Technique II.
Preparation: Solution a. 10% triethylamine in methylene chloride.
Solution b. 0.05% fluorescamine in acetone.
Procedure: Silica gel plates on which amino acids have been chromatographed are dried at 110°C for 10 min and cooled to room temperature. Spray with solution a and air dry for several seconds, spray with solution b and air dry for several seconds, finally respray with solution a. The plates are viewed using a longwave (360 nm) ultraviolet source.
Results: Detects amino acids. The fluorescence intensities of the amino acids are only slightly faded after 24 hr. When triethylamine is omitted from the spray the spots are less intense and fade almost completely after 24 hr. As little as 500 picomoles of amino acid can be detected.
Comments: It is claimed that a larger fluorescence intensity is obtained when the plates are dried only at room temperature for 30 mins using an ambient air current from a hair dryer. Spraying with solutions of fluorescamine in dimethylsulfoxide or dimethylformamide proved suitable in that no triethylamine stabilization was required.
References: Felix, A. M. and Jimenez, M. H., *J. Chromatogr.*, 89, 361, 1974.
Touchstone, J. C., Sherma, J., Dobbins, M. F., and Hansen, G. R., *J. Chromatogr.*, 124, 111, 1976.

Technique III.
Preparation: Solution a. 10% triethylamine in dichloromethane.
Solution b. 0.01% fluorescamine in acetone.
Procedure: Spray with solution a, solution b and solution a and observe under long-wavelength UV (366 nm).
Results: Detects picomole quantities of amino acids and peptides.
Comment: When peptides were recovered, hydrolyzed with acid and analyzed for constituent amino acids, o-phthalaldehyde and ninhydrin as spray reagents resulted in destruction of the amino-terminal residue, while fluorescamine did not.
References: Felix, A. M. and Jiminez, M. H., *J. Chromatogr.*, 89, 361, 1974.
Schiltz, E., Schnackerz, K. D., and Gracy, R. W., *Anal. Biochem.*, 79, 33, 1977.

Technique IV.
Assay procedure for secondary amines alone (method A).

A 50 $\mu\ell$ aqueous solution containing 0.5 to 5 nmol of secondary amine is transferred to a 12 × 75 mm glass test tube and 50 $\mu\ell$ of 0.05 M Kolthoff's buffer (NaOH-Na_2HPO_4) at pH 12.0 is added. The mixture is stirred vigorously on a vortex-type mixer and 100 $\mu\ell$ of a acetone solution of fluorescamine containing 20 mg/100 mℓ is added rapidly. After stirring for a further 10 sec the test tube is cooled in an ice bath. After 5 min 1.5 mℓ of 0.67 mM L-Leu-L-Ala dissolved in 0.2 M sodium phosphate buffer, pH 6.60 is added, and the mixture is heated immediately at 70°C for 10 min in a water bath and allowed to stand in an ice bath until the fluorescence is measured.

The fluorescence intensity is measured at 390 nm excitation and 480 nm emission, against a reagent blank containing no secondary amine.

Assay procedure for secondary amines in the presence of primary amines (method B). Determination of primary amines.

The amount of primary amine is determined as in method A, except that L-Leu-L-Ala is omitted from the reagent. When the sample contains both primary and secondary amines, the net fluorescence intensity obtained by method A represents the sum of their fluorescence intensities. The amount of secondary amine is obtained by subtracting the fluorescence intensity in method B from that obtained in method A. The method permits the determination of 0.5 nmol of secondary amines. The fluorescence of the blank limits the sensitivity of the assay.

Reference: Nakamura, H. and Tamura, Z., *Anal. Chem.*, 52, 2087, 1980.

Technique V.

Preparation:	Solution a. 1% v/v triethylamine in acetone.
	Solution b. 10 mg of fluorescamine in 100 mℓ acetone.
	Solution c. 0.1 M acetic acid in acetone.
	Solution d. 0.1 M N-chlorosuccinimide in acetone.
Procedure:	After electrophoresis or chromatography the papers are dried in air and heated at 55°C for at least 1 hr. They are then washed with acetone, wet with solution a and allowed to dry for 5 min at room temperature, wet with solution b, allowed to dry for 5 min, and then washed with acetone and dried. The fluorescent spots are detected by viewing under a longwave (336 nm) ultraviolet lamp.
Results:	Detects amino acids and peptides. To detect proline and hydroxyproline, after treatment with solution b, the paper is heated at 110°C for at least 2 hr or left at room temperature for at least 2 days. Alternatively the paper may be wet successively with solutions c and d and allowed to stand for 5 min at room temperature. After washing with acetone, the paper is heated for 5 to 10 min at 110°C. With either procedure, proline and hydroxyproline are visualized as fluorescent spots.
Comment:	It is suggested that the reaction of fluorescamine with peptides and amino acids on paper proceeds differently from the reaction in aqueous medium, stopping at an intermediate step to yield a fluorescent compound that can revert to the free amino acid on hydrolysis.
Reference:	Mendez, E. and Lai, C. Y., *Anal. Biochem.*, 65, 281, 1975.

FLUORESCAMINE-PERCHLORIC ACID

Preparation:	Solution a. 0.2 M sodium borate buffer pH 9.0.
	Solution b. Acetone-n-hexane (1:4) containing fluorescamine, 10 mg/100 mℓ.
	Solution c. 40% HClO$_4$.
	Solution d. 0.5 M sodium borate buffer, pH 9.0.
Procedure:	Method I: detection after chromatography of native compounds.
	After thin-layer chromatography the plates are dried at 110°C for 5 min, sprayed with solution a, re-dried at 110°C for 5 min and dipped for 30 min in solution b. The plates are then sprayed with solution c for 5 sec. The spraying time should be changed depending upon the plate size to obtain maximal fluorescence. Fluorescence is observed

within a few minutes under a UV lamp at 366 nm. Plates developed with acidic media are sprayed with solution d instead of solution a and sprayed with solution c for about 10 sec.

Method II: detection after chromatography of fluorescamine derivatives.

The compounds are derivatized at the origin by dipping the lower 2 cm of the plate in solution b for 15 min. The plate is dried without heating and then developed with the appropriate solvent. Air-dried plates are sprayed with solution c for 5 sec.

Results: Tryptophan, tryptamine, N-terminal tryptophan peptides and other 3-(2-aminoethyl)-indoles give positive results. Using method I, fluorescamine derivatives of tryptophans and N-terminal tryptophan peptides given an orange fluorescence, while those of tryptamines give a yellowish green or bluish green fluorescence. After spraying with solution b, all changed to a characteristic yellowish fluorescence which was more intense. Using method II, neither the underivatized compounds nor fluorescamine derivatives are displaced or extracted by the fluorescamine solution. The derivatization seems to be complete within 15 min. After chromatography most positive compounds can be detected at the 1-nmole level as their fluorescamine derivatives. The fluorescence is increased by spraying with solution c.

Comments: The detection limit in methods I and II varied from 2 to 800 pmole and 0.8 to 800 pmole, respectively.

Reference: Nakamura, H. and Pisano, J. J., *J. Chromatogr.*, 152, 153, 1978.

IODINE

Procedure: The dried commercial silica gel plate on which amines have been chromatographed is placed in a closed cell with several iodine crystals for 10 to 15 min.

Results: The detection of amines is based on the readily formed and intensely colored ion radicals. The spots of aniline, diphenylamine and phenyl-2-naphthylamine, which develop color only through adsorption, disappear after some time because of the desorption of iodine.

Comments: Bromine vapor has been used in place of iodine vapor as the oxidizing agent. The disadvantage of bromine for detection is that prolonged treatment with bromine vapor leads to further oxidation of the cation radicals formed and thus to the disappearance of the spots.

Reference: Uchytil, I., *J. Chromatogr.*, 93, 447, 1974.

PERCHLORIC ACID

Preparation: Solution a. 70% $HClO_4$.

Procedure: After ascending chromatography on precoated silica gel glass plates, the plate is briefly air-dried with a hair dryer. The plate is sprayed with the solution for 5 sec and the fluorescence observed in the dark under longwave UV light. After development with alkaline solvents, the plate is sprayed for 10 sec.

Results: Detects 3-substituted indoles including tryptophan, tryptamine, and peptides containing tryptophan. The indoles give an intense yellowish fluorescence immediately after spraying with the solution. The fluorescence is stable for at least 30 min. As little as 40-850 pmole of 3-substituted indoles can be detected.

Reference: Nakamura, H. and Pisano, J. J., *J. Chromatogr.*, 152, 167, 1978.

o-PHTHALALDEHYDE

Technique I.
Preparation: Solution a. 0.1% o-phthalaldehyde and 0.1% 2-mercaptoethanol in acetone.
Solution b. 1% triethylamine in acetone.
Procedure: Spray with solution a and 5 min later with solution b. After 10 min observe under longwave (350 nm) UV.
Results: Detects amino acids or peptides on thin layer plates. As little as 50-100 pmoles of many amino acids can be detected.
Reference: Lindeberg, E. G. G., *J. Chromatogr.*, 117, 439, 1976.

Technique II.
Preparation: Solution a. 10% triethylamine in dichloromethane
Solution b. 0.05% phthalaldehyde in methanol containing 0.2% 2-mercaptoethanol and 0.09% Brij-35.
Procedure: Spray with solution a, solution b and solution a. Observe under longwave UV light.
Results: Detects picomole quantities of amino acids and peptides on thin layer chromatograms.
Reference: Schiltz, E., Schnackerz, K. D., and Gracy, R. W., *Anal. Biochem.*, 79, 33, 1977.

Technique III.
Preparation: Solution a. 1% triethylamine and 0.05% 2-mercaptoethanol in acetone.
Solution b. o-phthalaldehyde, 0.3 mg/mℓ in acetone.
Procedure: After electrophoresis or chromatography the paper is dried at 50°C for 1 hr, is washed with acetone, and is then dipped in a tray containing solution a. After 5 min at room temperature the paper is dipped in a tray containing solution b and allowed to dry for 5 min at room temperature. Finally the paper is washed with acetone and dried. The fluorescent spots are detected under a longwave (366 nm) UV lamp.
Results: Detects amino acids and peptides. Amino acids are more readily visualized at the 500 pmole level with o-phthalaldehyde than with fluorescamine. While the fluorescent spots produced by fluorescamine remain visible for several weeks the spots produced by o-phthalaldehyde disappeared after several hours.
Reference: Mendez, E. and Gavilanes, J. G., *Anal. Biochem.*, 72, 473, 1976.

STANNOUS CHLORIDE-DIMETHYLAMINOBENZALDEHYDE

Preparation: Solution a. 5% stannous chloride in HCl.
Solution b. 1% p-N,N-dimethylaminobenzaldehyde in 1 *M* HCl (Ehrlich reagent).

Procedure: Spray with solution a to reduce nitro groups to amino groups and after drying the chromatogram spray with solution b.
Results: Detects the 2,4-dinitrophenyl (DNP) derivatives of aromatic amines.
Reference: Franc, J. and Koudelkova, V., *J. Chromatogr.*, 170, 89, 1979.

7,7,8,8-TETRACYANOQUINODIMETHANE (TCNQ)

Preparation: Solution a. 0.3% TCNQ in pyridine-acetone (1:1).
Procedure: Spray the chromatogram (Whatman 3MM paper or silica gel plates) with dimethylsulfoxide, dry with gentle heat from a hair dryer and spray with solution a. Air dry for 15 min at room temperature. Record the visible colors, put under UV light for 5 min, and record the fluorescent colors.
Results: Amino acids and amines give characteristic colors and fluorescent colors.
Comments: The preliminary spraying with dimethyl sulfoxide helps in the development of fluorescent colors in the reaction of TCNQ with amino acids, ethylenediamine and ethanolamine.
Reference: Guyer, M., Jr. and Sawicki, E., *Anal. Chim. Acta.*, 49, 182, 1970.

4,4'-TETRAMETHYLDIAMINO-DIPHENYLMETHANE (TDM)

Preparation: Solution a. Sodium hypochlorite containing 13 to 14% active chlorine diluted to 6 times its original volume.
Solution b. A mixture of the following amounts of solutions c and d with 1.5 ml of solution e added.
Solution c. 2.5 g 4,4'-tetramethyldiamino-diphenylmethane dissolved in 10 ml glacial acetic acid and diluted with 50 ml water.
Solution d. 5 g potassium iodide in 100 ml water.
Solution e. 0.3 g ninhydrin dissolved in 90 ml water and 10 ml glacial acetic acid added.
Procedure: Thin-layer plates are dried at 100°C for 15 min after chromatography and sprayed with solution a until damp. Silica gel plates are then dried for 10 min in warm air, cellulose plates for 45 min in warm air or for 5 min at 100°C, and are lightly sprayed with solution b.
Results: Amines, amides, peptides, and amino acids produce green spots, which slowly turn blue-green to dark blue. The reaction can be speeded up by a stream of warm air or by UV irradiation.
Reference: Von Arx, E., Faupel, M., and Brugger, M., *J. Chromatogr.*, 120, 224, 1976.

2,4,6-TRINITROBENZENE SULFONATE (TNBS)

Preparation: 100 mg of TNBS in 150 ml of dimethyl sulfoxide.
Procedure: 1 ml of TNBS reagent is added to 3 ml of eluate (pH 8.5 borate buffer) from the chromatographic column. Reaction is carried out at 50°C for 10 min and terminated by cooling the reaction mixture in water. The absorbance at 420 nm is then measured.
Results: Detects 10-20 nmol of polyamines.
Reference: Endo, Y., *J. Chromatogr.*, 205, 155, 1981.

Section III.II.

SUMMARY TABLES FOR DETECTION REAGENTS*

The following tables represent summary information on several classes of compounds, various color reagents, colors produced and sensitivity of detection. Procedures for the preparation of detection reagents are included in the tables and in the earlier parts of this section.

---

* Tables 1-6 were prepared by Z. Deyl.

Table 1
SENSITIVITY OF o-PHTHALALDEHYDE DETECTION

Detection limit (pmoles per spot)

Substance	Cellulose				Silica gel				After TLC
	10 min	1 hr	2 hr	1 hr, 100°	10 min	1 hr	2 hr	1 hr, 100°	10 min
Alanine	250	250	250	250	50	250	500	1000	100
Arginine	10	10	25	25	10	50	100	500	25
Asparagine	250	250	250	250	25	100	250	500	50
Aspartic acid	100	100	100	100	25	100	100	1000	100
Cystine	10	10	10	10	100	250	250	100	250
Glutamine	100	100	100	100	25	100	250	1000	50
Glutamic acid	100	100	100	100	25	100	250	1000	100
Glycine	100	100	100	100	25	250	1000	—	100
Histidine	25	50	50	50	25	100	100	100	25
Isoleucine	100	100	100	100	50	500	500	—	100
Leucine	100	100	100	100	50	500	500	—	100
Lysine	25	25	25	25	25	100	250	1000	50
Methionine	100	100	100	100	50	100	250	1000	100
Phenylalanine	50	50	50	50	50	250	500	1000	100
Proline	—	—	—	250	—	—	—	250	250[a]
Serine	100	100	100	100	10	250	250	1000	25
Threonine	100	100	100	100	25	250	250	1000	50
Tryptophan	25	25	25	50	50	100	100	100	250
Tyrosine	25	25	25	50	50	500	500	500	100
Valine	25	25	25	50	50	500	1000	—	100
Glycyl-glycine	250	250	250	250	100	500	1000	—	250
Bradykinin	50	50	50	50	50	5000	1000	250	100
Arginine-vasopressin	100	100	100	500	250	1000	1000	1000	500

[a] After heating at 100° for 1 hr. 1-uℓ aliquots containing 5-1000 pmoles were spotted on thin-layer plates, dried at 100° for 30 min, treated with the spray reagents and viewed under UV light (350 nm) after the time indicated. Thin-layer

chromatography (TLC) was carried out on silica gel with n-butanol-acetic acid-water (4:1:1); front migration, ca 10 cm; dried and visualized as above.

REFERENCE

1. Lindeberg, E. G. E., *J. Chromatogr.*, 117, 439, 1976.

Reproduced by permission of the Elsevier Scientific Publishing Co.

Table 2
MINIMUM DETECTABLE AMOUNTS (μg) OF TRYPTOPHAN, INDOLEACETIC ACID AND TRYPTOPHAN CONTAINING PEPTIDES BY FLUORESCENCE AT 365 nm (SILICA GEL) AND 405 nm (FILTER PAPER) AFTER TREATMENT WITH FORMALDEHYDE, FORMALDEHYDE-OZONE, OR FORMALDEHYDE HCl

Compound	Filter-paper			Silica gel		
	Formaldehyde	Formaldehyde-ozone	Formaldehyde HCl	Formaldehyde	Formaldehyde-ozone	Formaldehyde HCl
L-Tryptophan	0.3-1.0	0.03	0.03	0.03	0.1	0.03
Indoleacetic acid	—	—	0.03	—	—	0.03
L-Tryptophyl-L-alanine	0.3-1.0	0.03	0.03	0.03	0.1	0.03
L-Tryptophyl-L-glutamic acid	0.3-1.0	0.03	0.03	0.1	0.1	0.1
L-Tryptophyl-L-glycine	0.3-1.0	0.03	0.03	0.1	0.1	0.1
L-Tryptophyl-L-phenylalanine	0.3-1.0	0.03	0.03	0.1	0.1	0.1
L-Tryptophyl-L-tyrosine	0.3-1.0	0.03	0.03	0.1	0.1	0.1
L-Glycyl-L-tryptophan	—	—	0.03	—	—	0.1
L-Phenylalanyl-L-tryptophan	—	—	0.03	—	—	0.1
L-Prolyl-L-tryptophan	—	—	0.03	—	—	0.1
L-Arginyl-L-tryptophyl-L-glycine	—	—	0.1	—	—	0.3
Tetragastrin	—	0.1	0.3	0.1—0.3	0.3	0.3
Peptavlon	—	—	0.1—0.3	—	—	0.3
EAE	—	—	0.3—1.0	—	—	0.3
Glucagon	—	—	—[a]	—	—	—
Thyroglobulin	—	—	—	—	—	—

[a] With 10 μg of glucagon, weak fluorescence was observed.

Note: Volumes of 1 μℓ applied, results indicated by dashes represent amounts greater than 1 μg.

REFERENCE

1. Larsson, L.-I., Sundler, F., and Hakanson, R., *J. Chromatogr.*, 117, 355, 1976.

Reproduced by permission of the Elsevier Scientific Publishing Co.

Table 3
PTH-AMINO ACIDS, COLORS IN UV LIGHT

Compound	Color after Second development	Color after Alkaline treatment
Valine	Red	Red
Proline	Red	Red
Alanine	Red	Red
Glycine[a]	Red	Brownish red
Serine	Red	Brownish red (blue)
Dehydroserine	Green (blue)	Bluish green (blue)
Asparagine[a]	Red	Greenish brown (bluish green)[d]
Asparagine by-product	Colorless	White blue[d] (greenish blue)[d]
Aspartic acid	Red	Brownish red (dark brown)
S-Carboxymethylcysteine	Green[b] (blue)	Brownish red[e] (blue)
Methionine[a]	Red	Brownish red
Methionine sulfone[a]	Red	Brownish red (blue)
Leucine	Red	Brownish red
Isoleucine	Red	Red
Lysine	Red	Red
Tyrosine[a]	Red	Red (bluish green)[d]
Tyrosine by-product	Greenish yellow	Green (green)[d]
Threonine	Red	Bluish green (blue)
Dehydrothreonine	Green (blue)	Bluish green (blue)
Glutamine[a]	Red	Greenish brown (white yellow)[d]
Glutamine by-product[a]	Colorless	Green (white yellow)[d]
Glutamic acid	Red	Red
Phenylalanine[a]	Red	Greyish red (white blue)[d]
Tryptophan[a]	Red	Greyish red (white blue)[d]
Histidine[a]	Red	Blue[d] (light blue)[d]
Histidine by-product	Red (blue)[c,d]	Purple (bluish purple)[d]
Arginine	Red (blue)[c,d]	Purple (blue)[d]
Cysteic acid	Red	Brownish red (dark brown)

[a] Spots appear yellow to the naked eye. The glycine derivative is an exception. It is uniquely pink, especially after spraying with 1 M NaOH.
[b] Red after the first development. The red color sometimes persisted after the second development.
[c] No fluorescence after the first development.
[d] Fluorescent.
[e] Sometimes mixed with green color.

Note: The plates were viewed under a Pan UV lamp after two-dimensional chromatography in toluene-*n*-pentane-acetic acid (6:3:2) (first direction) and acetic acid-water (1:3). Unless otherwise specified, the colors were the same before and after chromatography. Alkaline treatment involved spraying with 0.05 M sodium hydroxide in methanol-water (1:1) and heating at 115° for 30 min. Colors seen under the longwave UV lamp are shown in parentheses.

REFERENCE

1. Nakamura, H., Pisano, J. J., and Tamura, Z., *J. Chromatogr.*, 175, 153, 1979.

Reproduced by permission of the Elsevier Scientific Publishing Company.

Table 4
LIMITS OF DETECTION OF PTH AMINO ACIDS AFTER DEVELOPMENT WITH CHLOROFORM-FORMIC ACID (20:1) AND REACTION WITH N-CHLORO-5-DIMETHYLAMINONAPHTHALENE-1-SULFONAMIDE

PTH-amino acid	Limit of detection (pmol)	PTH-amino acid	Limit of detection (pmol)
Ala	60	Lys	40
Asp	40	Met	40
Cys	60	Phe	60
Glu	40	Pro	60
Gly	60	Thr	40
Ile	60	Trp	40
Leu	60	Val	60

REFERENCE

1. Murayama, K. and Kinoshita, T., *J. Chromatogr.*, 205, 349, 1981.

Reproduced by permission of the Elsevier Scientific Publishing Company.

Table 5
RELATIVE FLUORESCENCE INTENSITIES (F) OF VARIOUS PTH-AMINO ACIDS (500 PMOL PER SPOT) ON A SILICA GEL G PLATE AND REACTION WITH N-CHLORO-5-DIMETHYLAMINONAPHTHALENE-1-SULFONAMIDE

PTH-amino acid	F	PTH-amino acid	F	PTH-amino acid	F
Met	130	Leu	100	Thr	80
Lys	126	Phe	101	Ileu	77
		Ala	98	Trp	66
		Gly	97	Pro	55
		Val	95	CysOH	32
				Asp	23

Note: The fluorescence intensity of PTH-Leu is expressed as 100.

REFERENCE

1. Murayama, K. and Kinoshita, T., *J. Chromatogr.*, 205, 349, 1981.

Reproduced by permission of the Elsevier Scientific Publishing Company.

Table 6
AROMATIC AMINES AND AMINOPHENOLS

Compound	Spray I initial color	After heat	Spray II after heat	HCl overspray
o-Aminophenol	Pale Yellow	Yellow	Dark yellow	Peach
m-Aminophenol	—	Dark brown	Rust-brown	Brown
p-Aminophenol	—	Purple	Blue	Red
o-Anisidine	—	Purple	—	Lilac
m-Anisidine	—	Purple-brown	—	Peach

Table 6 (continued)
AROMATIC AMINES AND AMINOPHENOLS

Compound	Spray I initial color	After heat	Spray II after heat	HCl overspray
p-Anisidine	—	Light brown	—	Red
o-Aminobenzoic acid	—	Rust-brown	—	—
m-Aminobenzoic acid	—	Light brown	—	—
p-Aminobenzoic acid	—	Lime-green	—	—
3,5-Diaminobenzoic acid	Buff	Dark brown-purple	Light pink	Buff
2,4-Dimethoxyaniline	—	Maroon	—	Purple
2,5-Dimethoxyaniline	—	Black-brown	—	Green-brown
3,5-Dimethoxyaniline	—	Dark brown, yellow halo	—	Yellow
2,4-Dichloroaniline	—	Blue-purple	—	Buff
2,5-Dichloroaniline	—	Cream-yellow	—	Mustard
3,4-Dichloroaniline	—	Mauve	Light blue	Buff
2-Amino-5-hydroxybenzoic acid	—	Light brown-purple	Light blue	Dark brown
3-Amino-2-hydroxybenzoic acid	—	Gold	—	Cream
3-Amino-4-hydroxybenzoic acid	Yellow	Yellow-black	Green	Brown
2-Amino-1,3-dimethyl-benzene	—	Purple	—	Light red
2-Amino-1,4-dimethyl-benzene	—	Green, purple halo	—	Light green
3-Amino-1,2-dimethyl-benzene	—	Rust-brown	—	Light green
4-Amino-1,2-dimethyl-benzene	—	Mustard green	—	Yellow
4-Amino-1,3-dimethyl-benzene	Light brown	Mustard, purple halo	—	Mauve
2-Amino-5-nitrotoluene	Bright yellow	Bright yellow	Bright yellow	Bright yellow
4-Amino-2-nitrotoluene	Yellow	Lime-green	Yellow	Cream-yellow
4-Amino-3-nitrotoluene	Dark yellow	Dark yellow	Orange-yellow	Deep yellow
Aniline	—	Brown-green	—	—
Phenol	—	—	—	—
Eugenol	—	Orange-peach	Buff	Buff
Orcinol	—	Rust-brown	Pink	Cream
Resorcinol	—	Brown	Khaki	Grey-buff
Metol	Light brown	Rust	Blue-grey	Brown
Pyrocatechol	Buff	Dark brown	Dark grey	Dark brown
Pyrogallol	Light brown	Brown	Brown-yellow	Brown-yellow
α-Naphthol	—	Green-grey	Grey	Brown
β-Naphthol	Beige	Yellow-grey	Brown	Brown
Cinchonine	—	Orange	—	Pink
Brucine	Orange-yellow	Orange yellow	—	—
Hydrazine	—	Tangerine	—	—
Phenylhydrazine	Light yellow	Mustard	Light yellow	Beige

Spray I.   A freshly prepared 3% aqueous solution (w/v) of selenium dioxide is sprayed on a dried silica gel TLC plate which is then heated at 120° for 15-20 min.

Spray II. A freshly prepared 1% solution of phenol in 20% Na_2CO_3 is sprayed on a silica gel plate which is heated for 20-25 min at 120°. Overspraying with concentrated HCl results in color change immediately or 48 hr later.

REFERENCE

1. **Mitchell, S. C.** and **Waring, R. H.**, *J. Chromatogr.*, 151, 249, 1978.

Reproduced by permission of the Elsevier Scientific Publishing Co.

Table 7
COLORS GIVEN BY PRIMARY ARYLMONOAMINES IN RESPONSE TO VARIOUS DETECTION AGENTS

Method of detection

Compound	1	2	3	4	5	6	7	8	9	10
4-Aminoazobenzene	Carmine	Red	Light Orange	NC	NC	Red	Olive green	Blue	Negligible	NC
2-Aminobenzoic acid	Red	Brown	Yellow	Yellow	Yellow	Faint green	Dirty grey	Blue	Faint brown	Pale grey
4-Aminobenzoic acid	Red	Brown	Yellow	NC	Pale yellow	Negligible	Dirty grey	Blue	Faint purple	NC
4-Aminobiphenyl	Red	Pale brown	NC	Yellow	NC	Negligible	Negligible	Faint blue	Purple	NC
4-Amino-4-di-methyl-aminoazobenzene	Brownish mauve	Pink	Orange	Pale pink	NC	Reddish brown	Dirty green	Negligible	NC	Light brown
4-Aminodiphenylamine	Black	Brownish grey	Orange	Brown	Brown	Blue	Dark blue	Blue	Blue	Brown
1-Amino-2-hydroxy-naphthalene-4-sulfonic acid	Negligible	Faint purple	Negligible	NC	NC	Negligible	Negligible	Blue	NC	Brown
2-Aminophenol	Brown	Light brown	Yellow	Negligible	Yellow	Grey	Dirty green	Blue	Purple	Brownish black
2-Aminopyridine	NC	NC	Very faint yellow	NC	Negligible	NC	Pale yellow	NC	NC	NC
4-Anisidine	Mauve-red	Light brown	Faint yellow	Yellow	Yellow	Blue	Dark blue	Purple	Light blue	Bluish brown
1-Naphthylamine	Brownish purple	Light brown	Negligible	Light brown	NC	Blue	Blue	Blue	Purple	NC
2-Naphthylamine	Brownish red	Light brown	Faint yellow	Light brown	Pale yellow	Steel blue	Dark blue	Blue	Brown	NC
2-Nitroaniline	Orange	Brown	NC	NC	NC	NC	NC	NC	Negligible	NC
3-Nitroaniline	Orange-yellow	Brown	NC	NC	NC	NC	NC	NC	Negligible	NC

	1	2	3	4	5	6	7	8	9	10
4-Nitroaniline	Orange-red	Brown	Negligible	Yellow	NC	NC	NC	NC	Negligible	NC
Sulfanilic acid	Orange-red	Brown	Yellow	Yellow	NC	NC	NC	NC	Faint purple	NC
4-Toluidine	Orange-red	Brown	Yellow	Grey	Blue	Blue	Purple	Pale pink		

Note: NC = no color.

Method of detection.

Compounds were chromatographed on Silica Gel F_{254} thin-layer plates before spraying with the reagent.

(1) The plates are sprayed liberally with a freshly-prepared 3% solution of pentyl nitrite in diethyl ether containing 3% of 98% formic acid. After drying at room temperature in a current of air for 10 min the second reagent, a 1% solution of 2-naphthol in 5% NaOH, is applied.

(2) Plates are sprayed with a solution of 0.2 g of glucose in a mixture of 4 ml of water, 3 ml of ethanol, 3 ml of n-butanol, 0.9 ml of H_3PO_4, S.G.1.75, before heating at 115° for 10 min.

(3) The initial spray reagent is 2% vanillin in n-propanol; after drying at room temperature the second reagent, 1% ethanolic KOH, is applied and the plates are heated at 110°C for 10 min.

(4) The plates are sprayed with a 1% solution of salicylaldehyde in ethanol containing 1% glacial acetic acid and heated at 115°C for 10 min.

(5) The plates are sprayed with a 1% solution of salicylaldehyde in ethanol containing 1% glacial acetic acid, are then sprayed with 1% ethanolic KOH and are heated at 115°C for 10 min.

(6) Plates are sprayed with the Folin-Ciocalteu reagent (BDH Chemicals Ltd., Poole, Dorset, Great Britain).

(7) Plates are sprayed with the Folin-Ciocalteu reagent and dried in air before spraying with 20% Na_2CO_3 solution.

(8) Equal volumes of 0.1 M $FeCl_3$ and 0.1 M potassium ferricyanide are mixed just before use and sprayed on the plates.

(9) The plates are sprayed with a 5% solution of potassium dichromate in 40% H_2SO_4.

(10) The plates are sprayed with 5% ethanolic phenol, followed by 0.2% sodium hypochlorite in 15% NaOH.

REFERENCE

1. Jones, G. R. N., *J. Chromatogr.*, 77, 357, 1973.

Reproduced by permission of the Elsevier Scientific Publishing Company.

Table 8
DETECTION OF AROMATIC SECONDARY AMINES ON FILTER PAPER

Amine	Color of spot	
	Test 1	Test 2
Color of spot/ Test 1 Test 2		
Diphenylamine	Bluish black	Red purple
p-Octyldiphenylamine	Dark yellow-green	—
p,p′-Dioctyldiphenylamine	Yellow	—
N-Phenyl-1-naphthylamine	Olive green	Blue-purple
N-Phenyl-2-naphthylamine	Yellowish brown	Light red
N-Octylphenyl-2-naphthylamine	Yellowish brown	Light yellowish red
Di-1-naphthylamine	Olive green	Blue-purple
Di-1,2-naphthylamine	Olive	Purple
Di-2-naphthylamine	Yellowish brown	Light red
Phenothiazine	Brown	Brown
3,7-Dioctylphenothiazine	Reddish yellow	Pale red-purple

Test 1: The paper is sprayed with concentrated HNO_3.
Test 2: The paper is sprayed with a diazo-reagent prepared by slowly adding a solution of 3 g of $NaNO_2$ in 24 ml of water to a solution of 1 g of sulfanilic acid and 8 ml of 36% HCl in 20 ml of water, the temperature being maintained below 5°C.

The tests are applied following separation of the amines by partition paper chromatography.

REFERENCE

1. Shimizu, I., *J. Chromatogr.*, 118, 96, 1976.

Reproduced by permission of the Elsevier Scientific Publishing Company.

Section IV
*Guide to Methods of Sample Preparation
Including Derivatizations*

Section IV

GUIDE TO METHODS OF SAMPLE PREPARATION INCLUDING DERIVATIZATION

This section describes procedures for extraction, hydrolysis, cleanup, and derivatization of samples prior to chromatography. It is hoped that the methods can, in many cases, be used directly by researchers for the preparation of samples for chromatography, but they should at least serve as a guide, illustrative of the kind of steps which are required, for workers designing their own procedures. The methods are classified according to chemical classes.

1. Hydrolysis of proteins and peptides for amino acid analysis.................198
2. The modification of cysteine and cystine in proteins for amino acid analysis...200
3. Hydrolysis and cleanup procedure for the GC analysis of amino acids in food samples...201
4. Extraction of minute amounts of free amino acids in natural waters prior to GC analysis...201
5. Sample preparation for the analysis of amino acids from soil or ground rock samples...202
6. N-TFA amino acid isobutyl esters, the preparation of for GC.............203
7. N-isobutyloxycarbonyl amino acid methyl esters, the preparation of for GC...204
8. Amino acid enantiomers, the preparation of derivatives for the resolution of by GC...204
9. Pentafluoropropionyl-amino acid (+)-3-methyl-2-butyl esters, the preparation of for the separation of amino acid enantiomers by GC.........205
10. L-α-Chloroisovaleryl derivatives of DL-amino acid methyl esters and of amines, the preparation of for separation of enantiomers by GC.............205
11. Derivatives of dicarboxylic amino acids and their amides, the preparation of for GC...205
12. N-Trifluoroacetyl-amino acid trimethylsilyl esters, the preparation of for GC...206
13. Dansyl-amino acids, the preparation of for HPLC......................207
14. Optically active terpene derivatization reagents applied to the resolution of amino acid enantiomers by HPLC...207
15. N-d-10-Camphorsulfonyl amino acid p-nitrobenzyl esters, the preparation of for HPLC...208
16. N-(−)-α-Methoxy-α-methyl-1-naphthalene acetylamino acid methyl esters, the preparation of for HPLC...208
17. o-Phthalaldehyde/ethanethiol derivatives of amino acids, the preparation of for HPLC...209
18. Thiourea derivatives of amino acids, the preparation of for the separation of enantiomers by LC...209
19. 4-Dimethylaminoazobenzene-4'-sulfonyl chloride (dabsylchloride), the chromophoric labeling of amino acids with for TLC.....................209
20. 4-N,N-Dimethylaminonaphthylazobenzene-4'-thiohydantoin (DANABTH) amino acids, the preparation of for TLC..................210
21. Histidine and histamine, fluorogenic reaction using fluorescamine for the detection of...210
22. N-2,6-Dinitro-4-trifluoromethyl, O-trimethylsilyl derivatives of amines, the preparation of for GC...211

23. N-Dinitrophenyl, O-trimethylsilyl derivatives of amines, the preparation of for GC 211
24. Trifluoroacetyl derivatives of primary and secondary amines, the preparation of for GC 212
25. Control of trimethylsilylation reactions by the use of color indicators 212
26. Schiff base of amines, the preparation of for GC 213
27. Dimethylalkylamines, the preparation of for GC 214
28. Sulfonamides of secondary amines, the preparation of for GC 214
29. Polyamines, the N-permethylation of for GC 215
30. N-Trifluoroacetyl derivatives of amines, the preparation of for GC 215
31. Trimethylsilyl derivatives of amines and amino acids, the preparation of for GC 215
32. m-Toluoyl derivatives of amines, the preparation of for HPLC 216
33. Tosyl derivatives of polyamines, the preparation of for HPLC 216
34. 4-Methoxybenzamide derivatives of amines, the preparation of for HPLC ... 216
35. Fluorescamine derivatives of aliphatic diamines and polyamines, the preparation of for HPLC 217
36. o-Phthalaldehyde derivatives of amines, the preparation of for HPLC 217
37. The separation of polyamines in serum and their conversion to fluorescamine derivatives for HPLC 217
38. N,N'-Disubstituted ureas of aliphatic and aromatic primary and secondary amines, the preparation of for HPLC 218
39. Aliphatic amines, reaction with 7-chloro-4-nitrobenzo-2-oxa-1,3-diazole (NBD-Cl) for TLC 219

1. HYDROLYSIS OF PROTEINS AND PEPTIDES FOR AMINO ACID ANALYSIS

The most frequently used method for hydrolyzing proteins or peptides to amino acids is heating with 6 N HCl. As most proteins contain tryptophan, which is readily destroyed in HCl by oxidation, an alternative method of hydrolysis applicable to all the amino acids is desirable. Sulfonic acids, which are nonoxidizing strong acids, may be used to replace HCl. Suggested methods are (1) hydrolysis with 3 N p-toluenesulfonic acid for 22 hr, which gives more than 90% recovery of tryptophan.[1] However, p-toluenesulfonic acid is a solid recrystallized from concentrated HCl and is difficult to prepare free from HCl. The method also does not permit the analysis of half-cystine. (2) Hydrolysis using 4 N methanesulfonic acid containing 0.2% 3-(2-aminoethyl) indole as a catalyst permits the determination of the complete amino acid composition of proteins from a single hydrolysate.[2] Half-cystine is determined as S-sulfocysteine. A stock solution of 4 N methanesulfonic acid is prepared by diluting concentrated acid with glass-distilled water. Solid 3-(2-aminoethyl) indole is added to give a final concentration of 0.2% and the solution is immediately flushed with nitrogen. The solution is stored at 4° and thoroughly flushed with nitrogen after each use.

Hydrolysis of proteins or peptides is performed in heavy walled ignition tubes which have been washed with a 3 to 1 mixture of H_2SO_4 and HNO_3, rinsed in deionized water and oven-dried. 0.2 to 2 mg of sample is hydrolyzed in vacuo (20 μ) at 115° for 22, 48 or 72 hr with 1.0 ml of 4 N methanesulfonic acid containing 0.2% 3-(2-aminoethyl) indole. For samples weighing less than 1 mg, 0.3 to 0.5 ml of acid is used. The hydrolysate is partially neutralized with 1.0 ml of 3.5 N NaOH and centrifuged or filtered for amino acid analysis on the ion-exchange column.

For the determination of half-cystine, 1.0 ml of the hydrolysate is cooled to 4° and 0.3 ml of pyridine is added, followed by 0.9 ml of 4 N NaOH to bring the pH to 6.8. An aqueous solution of dithiothreitol (1 ml, 4 μmol) is added and the hydrolysis tube immediately flushed with nitrogen for 1 min, sealed with a layer of Parafilm and incubated at 37° for 1 hr. After reduction, 60 mg of solid sodium tetrathionate is added to the hydrolysate, and the mixture allowed to stand at 25° for at least 5 hr. The mixture is then rotary-evaporated to dryness in vacuo at 30°. To remove pyridine completely, the residue is redissolved in 0.5 mℓ of water and redried. The residue is dissolved in 1 to 5 mℓ of pH 2.2 buffer, filtered through a millipore filter (0.45 μ) and analyzed for amino acids. It is imperative that the pH of the sample is brought to pH 2.2 ± 0.5. (3) Hydrolysis with 6 N HCl containing 0.5 to 6% of thioglycolic acid.[3] This method is stated, however, to give low recoveries for tryptophan. (4) Hydrolysis with mercaptoethane sulfonic acid.[4] Hydrolysis is carried out in thick-walled sealed Pyrex tubes, 80 × 11 mm, which have been previously washed with chromic acid and water, then dried. 0.5 to 2 mg of peptide or protein is hydrolyzed with 1 mℓ of 3 N mercaptoethane sulfonic acid at 110 ± 2°C for 24 or 72 hr. The tubes are then cooled, opened and 2 mℓ of N NaOH added to the solution. The contents of the tube are washed into a 5.00 mℓ volumetric flask and 1.00 mℓ aliquots used for amino acid determination.

THE PURITY OF REAGENTS

The availability of fluorescent reagents for quantitative amino acid analysis allows existing equipment to be upgraded from the micromolar to the nanomolar level of sensitivity. This is often accompanied by baseline noise and extraneous emission peaks. Constant boiling HCl may contain an impurity which causes problems of this nature.[5] The impurity, assumed to be a primary amine, distills with the acid, does not react with ninhydrin, and bleeds through sulfonated polystyrene columns even at acid pH. Removal of the impurity is achieved by distilling 2 ℓ of 6 N HCl, prepared from reagent grade concentrated acid and deionized water, from 10 g of sodium dichromate. The impurity, possibly a low molecular weight primary amine, is presumably oxidized to a compound containing a nitro group which no longer reacts with fluorescamine or o-phthalaldehyde and does not interfere with the chromatographic procedure.

A problem related to that of ensuring the purity of reagents is that of avoiding contamination.[6] A serious source of amino acid contamination is the fingerprints of individuals involved in handling samples and glassware. Features of contamination by free amino acids derived from hands are the presence of nearly all the usual protein amino acids and of two metabolic intermediates, citrulline and ornithine. Precautions must therefore be taken to avoid hand contamination when determining very small amounts of amino acids.

Cleaning procedures have been devised to avoid contamination of samples by glassware.[7] The glassware is initially cleaned with soap and water and is then soaked in 5% NaHCO$_3$ solution to saturate all active negative sites on the surface with sodium ions. After drying at 160°C the glassware is wrapped in aluminum foil and fired at 625°C for a minimum of 4 hr. Both glass and wrapping are decontaminated. Storage for some days before use is possible.

Hamilton considered that in view of conflicting findings and experimental difficulties the presence of amino acids in lunar samples was highly questionable.[8] If amino acids are present in lunar samples the majority of them do not have the distribution pattern, type or variety usually associated with biological origin, i.e., they are not contaminations from terrestrial biological sources. In Hamilton's opinion it was not possible with the amounts of lunar samples that had been made available to determine if amino acids were indigenous to the moon either in free, in complexed, or in precursor forms.

The elution of proteins from polyacrylamide gels has been used in a number of procedures, including comparative amino acid composition determination and peptide mapping. Experiment has shown that there is a reproducible and measurable "background" present in eluted polyacrylamide gels and that when it is applied appropriately to amino acid composition data the accuracy of the analysis is increased by the correction.[9]

REFERENCES

1. Liu, T.-Y. and Chang, Y. H., *J. Biol. Chem.*, 246 2842, 1971.
2. Simpson, R. J., Neuberger, M. R., and Liu, T.-Y., *J. Biol. Chem.*, 251, 1936, 1976.
3. Matsubara, H. and Sesaki, R. M., *Biochem. Biophys. Res. Commun.*, 35, 175, 1969.
4. Penke, B., Ferenczi, R., and Kovacs, K., *Anal. Biochem.*, 60, 45, 1974.
5. Schwabe, C. and Catlin, J. S., *Anal. Biochem.*, 61, 302, 1974.
6. Hamilton, P. B., *Nature (London)*, 205, 284 (1965)
7. Hamilton, P. B. and Nagy, B., *Space Life Sci.*, 3, 432, 1972.
8. Hamilton, P. B., *Anal. Chem.*, 45, 1718, 1975.
9. Brown, W. E. and Howard, G. C., *Anal. Biochem.*, 101, 294, 1980.

2. THE MODIFICATION OF CYSTEINE AND CYSTINE IN PROTEINS FOR AMINO ACID ANALYSIS

Automatic amino acid analysis of proteins after HCl hydrolysis is unsatisfactory for cysteinyl and cystinyl residues, which are unstable to this treatment. This disadvantage may be overcome by suitable prior modification of cysteinyl groups. Suggested procedures are

(1) The protein is treated with tri-n-butylphosphine and 1,3-propane sultone.[1] Tri-n-butylphosphine reduces cystine to cysteine and 1,3-propane sultone alkylates cysteine to S-3-sulfopropylcysteine; these reactions may be performed simultaneously. One nmol of protein that has been lyophilized in an ignition tube is dissolved in 0.1 mℓ of 0.5 N NaHCO$_3$ mixed with n-propanol (1:1). Ten mℓ of 50 mM tri-n-butylphosphine in 100% n-propanol and 6 $\mu\ell$ of 240 mM 1,3-propane sultone in 60% n-propanol are added and the mixture kept under N$_2$ at 25°C for 2 hr. Both of these solutions are prepared immediately before use. The reaction is terminated by addition of 50 $\mu\ell$ of constant boiling HCl and the mixture is diluted with 0.5 mℓ of deionized water, frozen and lyophilized.

Since 1,3-propane sultone is a suspected carcinogen, 3 vapor traps cooled with dry ice in acetone are connected in series between the lyophilization flask and the lyophilizer. The sample tube, insulated with a piece of Styrofoam, is placed inside the flask. One mℓ of 10% NaOH is placed inside each trap; the 1,3-propane sultone is thus condensed and hydrolyzed to noncarcinogenic products. One-nmol protein samples are hydrolyzed with 100 $\mu\ell$ of 4 N methanesulfonic acid containing 0.2% tryptamine at 115°C for 24 hr. After hydrolysis the samples are partially neutralized with 100 $\mu\ell$ of 3.5 N NaOH. The method gives good analysis of tryptophan (about 80%) and excellent analyses of the other amino acids.

(2) Cysteinyl residues are determined as S-β-(4-pyridylethyl)cysteine and cystinyl residues as S-sulfocysteine after hydrolysis of the pyridylethyl protein with 4 M methanesulfonic acid.[2] Tryptamine must be present during hydrolysis for quantitative tryptophan recovery. Soluble proteins (300-500 mg) are dissolved in 20 to 30 mℓ of Tris buffer solution, 8 M urea, pH 7.5, under a nitrogen atmosphere. 4-vinylpyridine (20 M excess over sulfhydryl residues) is added and the solution is stirred for 2 to 3 hr. The solution is acidified to pH 3 with glacial acetic acid, dialyzed against 0.01 N acetic acid and freeze dried. The recommended hydrolysis procedure uses 4 M methanesulfonic acid.

REFERENCES

1. Lee, K. S. and Drescher, D. G., *J. Biol. Chem.*, 254, 6248, 1979.
2. Inglis, A. S., McMahon, D. T. W., Roxburgh, C. M., and Takayanagi, H., *Anal. Biochem.*, 72, 86, 1976.

3. HYDROLYSIS AND CLEANUP PROCEDURE FOR THE GC ANALYSIS OF AMINO ACIDS IN FOOD SAMPLES

HYDROLYSIS[1]

The samples are weighed into hydrolysis tubes, 15 × 160 mm, and 5 mℓ of 6 N HCl containing 10 g of phenol per liter are added. After adding an appropriate amount of the internal standard, α-aminocaprylic acid, the tubes are frozen in a mixture of solid CO_2 and ethanol and the frozen samples placed in a desiccator. The dissolved air and oxygen are removed by applying a vacuum for 20 min. Dry nitrogen is then admitted to release the vacuum and the tubes are sealed and rotated constantly in a hot-air oven at 110°C for 24 hr.

ION-EXCHANGE CLEANUP[1]

To remove impurities which might interfere with the GC analysis, the hydrolyzate is passed through an ion-exchange resin column. Amberlite® IR-120, a strong cation-exchange resin, is stirred for 1 hr with N NH_4OH using a magnetic stirrer. This procedure is repeated twice before the resin is washed with deionized water until it is neutral to litmus paper. An excess of 3 N HCl is added and the mixture is stirred very slowly with a magnetic stirrer. After 1 hr the HCl is discarded and the resin is washed with deionized water to neutral pH and packed into a column. The hydrolyzate is evaporated, dissolved in 0.1 M HCl and added to the column without disturbing the resin bed. Impurities are washed out by passing 5 mℓ of deionized water through the column and the amino acids are eluted with 5 mℓ of 7 N NH_4OH, followed by 5 mℓ of deionized water. The column is regenerated with 3 N HCl before further use. The amino acids are converted to their N-TFA n-butyl esters for GC analysis.

Boila and Milligan found that when NH_4OH at concentrations greater than 3 N was used to elute amino acids from the ion-exchange resin, recoveries of lysine and arginine were less than 50% and highly variable on subsequent GC measurement.[2] By using 1 N NH_4OH to elute amino acids from the resin the recovery of the amino acids was maximized and the variability of the results was reduced.

REFERENCES

1. Nair, B. M., *J. Agr. Food Chem.*, 25, 614, 1977.
2. Boila, R. J. and Milligan, L. P., *J. Chromatogr.*, 202, 283, 1980.

4. EXTRACTION OF MINUTE AMOUNTS OF FREE AMINO ACIDS IN NATURAL WATERS PRIOR TO GC ANALYSIS

An ion-exchange column is first prepared. Dowex® 50 W (H), 100 to 200 mg, is placed in a Pasteur pipet with a glass wool plug in the stem. Distilled and deionized

(D2) water is used to wash the resin. The resin is loaded with 4 mℓ of 3 M HCl and then washed with (D2) water until the eluate is neutral. After elution of the sample, the resin is washed with 15 mℓ of (D2) water to pH 5 to 6 and regenerated with 4 mℓ of 3 M HCl.

Aqueous samples, ranging from 1000 mℓ for field samples to 3 to 500 mℓ for laboratory experiments, are filtered through a membrane filter (Millipore, mixed cellulose acetate and nitrate, pore size 0.45 μm) to remove particulate matter. The filtrate is evaporated in a rotary evaporator at 80°C and reduced pressure. The volume is reduced to 2 to 10 mℓ, depending on the amount of salts and organic matter, and transferred to a 50 mℓ test tube. The evaporator flask is rinsed with 3 × 2 mℓ of (D2) water which is added to the sample.

Three M HCl is added to the tube to bring the pH to 2.5 to 3.0, followed by 2 mℓ of chloroform which has been previously equilibrated with water to remove ethanol. The sample is shaken vigorously for 10 min, the tube centrifuged and the aqueous phase transferred to another tube. 2 × 1 mℓ of (D2) water is carefully added to the chloroform surface, sucked off, and added to the sample. The combined aqueous phases are extracted with 2 mℓ of chloroform and transferred to a glass funnel attached to the resin column. The sample is allowed to pass through the ion-exchange column at about 1.0 mℓ/min and the resin is washed with 5 mℓ of (D2) water to remove neutral molecules and anions. The amino acids are then eluted with 3 mℓ of 3 M NH_4OH. The eluate is collected in a Teflon-lined screw-cap culture tube, 14 × 90 mm, and evaporated almost to dryness in a rotary evaporator at 70°C. The aqueous sample is transferred to a Pyrex glass capillary tube, 80 × 2 mm i.d. The culture tube is washed twice with 50 μℓ portions of (D2) water, each of which is added to the sample.

The aqueous column is sucked up the tube to a distance of about 1.5 cm from the end and the capillary tube sealed in a micro-flame. The sample is then evaporated to dryness overnight in a desiccator at 10 to 20 mm pressure, before conversion of the amino acids to their N-heptafluorobutyryl isobutyl esters and gas chromatographic analysis. An alternative procedure using Amberlite® IR-120 (H⁺) resin in place of Dowex® 50 W (H⁺) resin is described in the Reference.

REFERENCE

1. Bengtsson, G. and Odham, G., *Anal. Biochem.*, 92, 426 (1979).

5. SAMPLE PREPARATION FOR THE ANALYSIS OF AMINO ACIDS FROM SOIL OR GROUND ROCK SAMPLES

Soil scientists and geochemists frequently wish to study the amino acid composition of soils or rocks. To do this it was often necessary to process large samples to obtain adequate material for study. Amino acids were generally isolated by hydrolysis in 6 N HCl followed by evaporation to dryness and desalting. Strong cation exchangers were the most frequent choice for desalting hydrolysates; the hydrolysates however required large columns of resin and large volumes for elution of amino acids. Long and cumbersome evaporations, losses, and contamination could be involved.

Pollock and Miyamoto found that the yields of amino acids were improved and desalting operations simplified by using HF to precipitate calcium, magnesium, and aluminum, while simultaneously reducing the concentrations of sodium and potassium.[1] The iron fluoride was removed by adjustment to pH 9 when it precipitated as mixed fluorohydroxides.

Pollock et al. determined the D and L isomers of protein amino acids in soils by gas chromatographic techniques.[2] The desalting procedure they adopted utilized fluoride precipitation in conjunction with a Biorad AG 1-X8 (F⁻) column. The D-amino acid concentrations were determined by direct derivatization after complete desalting, or by thin-layer chromatography of the desalted amino acids, followed by gas chromatography. The second method was considered to give the more accurate values.

A recommended procedure combined HF treatment after hydrolysis with a concentrated formic acid extraction.[3] The cations present in a 6 N HCl hydrolysate and in the formic acid extract after conversion of the salts from chlorides to fluorides were determined, as well as the amino acid nitrogen extracted by the formic acid. After a lengthy study of different extraction times temperatures and the amount of racemization taking place, the following processing sequence was adopted.

1. Place 1 g of soil in the hydrolyzer and add 5 ml of 6 N HCl.
2. Heat the closed system at 100°C for 16 hr.
3. Evaporate to dryness. Allow the temperature to fall to about 70°C, apply a vacuum, and heat the vessel slowly to 100°C.
4. Add 5 ml of 10 N HF to the residue and mix for 2 hr.
5. Evaporate to dryness, ensuring that the residues are dry. Apply vacuum to the vessel at ambient temperature and heat is applied until 100°C is attained.
6. Add 5 ml of 97 + % formic acid to the residue, heat and agitate for 1 hr at 70°C.
7. Filter and evaporate the filtrate to dryness in a second vessel.
8. Prepare an ion-exchange column before proceeding to step 9. The 10 ml Dowex® 50 H⁺ column is 1.3 cm internal diameter and 7.6 cm long. Add 40 ml of 4 N NaOH, followed successively by 40 ml of water, 40 ml of 6 N HCl, and 30 ml of water directly into the column.
9. Add 5 ml of water to the residue from step 7, heat gently to ensure solution of the amino acids, allow to cool and add to the freshly regenerated ion-exchange column. Immediately add 30 ml of water to the column to wash through anions and neutral organics, followed by 15 ml of 4 N NH₄OH. Start to collect the amino acids when the ammonia begins to break through the column. Collect only the first 3 ml.
10. Evaporate the amino acid containing solution to dryness at 100°C.

The amino acids are then converted to the N-TFA D(+)-2-butyl esters prior to gas chromatographic determination. Pollock et al. have described semiautomated breadboard instrumentation for this processing scheme.[3]

REFERENCES

1. Pollock, G. E. and Miyamoto, A. K., *Agric. Food. Chem.*, 19, 104 (1971).
2. Pollock, G. E., Cheng., C-N and Cronin, S. E., *Anal. Chem.*, 49, 2 (1977).
3. Pollock, G. E., Day, R. Kinsey, S. and Miller, S. L., *COSPAR, Life Sciences and Space Research*, Volume 15, Holmquist R. and Stickland, A. C. Eds., Pergamon Press, New York, 1977, 27.

6. THE PREPARATION OF N-TFA AMINO ACID ISOBUTYL ESTERS FOR GC

In this procedure samples are heated using a TNCAM, Model DB-3H Driblock heater. Bath wax is added so that each hole in the block forms a miniature oil bath. Vials are inserted in the oil to the level of the fluid inside the vial; a vertical temperature gradient is thus established in the vial and the reactants reflux.

The amino acid mixture, 25 μℓ containing about 2.5 μmol/mℓ of each amino acid, is added to a 1-mℓ Reactivial (Pierce, Rockford, Ill.) and excess solvent is evaporated at 50°C in a stream of dry N_2. Norleucine and pipecolic acid are added as internal standards in amounts equivalent to the other amino acids and the solvent is again evaporated under N_2. 100 μℓ of 3 MHCl in isobutanol are added and the solution heated at 120°C. After about 5 min the vial is removed from the heater and, while hot, the contents are shaken for 15-30 sec. The vial is again heated for a total period of 30 min. The vial is cooled to room temperature, is opened and excess reagent evaporated at 50°C using a stream of dry N_2. Heptafluorobutyric anhydride (HFBA), 50-75 μℓ, is added and the vial is heated at 150°C for 10 min. The vial is cooled to room temperature and evaporated just to dryness using a stream of dry N_2. Hexadecane dissolved in ethyl acetate is added in amount equivalent to half the molar concentration of the amino acids. Ethyl acetate is added to give a total volume of 25 μℓ and the whole is agitated. Appropriate aliquots of this solution are applied to the chromatographic column.

REFERENCES

1. MacKenzie, S. L. and Tenashuk, D., *J. Chromatogr.*, 97, 19, 1974; 111, 413, 1975; 171, 195, 1979; 173, 53, 1979.

7. THE PREPARATION OF N-ISOBUTYLOXYCARBONYL AMINO ACID METHYL ESTERS FOR GC

To an aqueous aliquot containing up to about 2 mg of total amino acids (from 10 to 100 μg of each amino acid) in a 10 mℓ polyethylene-stoppered vial is added 0.5 mℓ of 10% NA_2CO_3 and the volume made up to 2 mℓ with distilled water. Isobutyl chloroformate (0.1 mℓ) is added immediately, and the mixture is shaken mechanically for 10 min at room temperature. The product is extracted twice with 3 mℓ of diethyl ether to remove excess reagent, the aqueous layer is saturated with NaCl, acidified to pH 1-2 with 10% orthophosphoric acid, using thymol blue test paper and then extracted 3 times with 3 mℓ of diethyl ether, being shaken by hand for 1 min. One mℓ of methanol is added to the combined ethereal extracts and methylation is performed by bubbling diazomethane through the solution until a yellow tinge is visible. After standing at room temperature for 5 min the solvents are quickly evaporated to dryness at 50°C under nitrogen. The residue is dissolved in 0.1 to 0.2 mℓ of ethyl acetate, the solution is dried over anhydrous Na_2SO_4 and 4 μℓ of the solution used for gas chromatography.

REFERENCE

1. Makita, M., Yamomoto, S., and Kono, M., *J. Chromatogr.*, 120, 129, 1976.

8. THE PREPARATION OF DERIVATIVES FOR THE RESOLUTION OF AMINO ACID ENANTIOMERS BY GC

Procedure A. To a solution of about 1 mg of amino acid ester in 0.8 mℓ of tetrahydrofuran containing 0.2 mℓ of pyridine is added about 4 mg of acid chloride (d-isoketopinyl chloride, 1-dihydroteresantalinyl chloride or 1-teresantalinyl chloride). The acid chlorides should be freshly prepared by treatment of the appropriate acid with thionyl chloride prior to use. The reaction mixture, consisting of the N-acyl derivative and excess reagent, is injected directly into the gas chromatograph.

Procedure B. About 1 mg of the amino acid ester in 0.5 m*l* of acetonitrile is treated with about 4 mg of the acid chloride in the presence of 1 drop of triethylamine. The reaction product is analyzed without further purification.

REFERENCE

1. Nambara, T., Goto, J., Taguchi, K., and Iwata, T., *J. Chromatogr.*, 100, 180, 1974.

9. THE PREPARATION OF PENTAFLUOROPROPIONYL-AMINO ACID (+)-3-METHYL-2-BUTYL ESTERS FOR THE SEPARATION OF AMINO ACID ENANTIOMERS BY GC

About 100 µg of a DL-amino acid are heated in 150 µ*l* of a 7 *N* solution of HCl in (+)-3-methyl-2-butanol of about 99% optical purity. Basic amino acids are esterified previously with a 1.25 *N* solution of HCl in methanol for 60 min at 100°C. After removal of excess of reagent under reduced pressure, the sample is acylated with a mixture of 200 µ*l* of dichloromethane and 50 µ*l* of pentafluoropropionic anhydride for 30 min at room temperature. After removing excess of reagent, the sample is dissolved in 100 µ*l* of ethyl acetate and the solution used for gas chromatography.

REFERENCE

1. König, W. A., Rahn, W., and Eyem, J., *J. Chromatogr.*, 133, 141, 1977.

10. THE PREPARATION OF L-α-CHLOROISOVALERYL DERIVATIVES OF DL-AMINO ACID METHYL ESTERS AND OF AMINES FOR SEPARATION OF ENANTIOMERS BY GC

Samples of about 10^{-4} g of amino acid methyl esters or amines are treated with a mixture of 80 µ*l* CH_2Cl_2 and 20 µ*l* of L-α-chloroisovaleryl chloride at room temperature in a screw cap vial with Teflon® lining in the cap. After removal of excess reagent in a nitrogen stream the samples are injected into the gas chromatograph. Derivatives of hydroxy amino acids are dissolved in 50µ*l*, CH_2Cl_2 and 50µ*l* N-methyl-N-trimethylsilyl-trifluoroacetamide (MSTFA) are added. Th temperature for 30 min. Conversion to the O-TMS derivative is thus brought about.

REFERENCE

1. König, W. A., Stölting, K., and Kruse, K., *Chromatographia*, 10, 444, 1977.

11. THE PREPARATION OF DERIVATIVES OF DICARBOXYLIC AMINO ACIDS AND THEIR AMIDES FOR GC

Condensation of the amino acids with DCTFA (1,3-dichlorotetrafluoroacetone) gives 2,2-bis (chlorodifluoromethyl)-1,3-oxazolidinones. Treatment of the ketone with a small amount of methanol produces a halogenated alcohol which esterifies the second carboxyl group of glutamic and aspartic acids after a reactive anhydride such as HFBA (heptafluorobutyric anhydride) is added. The amide derivatives formed from glutamine and asparagine during the first acylation are destroyed during the extraction procedure. The second acylation is necessary to restore these derivatives.

Condensation — 10 to 100 nmoles of each amino acid in the mixture are condensed with 100 μl of mixed solvent, consisting of benzene, acetonitrile and pyridine in volume ratio 60:30:28 and 20 μl of DCTFA. Phenylalanine is used as an internal standard.

First acylation — The sample is treated with 20 μl of 20% of methanol in benzene followed by 12 μl of HFBA and after at least 10 sec by an additional 24 μl of HFBA.

Extraction — 500 μl of light petroleum-dichloromethane mixture (4:1,v/v) are added to the sample and the contents shaken for 10 to 15 sec with 400 μl each of 1 M Na_2CO_3, 1 M HCl (twice) and 1 M $NaHCO_3$. The water phase is removed using a Pasteur pipet, and after drying the organic phase with anhydrous Na_2SO_4 the extraction medium is transferred to another tube, leaving the sulfate in the original tube, and evaporated just to dryness. Excessive blowing of gas into the dry residue is avoided.

Second acylation — 100 μl of hexane and 5 μl of HFBA are added to the evaporated residue. After heating the sample at 70°C for 10 min, an aliquot of 2 to 3 μl is injected into the chromatographic column.

REFERENCE

1. Husek, P. and Felt, V., *J. Chromatogr.*, 152, 363, 1978; 152, 546, 1978.

12. THE PREPARATION OF N-TRIFLUOROACETYL-AMINO ACID TRIMETHYLSILYL ESTERS FOR GC

N-Trifluoroacetyl-amino acid trimethylsilyl esters are prepared by reaction of the amino acid hydrotrifluoroacetate with HMDS (hexamethyldisilazane). The procedure can also be used for the preparative isolation of the N-TFA amino acid, since alcoholysis splits off the TMS residue giving the alcohol soluble N-TFA amino acid.

0.01 mol of amino acid is dissolved in 0.1 mol (8 ml) of trifluoroacetic acid and the viscous solution diluted with 10 ml of dioxan. This solution is added dropwise with stirring to 0.06 mol (13 ml) of ice-cold HMDS and boiled under reflux with exclusion of moisture until the reaction mixture is clear. The ammonium trifluoroacetate that is formed sublimes into the condenser. The resulting solution can be used directly for gas chromatographic experiments or the N-TFA amino acid may be isolated. For this purpose the excess HMDS and dioxan are distilled off and the oily residue treated with alcohol. Free amino acid which precipitates out is filtered off. The alcohol is distilled off from the clear solution and the residue taken up in ether. Still unaltered amino acid hydrotrifluoroacetate remains undissolved and is filtered off. By distilling the ether off from the clear solution the N-TFA amino acid can be isolated in a pure state.

In preparative experiments yields of 95% of the N-TFA amino acid TMS ester are obtained. Tryptophan, tyrosine, lysine and ornithine readily form the corresponding derivatives. Arginine and histidine hydrotrifluoroacetates do not react with HMDS. Cysteine and cystine give no defined reaction products.

REFERENCE

1. Michael, G., *J. Chromatogr.*, 196, 160, 1980.

13. THE PREPARATION OF DANSYL-AMINO ACIDS FOR HPLC

Dansyl chloride (1-dimethylaminonaphthalene-5-sulfonyl chloride) is a well-known derivatizing agent for amino acids. The analysis of dansyl-amino acids by HPLC is an attractive approach because of the speed, automation, and low detection limits possible. The percentage conversion of some amino acids to their dansyl derivatives depends on the concentration ratio of dansyl chloride to amino acid. This variation in yield is explicable in terms of three competing reactions producing (1) dansyl-amino acid, (2) dansyl sulfonic acid, and (3) dansyl amide. The desirable reaction (1) is accelerated by high pH, which, however, also favors reaction (2). Conversion of appropriate amino acids to didansyl derivatives also depends on the pH. Reaction conditions have been reported that give high yields of the mono- and didansyl derivatives, and where the yield is independent of the ratio of dansyl chloride to amino acid over a 1000-fold range. The dansylation procedure thus provides a means of quantitative analysis for amino acids.

For derivatization, a solution of dansyl chloride in acetonitrile (1.5 mg/mℓ) is prepared. For trace detection of amino acids, HPLC grade acetonitrile is distilled over dansyl chloride. Partially hydrolyzed dansyl chloride should not be used, since it will not totally dissolve in acetonitrile. The amino acids are dissolved in 40 mM lithium carbonate buffer (pH 9.5, with HCl). For room temperature reaction, 1 mℓ of dansyl chloride solution is rapidly added to 2 mℓ of amino acid solution (0.001-1 mM in the buffer). The mixture is gently shaken for 2 min and then allowed to stand at room temperature. The precipitate which appears initially on mixing dissolves immediately. The reaction is terminated by adding 100 μℓ of 2% ethylamine or methylamine hydrochloride solutions. For reaction at higher temperatures, 2 mℓ of the amino acid solution is preheated to the required temperature in a heating module and 1 mℓ of dansyl chloride solution added. After brief shaking, the reaction vial is maintained at the given temperature in the heating block. The reaction is terminated by adding 100 μℓ of 2% ethylamine solution. Reaction vials are wrapped with aluminum foil to exclude light. At room temperature all the amino acids are derivatized in high yields after 35 min reaction with dansyl chloride. Derivatization at 60°C yielded reasonable results even in 5 min.

REFERENCE

1. Tapuhi, Y., Schmidt, D. E., Lindner, W., and Karger, B. L., *Anal. Biochem.*, 115, 123, 1981.

14. OPTICALLY ACTIVE TERPENE DERIVATIZATION REAGENTS APPLIED TO THE RESOLUTION OF AMINO ACID ENANTIOMERS BY HPLC

Two chiral reagents, (−)-1,7-dimethyl-7-norbornyl isothiocyanate (I) and (+)-neomenthyl isothiocyanate (II) may be used.

DERIVATIZATION

Methyl ester method — to a solution of about 100 μg of amino acid methyl ester in 0.2 mℓ of acetonitrile is added about 0.2 mg of reagent I or II and about 2 mg of sodium acetate. The solution is allowed to stand at 37°C for 1 hr, the reaction mixture is diluted with 0.5 mℓ of ethyl acetate, is washed successively with 5% HCl, 5% NaHCO$_3$, and water, and dried over Na$_2$SO$_4$. A sample is injected into the chromatograph.

tert-Butyldimethylsilyl ester method — to a solution of about 100 μg of amino acid in 0.2 ml of aqueous 50% pyridine is added about 1 mg of reagent I or II and 10 μl of 2 M NaOH. The solution is heated at 90°C for 1 hr, the reaction mixture then being diluted with water and extracted with ether to remove excess reagents. The organic layer is washed with water, dried over Na_2SO_4 and evaporated. A silylating agent is freshly prepared by shaking a solution of 200 mg of tert-butyldimethylchlorosilane and 100 mg of imidazole in 0.1 ml of dimethylformamide with 0.3 ml of hexane. The upper layer is used for silylation. To the residue from evaporation is added 50 μl of silylating agent and the solution is allowed to stand at room temperature for 1 hr. The reaction mixture is diluted with 0.5 ml of ethyl acetate, is washed successively with 5% $NaHCO_3$ and water and dried over Na_2SO_4. A 5 μl aliquot is injected into the chromatograph.

REFERENCE

1. Nambara, T., Ikegawa, S., Hasegawa, M., and Goto, J., *Anal. Chim. Acta.*, 101, 111, 1978.

15. THE PREPARATION OF N-d-10-CAMPHORSULFONYL AMINO ACID p-NITROBENZYL ESTERS FOR HPLC

Thirty ml of a solution of 2.0 mmole of d-camphorsulfonyl chloride in anhydrous diethyl ether is added drop by drop to a solution of 1.0 mmole amino acid in 10 ml of diethyl ether plus 20 ml of 1 N NaOH. The mixture is stirred vigorously at 0°C, stirring then being continued for 3 hr at room temperature The aqueous layer is separated, washed twice with diethyl ether, acidified with concentrated HCl and then extracted with diethyl ether. The ethereal solution is dried over anhydrous Na_2SO_4 and evaporated to dryness. The residue is dissolved in 10 ml of N,N-dimethylformamide, and one drop of trimethylamine and 1.1 mmole of p-nitrobenzyl bromide are added. The reaction mixture is heated at 55°C for 2 hr, diluted with 40 ml of chloroform, dried over anhydrous Na_2SO_4 and evaporated to dryness. A chloroform solution of the derivative is used for HPLC.

REFERENCE

1. Furukawa, H., Mori, Y., Takeuchi, Y., and Ito, K., *J. Chromatogr.*, 136, 428, 1977.

16. THE PREPARATION OF N-(−)-α-METHOXY-α-METHYL-1-NAPHTHALENE ACETYLAMINO ACID METHYL ESTERS FOR HPLC

To a solution of amino acid methyl ester, about 100 μg in 0.2 ml of pyridine, is added about 2 mg of (−)-α-methoxy-α-methyl-1-naphthaleneacetic acid and about 2 mg of N,N-dicyclohexylcarbodiimide. The mixture is allowed to stand at room temperature for 30 min, is then diluted with 0.5 ml of ethyl acetate, successively washed with 5% HCl, 5% $NaHCO_3$ and water and dried over anhydrous Na_2SO_4. A 5 μl aliquot is injected into the chromatograph. The derivatization procedure has been used in the resolution of amino acid enantiomers.

REFERENCE

1. Goto, J., Hasegawa, M., Nakamura, S., Shimada, K., and Nambara, T., *J. Chromatogr.*, 152, 413, 1978.

17. THE PREPARATION OF o-PHTHALADEHYDE/ETHANETHIOL DERIVATIVES OF AMINO ACIDS FOR HPLC

A borate buffer is prepared by adding enough boric acid with heat to 1 ℓ of water to form a saturated solution.[1] After cooling the solution is filtered and brought to pH 9.5 with NaOH. 50 mg of o-phthalaldehyde are dissolved in 4.5 mℓ of methanol, 50 $\mu\ell$ of ethanethiol and 0.5 mℓ of the borate buffer are added and the solution mixed. The solution should be protected from light and used for only 1 day. To a 5.0 mℓ volumetric flask are added 1.00 mℓ of a 5.0 μM amino acid solution, 1.00 mℓ of borate buffer and 0.50 mℓ of the o-phthalaldehyde/ethanethiol solution. The mixture is diluted to volume with methanol and allowed to react at room temperature for 1 min. For derivatization of serum samples, 1.00 $\mu\ell$ of serum is added to 2.00 mℓ of methanol, the mixture mixed and centrifuged. 1.00 mℓ of the clear supernatant is derivatized. A 5.0 $\mu\ell$ aliquot of the derivatized amino acid solution or serum is analyzed by HPLC.

ALTERNATE PROCEDURE[2]

3.0 g of boric acid is added to purified water and the pH adjusted to 10.5 with KOH. In a separate container 0.05 g of o-phthalaldehyde is dissolved in 1.0 mℓ of ethanol and mixed with 0.05 mℓ of 2-mercaptoethanol. The solutions are mixed and diluted to 100 mℓ, 0.3 mℓ of a 30% solution of Brij being added to enhance the fluorescence of lysine. Three to four minutes before addition to the column, amino acids, dry or dissolved in aqueous solution, are mixed with the o-phthalaldehyde reagent to form fluorescent derivatives. Decomposition of derivatives and gradual loss of fluorescence occurs after reaction times longer than 5 min.

REFERENCES

1. Hill, D. W., Walters, F. H., Wilson, T. D., and Stuart, J. D., *Anal. Chem.*, 51, 1338, 1979.
2. Gardner, W. S. and Miller, W. H., III, *Anal. Biochem.*, 101, 61, 1980.

18. THE PREPARATION OF THIOUREA DERIVATIVES OF AMINO ACIDS FOR THE SEPARATION OF ENANTIOMERS BY LC

Five mg of amino acid ethyl ester is dissolved in acetonitrile containing 0.2% (w/v) triethylamine to give a volume of 5 mℓ. To 50 $\mu\ell$ of this solution are added 50 $\mu\ell$ of 0.5% (w/v) 2,3,4,6-tetra-O-acetyl-β-D-glucopyranosyl isothiocyanate in acetonitrile. The mixture is allowed to stand at room temperature for 60 min and a 5 $\mu\ell$-aliquot is injected directly into the chromatograph.

REFERENCE

1. Nimura, M., Ogura, H., and Konoshita, T., *J. Chromatogr.*, 202, 375, 1980.

19. THE CHROMOPHORIC LABELING OF AMINO ACIDS WITH 4-DIMETHYLAMINOAZOBENZENE-4'-SULFONYL CHLORIDE (DABSYL CHLORIDE) FOR TLC

Dabsyl chloride is synthesized by reaction of sodium 4-dimethylaminoazobenzene-4'-sulfonate with PCl_5. It is stable at room temperature and reacts readily with amino acids to form dabsyl amino acids. Dabsyl amino acids can be visualized on thin-layer plates and are photo-stable.

Since dabsyl chloride has the same functional group as dansyl chloride, conditions for dabsylation are similar to those of dansylation: 600 nmol of each amino acid is dissolved in 0.3 mℓ of pH 8.9 buffer made from 0.27 mℓ of 0.1 M NaHCO$_3$ plus 0.03 mℓ of 0.1 M Na$_2$CO$_3$. 0.3 mℓ of a solution containing 3.3 mg/mℓ of dabsyl chloride is added and the mixture tightly stoppered with a glass stopper. Reaction is allowed to proceed at 70°C in a water bath with constant shaking for 6 min. After the precipitate has redissolved and the color changed from red-orange to orange, the reaction mixture is ready for TLC. The volume of the reaction mixture can be reduced to 20 µℓ by using only 20 nmol of each amino acid. Higher pH values and lower ratios of dabsyl chloride to amino acid give rise to multiple derivatives of some amino acids.

REFERENCE

1. Lin, J-K. and Chang, J-Y., *Anal. Chem.*, 47, 1634, 1975.

20. THE PREPARATION OF 4-N,N-DIMETHYLAMINONAPHTHYLAZOBENZENE-4-′-THIOHYDANTOIN (DANABTH) AMINO ACIDS FOR TLC

A pH 9.65 buffer is prepared from 50 mℓ acetone + 50 mℓ distilled water + 0.2 mℓ triethylamine + 5 mℓ 0.2 M acetic acid. A solution of 150 nmoles amino acid in 40 µℓ buffer is treated with 40 nmoles of 4-N,N-dimethylaminonaphthyl-azobenzene-4′-isothiocyanate (DANABITC) in 20 µℓ of acetone and heated at 50°C for 75 min. The mixture is dried in vacuo over P$_2$O$_5$ and taken up in a mixture of 20 µℓ of distilled water and 40 µℓ of acetic acid saturated with HCl. After standing at 50°C for 45 min the solution is again dried in the desiccator. The residue is redissolved in 40 µℓ of ethanol. A sample (0.01-0.02 µℓ of each DANABTH- amino acid is applied to a polyamide sheet for TLC. A buffer with a higher pH value, 10.4, is used for the preparation of the derivatives of glutamic acid, aspartic acid and cysteic acid due to the higher pK values of these amino acids. This buffer is prepared from 0.4 mℓ triethylamine + 5 mℓ 0.2 M acetic acid + 50 mℓ acetone + 50 mℓ distilled water.

REFERENCE

1. Chang, J. Y. and Creaser, E. H., *J. Chromatogr.*, 132, 303, 1977.

21. FLUOROGENIC REACTION USING FLUORESCAMINE FOR THE DETECTION OF HISTIDINE AND HISTAMINE

A thin-layer chromatographic technique has been described for detecting histidine, histamine, histidyl peptides and related imidazole compounds. A fluorogenic reaction is employed in which compounds are derivatized with fluorescamine, are converted into different fluorescent products by heating with acid and separated on silica gel plates with a suitable solvent system. As little as 4 to 60 pmole of a histidine derivative can be detected under ultraviolet (360 nm) radiation.

PREPARATION OF FLUOROPHORES

Dispensing procedures are carried out with Eppendorf pipets. To 10 µℓ of 10 mM

aqueous solution of histidine or related compound in a disposable polyethylene micro test tube (capacity 1.5 mℓ), plus 40 μℓ of 0.2 M sodium borate buffer of pH 9.0 is added rapidly 50 μℓ of a solution of 20 mg of fluorescamine in 100 mℓ of acetonitrile, while the mixture is shaken vigorously on a Vortex type mixer. After standing for 5 min at room temperature, 50 μℓ of 2 N HCl are added and the tube is tightly capped. The mixture is incubated in a water bath at 80°C for 1 hr and 1 μℓ of the cooled reaction mixture applied to a precoated silica gel TLC plate for chromatography.

MICRO-PROCEDURE TO GIVE BETTER SENSITIVITY

A 4-μℓ volume of 0.2 M sodium borate buffer of pH 9.0 is applied to the wall of a polypropylene micro centrifuge tube (capacity 0.4 mℓ) with a 10-μℓ Hamilton syringe, 1 μℓ of aqueous sample is then added with a 1-μℓ pipet. After the buffered sample has fallen to the bottom of the tube, 5 μℓ of the fluorescamine solution is rapidly added with a 10-μℓ Hamilton syringe while the tube is shaken vigorously on a Vortex mixer. After standing for 5 min at room temperature, 5 μℓ of 2 N HCl is added, the tube is tightly capped and heated at 80°C for 1 hr in a water bath. The entire volume is then applied to a precoated silica gel plate for chromatography.

REFERENCE

1. Nakamura, H., *J. Chromatogr*, 131, 215, 1977.

22. THE PREPARATION OF *N*-2,6-DINITRO-4-TRIFLUOROMETHYL, *O*-TRIMETHYLSILYL DERIVATIVES OF AMINES FOR GC

A solution of the amine at a concentration of 1 mg/mℓ of free base is prepared in 0.001 N HCl. 10 μℓ of the amine solution is reacted with 100 μℓ of 0.11 M 2,6-dinitro-4-trifluoromethyl benzene sulfonic acid (DNTS) in 50% saturated sodium borate for 10 min at room temperature. The reaction mixture turns yellow on addition of DNTS. The DNT-amine is extracted twice with 400 μℓ of benzene, or with ethyl acetate in the case of dopamine, norepinephrine, diamine and polyamine derivatives. After centrifuging the organic layer is transferred to an acid-washed, silanized 0.3 mℓ vial, (Reacti-Vials, Pierce) and evaporated to dryness under nitrogen. Hydroxylated amine derivatives are converted to the corresponding trimethylsilyl (TMS)-ethers by adding 10 μℓ of *N,O*-bis(trimethylsilyl) acetamide (BSA), 40 μℓ of benzene or ethyl acetate and heating to 60°C for 15 min. *N*-DNT,*O*-TMS derivatives are evaporated to dryness under nitrogen.

REFERENCE

1. Dochi, P. S. and Edwards, D. J., *J. Chromatogr.*, 176, 359, 1979.

23. THE PREPARATION OF *N*-DINITROPHENYL, *O*-TRIMETHYLSILYL DERIVATIVES OF AMINES FOR GC

0.05 mℓ of a 0.25 M solution of 2,4-dinitrobenzenesulfonic acid in saturated sodium borate is added to a dried amine extract in 0.05 mℓ water or to 0.05 mℓ of an amine solution containing 100 or 200 ng of each amine. The centrifuge tubes are stoppered and heated in a boiling water bath for 15 min, the tubes are cooled and the 2,4-dinitrophenyl (DNP)-amines are extracted with 0.4 and 0.2 mℓ portions of benzene. After centrifuging, the benzene layers are transferred to a 0.3 mℓ Microflex tube (Kontes Glass Co.) whose conical bottom facilitates handling the small volumes of reagents used. The samples are evaporated to dryness in a stream of dry nitrogen, the tubes are closed with a screw cap containing a Teflon®-lined septum and 5 µℓ of O,N-bis(trimethylsilyl) acetamide (BSA) added through the septum. The tubes are then heated at 60°C for 30 min, are cooled and opened. Excess BSA and other volatile material are evaporated in vacuo for 1 hr before the tubes are recapped. To prevent hydrolysis of the TMS-ethers 1 µℓ of BSA is added and the solution made up to 10 µℓ with benzene. A 1 µℓ aliquot of the solution is used for gas chromatography.

REFERENCE

1. Edwards, D. J. and Blau, K., *Anal. Biochem.*, 45, 387, 1972.

24. THE PREPARATION OF TRIFLUOROACETYL DERIVATIVES OF PRIMARY AND SECONDARY AMINES FOR GC

To 10 mg of sample in a 1 mℓ vial is added 0.2 mℓ of N-methyl-bis-trifluoroacetemide and 0.5 mℓ of solvent (dimethylformamide, tetrahydrofuran or acetonitrile). The vial is capped and heated at 60°C to 100°C for 15 to 30 min, and the mixture is injected into the chromatograph. Hindered compounds may require additional heating.

REFERENCE

1. Donike, M., *J. Chromatogr.*, 78, 273, 1973.

25. CONTROL OF TRIMETHYLSILYLATION REACTIONS BY THE USE OF COLOR INDICATORS

Donike has described the selective N-trifluoroacylation-O-trimethylsilylation of phenol alkylamines, hydroxyamines and amino acids. The trimethylsilylating agent used was MSTFA (N-methyl-N-trimethylsilyltrifluoroacetamide), followed by MBTFA [N-methyl-bis(trifluoroacetamide)], as trifluoroacylating agent. Several advantages are claimed for this reaction sequence: (1) MSTFA and MBTFA are highly volatile. The reaction mixture alone or together with a suitable solvent may be injected directly into the gas chromatographic column. (2) N-TFA-O-TMS derivatives are stable in solution for long periods, even at the part per billion level. (3) The derivatives have excellent gas chromatographic properties. (4) The derivatives give rise to a characteristic fragmentation pattern in mass spectrometry.

The application of this derivatization reaction, however, is restricted by the equilibrium which exists between excess MSTFA and the resulting secondary N-trifluoroacetamides.

$$R{-}NH{-}TFA + MSTFA \rightleftharpoons R{-}N{\genfrac{}{}{0pt}{}{\diagup TMS}{\diagdown TFA}} + MTFA. \quad (MTFA = \text{N-trimethylfluoroacetamide}).$$

Thus, starting with primary amino groups, two gas chromatographic signals may be obtained for each compound. This occurs if the concentration of MSTFA in the reaction medium is high compared to that of MTFA. Steps may be taken to cause the equilibrium to shift to the left and maintain a constant trimethylsilylation potential as defined by the equilibrium constant of the equation.

Difficulties may be avoided, however, if small amounts of indicators of the azo-dye type, for example, methyl orange, are added. Methyl orange changes color in trimethylsilylation mixtures according to the silylation potential, as it does in aqueous solution according to the pH value.

When sufficient MSTFA is present to react with the anion of the azo dye, the color changes from red to yellow. If the color of the reaction mixture turns yellow, there are three conclusions that can be drawn. (1) The amount of N-TMS-amide is sufficient for all functions having relative low trimethylsilylation potentials, e.g., HO-, HS- and HOOC- groups. (2) Primary or secondary amines are not or not yet fully trimethylsilylated, so that they react rapidly with MBTFA. (3) The silylation potential, determined in trace analysis by the ratio of MSTFA to MTFA is low, so that the equilibrium in the equation is shifted to the left.

When the trimethylsilylation system contains N-TMS-amides and an acidic catalyst, the trimethylsilylation potential may be visually controlled. Suggested systems are MSTFA-TMSCl-acetonitrile: MSHFB-TMSCl-acetonitrile and MSTFA-trifluoroacetic acid (TMSCl = trimethylchlorosilane, MSHFB = N-methyl-N-trimethylsilylheptafluorobutyramide). The observation that indicators change their color according to the amount of MSTFA present means that it is not necessary to use a large excess of trimethylsilylating agent. The extent of trimethylsilylation of a sample can be evaluated from the color of the reaction mixture.

Procedure — Selective N-acylation O-trimethylsilylation. To an aqueous solution of the amines or amino acids acidified with HCl is added 10 μg of the sodium salt of methyl orange (10 μℓ of a 1 mg/mℓ methanolic solution) and the sample brought to dryness by evaporation in vacuo or freeze drying. The residue is dissolved in the minimum amount of trifluoroacetic acid or heptafluorobutyric acid with a microliter syringe. The amount of N-TMS-amide needed to reach the equivalence point is noted and an excess of N-TMS-amide, usually 10 vol %, is added. Alternatively the residue is dissolved in acetonitrile-MSTFA-TMSCl or acetonitrile-MSHFB-TMSCl. The amount of trimethylsilylation mixture is increased stepwise until the yellow color persists; a 10% excess of the mixture is then added. Heat in trimethylsilylation is often unnecessary and should only be used if it is not deleterious to the compounds to be analyzed. Acylation. 5-25 μℓ of MBTFA are added to the trimethylsilylated samples at room temperature. If during the reaction or on opening the vessels the color changes from orange to red, MSTFA (or MSHFB) must be added to protect hydroxyl, thiol, or carboxyl groups from attack by MBTFA.

REFERENCE

1. Donike, M., *J. Chromatogr.*, 115, 591, 1975; 103, 91, 1975; 78, 273, 1973; 85, 1, 1973; *Chromatographia*, 7, 651, 1974.

26. THE PREPARATION OF THE SCHIFF BASE OF AMINES FOR GC

(a) Using benzaldehyde: 0.1 to 0.3 mole of amine is mixed with 0.1 to 0.3 mole of benzaldehyde and nitrogen is bubbled through the mixture at room temperature.[1] The mixture is then distilled. Alternatively 1 to 6 × 10^{-4} mol of amine and 5 × 10^{-3} mol of benzaldehyde are mixed in 2 mℓ of n-hexyl alcohol at room temperature.[2] The reaction is rapid and exothermic. (b) Using 2-thiophene aldehyde: 1 × 10^{-5} mol of amine and 1 to 10×10^{-5} mol of 2-thiophene aldehyde are added to 1 ml of ethanol and the reaction is allowed to proceed at temperatures from room temperature to 60°C for 0.5 to 4 hr.[3] The rate of reaction can be evaluated by measuring the disappearance of the amine or 2-thiophene aldehyde by gas chromatography. (c) Using pentafluorobenzaldehyde: 10 or 100 μl of an amine-ethanol solution (10^{-5} mol/ml concentration) is added to 1 ml of n-hexane and then immediately 10 to 100 μl of a pentafluorobenzaldehyde-ethanol solution (10^{-4} mol/ml concentration) is added to the amine-hexane solution.[4] Reaction is allowed to proceed at room temperature or 60°C for 1 to 3 hr. Excess pentafluorobenzaldehyde is removed by adding 5 ml of 0.1 N NaOH and shaking vigorously for 1 min. The extent of reaction is estimated from the amount of Schiff base produced.

REFERENCES

1. Hoshika, Y., *J. Chromatogr.*, 115, 596, 1975.
2. Hoshika, Y., *Anal. Chem.*, 48, 1716, 1976.
3. Hoshika, Y., *J. Chromatogr.*, 136, 253, 1977.
4. Hoshika, Y., *Anal. Chem.*, 49, 541, 1977.

27. THE PREPARATION OF DIMETHYLALKYLAMINES FOR GC

The conversion of long chain primary amines to dimethylalkylamines to facilitate their separation by gas chromatography is easily achieved using the Leukart reaction. 0.5 to 1.0 ml of the primary amine is added to a test tube, and 2 ml of 37% formaldehyde and 2 ml of 90% formic acid added. The test tube is then heated in a steam bath until the vigorous bubbling ceases, a process which takes about 5 min. The test tube is cooled and phenolphthalein indicator added. 10% NaOH is added until the indicator turns red. Enough saturated NaCl solution is added to cause the tertiary amine to float to the top. The tertiary amine is dissolved in hexane for injection into the gas chromatograph.

REFERENCE

1. Metcalfe, L. D. and Martin, R. J., *Anal. Chem.*, 44, 403, 1972.

28. THE PREPARATION OF SULFONAMIDES OF SECONDARY AMINES FOR GC

A solution of the amine is prepared in 0.1 M HCl at a concentration of 20 μg/ml. One ml of the solution is pipetted into a 30 ml screw-capped test tube containing 4 ml of 10 M NaOH, followed by 0.2 ml of benzene sulfonylchloride. The test tube is capped, shaken vigorously for 30 sec and allowed to stand at room temperature for 30 min with occasional shaking. 5 ml of 10 M NaOH are added and the tube is immersed in a water bath at 80°C for 30 min to hydrolyze excess reagent. The solution is cooled, transferred to a 100 ml separatory funnel using deionized water, and extracted twice with 10 ml of diethyl ether. The diethyl ether layer is washed with 0.05 M Na_2CO_3, is transferred to a volumetric test tube and evaporated to 0.5 ml under a stream of N_2 and 60°C. A volume of 1-4 μl is injected into the gas chromatograph. Almost complete derivatization of amines is achieved.

REFERENCES

1. Hamano, T., Hasegawa, A., Tanaka, K., and Matsuki, Y., *J. Chromatogr.*, 179, 346, 1979.
2. Hamano, T., Mitsuhashi, Y., and Matsuki, Y., *J. Chromatogr.*, 190, 462, 1980.

29. THE N-PERMETHYLATION OF POLYAMINES FOR GC

A solution of the amine (1 equiv NH) in 3 M H_2SO_4 (0.55 mol/NH) and 40% aqueous formaldehyde (3.5 mol/NH) is treated with solid sodiumborohydride (2.1 mol/NH) added in small amounts with accurate temperature control (0-20°C) and efficient stirring in an open flask. Part of the sodium borohydride decomposes to hydrogen, which may cause some foaming. If the reaction mixture becomes too thick to be stirred it is diluted with a minimum of cold water. Excess water is to be avoided, since it depresses the yield, as does an increase in temperature. HCl should be avoided because of the danger of the formation of the very toxic and volatile α, α'-dichlorodimethyl ether on reaction with formaldehyde. When all the borohydride has been added, the mixture is strongly acidified by carefully adding concentrated H_2SO_4, is extracted with ether, made strongly alkaline with a large excess of solid KOH and re-extracted with ether. The ether solution contains the N-permethylated amine free from unmethylated precursors.

REFERENCE

1. Giumanini, A. G., Chivari, G., and Scarponi, F. L., *Anal Chem.*, 48, 484, 1976.

30. THE PREPARATION OF N-TRIFLUOROACETYL DERIVATIVES OF AMINES FOR GC

N-trifluoroacetyl derivatives may be formed directly from C_1 to C_6 aliphatic amine hydrochloride salts, thus obviating the regeneration of the very volatile free amines for derivatization purposes.

A sample containing 1 mg/mℓ each of the free amines in water, is acidified to pH 1 with HCl and the water removed in a rotary evaporator at 40°C. The amine hydrochlorides remain as a viscous yellow liquid which is made more crystalline by adding 10 mℓ of ethyl ether and evaporating, the process being repeated three times. Approximately 2 mg of the amine hydrochloride mixture is added to a PTFE-lined screw cap vial followed by 2 mℓ of trifluoroacetic anhydride. The vial is tightly capped and placed in a boiling waterbath for 5 min. The vial is then cooled and 3 mℓ of dichloromethane is added. Excess trifluoroacetic anhydride, trifluoroacetic acid, HCl and dichloromethane are removed at 0°C in a rotary evaporator, leaving a clear or slightly yellow liquid. A few milliliters of dichloromethane are added and the solution is dried by submerging the vial for several seconds in a liquid nitrogen bath to freeze out water or by adding a few milligrams of anhydrous Na_2SO_4, which has been previously heated at 550°C. The sample is then analyzed by gas chromatography.

REFERENCE

1. Jungclaus, G. A., *J. Chromatogr.*, 139, 174, 1977.

31. THE PREPARATION OF TRIMETHYLSILYL DERIVATIVES OF AMINES AND AMINO ACIDS FOR GC

One hundred $\mu\ell$ of pyridine and 100 $\mu\ell$ of bis(trimethylsilyl)-trifluoroacetamide containing 1% of trimethylchlorosilane are added to 1 to 5 mg of sample. A 5.0 $\mu\ell$ aliquot is injected into the gas chromatograph.

REFERENCE

1. Butts W. C., *Anal. Biochem.*, 46, 195, 1972.

32. THE PREPARATION OF m-TOLUOYL DERIVATIVES OF AMINES FOR HPLC

Twenty mℓ of the sample or standard containing the amine is taken; 5 mℓ of 20% NaOH and 2 mℓ of toluoyl chloride are added and the mixture shaken for 10 min. Two mℓ of concentrated NH_4OH are added to destroy excess reagent and the mixture is shaken for a further 10 min. If ammonia is to be determined, 1 mℓ of ethylene diamine is substituted for the NH_4OH. The mixture is extracted with 20 mℓ of dichloromethane, which is washed with 10% K_2CO_3 and distilled water. For maximum sensitivity the dichloromethane is evaporated and the derivatives dissolved in the mobile phase for column chromatography.

REFERENCE

1. Chen, E. C. M. and Farquharson, R. A., *J. Chromatogr.*, 178, 358, 1979.

33. THE PREPARATION OF TOSYL DERIVATIVES OF POLYAMINES FOR HPLC

An organic solvent is added to the reaction mixture to aid the dissolution of tosyl chloride. A water-acetone (1:1) mixture gives considerably higher yields in the tosylation reaction than other solvent mixtures. Water-acetone, however, gives a substantial interfering background on the chromatogram, thus affecting the separation of tosylated putrescine. The interfering background can be eliminated by warming the reaction mixture at about 70°C for more than 30 min.

To 1.00 mℓ of an aqueous solution containing putrescine, spermidine and spermine, 1 mℓ of 5 M $NaHCO_3$ and 20 mg of tosyl chloride dissolved in 2 mℓ of acetone are added. The mixture is warmed at about 70°C for 1 hr. After cooling, 100 μℓ of internal standard (1 mg of tosylated 1,10-diaminodecane in 1 mℓ of methanol) and 10 mℓ of 1 N NaOH are added. The mixture is washed with four 5-mℓ volumes of n-hexane and 15 mℓ of 1 N HCl are added. The tosylated polyamines are extracted with 10 mℓ of chloroform, the organic phase is dried over Na_2SO_4 and chloroform evaporated on a rotary evaporator. The residue is re-dissolved in a few drops of methanol and 10 μℓ of the solution is subjected to HPLC.

REFERENCE

1. Sugiura, T., Hayashi, T., Kawai, S., and Ohno, T., *J. Chromatogr.*, 110, 385, 1975.

34. THE PREPARATION OF 4-METHOXYBENZAMIDE DERIVATIVES OF AMINES FOR HPLC

To a 5 mℓ reaction vial is added 2.0 mℓ of tetrahydrofuran, 1.0 mℓ of a freshly prepared solution of 4-methoxybenzoyl chloride (4.1×10^{-2} M) in tetrahydrofuran, 1.0 mℓ of aqueous amine solution (12 μg/mℓ or less) and 0.1 mℓ of 5 M NaOH. The vial is sealed with a Teflon-lined screwcap and placed in a waterbath at 65°C for 1 hr. The pH of the solution is brought to 12 by adding 1 M NaOH, and the solution is extracted with chloroform (2×10 mℓ). The extracts are combined, washed with 10% K_2CO_3 (2×20 mℓ), 10 mℓ of 10% (v/v) H_2SO_4 and 10 mℓ of water. The organic phase is dried with $MgSO_4$, evaporated under a stream of air and the residue taken up in a known amount of methanol for analysis by HPLC.

REFERENCE

1. Clark, C. R. and Wells, M. M., *J. Chromatogr. Sci.*, 16, 332, 1978.

35. THE PREPARATION OF FLUORESCAMINE DERIVATIVES OF ALIPHATIC DIAMINES AND POLYAMINES FOR HPLC

To a mixture of one volume of sample solution containing less than 1 nmol amine/μl and one volume of 0.1 M borate buffer, pH 8.0, is added 1 volume of a solution of 20 mg of fluorescamine in 10 ml of acetone. Rapid addition and mixing are essential to achieve optimal fluorescence. An aliquot of the reaction mixture is applied to the column.

REFERENCE

1. Samejima, K., *J. Chromatogr.*, 96, 250, 1974.

36. THE PREPARATION OF o-PHTHALALDEHYDE DERIVATIVES OF AMINES FOR HPLC

The o-phthaldehyde reagent is prepared by dissolving 0.50 g of boric acid in 19 ml of water in a 50 ml beaker, adjusting the pH to 10.40 ± 0.02 with a 45 g/100 ml KOH solution and transferring the solution to a dark glass bottle with a PTFE-lined screw cap. o-Phthalaldehyde (17.5 mg) is then dissolved in 200 μl of glass-distilled methanol in a 5 ml beaker. This solution is added to the borate solution together with 40 μl of fresh 2-mercaptoethanol and stored under nitrogen at 5°C. The reagent is stable for 7 days.

In a cleanup procedure for HPLC analysis, frozen plasma samples are thawed in flowing water at room temperature, mixed thoroughly and a 2 ml aliquot pipetted into a 12 ml culture tube. The deproteinated sample is brought to pH 7.0 ± 0.2 with 0.5 mol/l KOH and immediately derivatized with 400 μl (350 μg) of o-phthalaldehyde reagent at pH 10.40 ± 0.02. Two grams of NaCl are added and the reaction mixture extracted twice with 2 ml of ethyl acetate. The mixture is shaken for 1 min during each extraction and centrifuged to separate the phases. The ethyl acetate is then partitioned twice with 2 ml of 50 nmol/l dibasic sodium phosphate buffer, pH 10.0 ± 0.1, shaken for 1 min and centrifuged. The final ethyl acetate solution is reduced to 100 μl in a stream of ultrapure, oxygen-free dry nitrogen and stored at 4°C until analysis.

REFERENCE

1. Davis, T. P., Gehrke, C. W., Gehrke, C. W., Jr., Cunningham, T. D., Kuo, K. C., Gerhardt, K. O., Johnson, H. D., and Williams, C. H., *J. Chromatogr.*, 162, 293, 1979.

37. THE SEPARATION OF POLYAMINES IN SERUM AND THEIR CONVERSION TO FLUORESCAMINE DERIVATIVES FOR HPLC

The polyamines, obtained after cleanup of deproteinized serum by Cellex P chromatography, are converted to their fluorescamine derivatives in the presence of nickel ion which inhibits the reaction of interfering amines with fluorescamine.

Procedure. Cellex P (H$^+$ form, 0.9 mequiv/g), is washed to remove contaminant before use. 20 g of Cellex P is washed successively with about 500 ml of 0.2 M NaOH, about 400 ml portions of water, 0.5 M HCl, 1 M NaCl, water, 1 M Na$_2$CO$_3$, water, ethanol and water, about 800 ml portions of 0.5 M NaCl and water and finally about 400 ml portions of 0.2 M and 0.01 M sodium phosphate buffers, pH 6.0. The Cellex

P is then suspended in about 500 mℓ of 0.01 M sodium phosphate buffer, pH 6.0 and stored in a refrigerator. When required for use, 0.01 M sodium phosphate buffer, pH 6.0, is added to three times the settling volume of Cellex P. After standing for 1 hr and mixing, the slurry is poured into a glass column, 15 × 0.5 cm I.D., to be 3.0 cm high after settling.

To 0.5 mℓ of serum in a centrifuge tube is added 0.1 mℓ of 4 nmole/mℓ 1,6-hexanediamine as internal standard, and the mixture is diluted with water to 1.0 mℓ. After adding 0.1 mℓ of 3 M HClO$_4$ and centrifuging, the pH of the supernatant is adjusted to 7.0 ± 0.5 with 1.5 M KOH, the mixture is cooled in ice-water and the KClO$_4$ produced is removed by centrifuging. For a sample solution of free polyamines, 1.0 mℓ of 0.1 M sodium phosphate buffer, pH 7.0, is added to the supernatant, which is then mixed with 0.5 mℓ of chloroform and 0.3 mℓ of methanol and centrifuged. The aqueous layer is used as the sample solution. For a sample solution of total polyamines, the supernatant obtained after removal of KClO$_4$ is placed in a screw-cap culture tube; 1.0 mℓ of concentrated HCl is added. The tube is heated at 115°C in an oil-bath for 12 hr and the mixture dried in vacuo at room temperature to remove excess HCl. The residue is dissolved in 1.0 mℓ of water and treated in the same way as in the preparation of the sample solution for free polyamines.

The sample solution is loaded on the Cellex P column, which is then washed successively with 2.0 mℓ of 0.01 M sodium phosphate buffer, pH 6.0, 1.0 mℓ of water and 1.5 mℓ of 0.05 M NaCl. The polyamines and internal standard are eluted with 2.0 mℓ of 3 M NaCl. To the eluate is added 0.5 mℓ of 0.4 M borate buffer, pH 9.0 and 0.2 mℓ of 20 mM nickel (II) sulfate, followed by 0.5 mℓ of 1 mM fluorescamine solution in anhydrous acetone with vigorous mixing. To the reaction mixture is added 1.0 mℓ of 0.3 M succinic acid, 1 g of NaCl and 0.4 mℓ of ethyl acetate, the mixture is shaken vigorously for about 30 sec on a Vortex-type mixer and allowed to stand for about 1 min. The organic layer is transferred to a test tube, 3.0 mℓ of cyclohexane added and fluorescamine derivatives extracted with 0.2 mℓ of 0.4 M borate buffer, pH 10.0, with mixing on a Vortex-type mixer for about 30 sec. Reagent blanks are prepared by treating 0.5 mℓ of water instead of serum in the same way as for free and total polyamines. 100 μℓ each of the extracts of the test samples and blanks are injected into the chromatograph.

REFERENCE

1. Kai, M., Ogata, T., Haraguchi, K., and Ohkura, Y., *J. Chromatogr.*, 163, 151, 1979.

38. THE PREPARATION OF N,N'-DISUBSTITUTED UREAS OF ALIPHATIC AND AROMATIC PRIMARY AND SECONDARY AMINES FOR HPLC

The samples are dissolved in N,N-dimethylformamide (DMF) which seems to catalyze the reaction and serves as a good solvent for the disubstituted ureas. A 0.5 mℓ volume of the amine (usually less than 10 mg) dissolved in DMF is pipetted into a glass-stoppered test tube containing a short magnetic rod. 0.5 mℓ of phenyl isocyanate (PHI) is added and the tube placed on a magnetic stirrer for 5 min. 0.5 mℓ of an aliphatic alcohol, e.g., methanol is then added to destroy excess PHI. This alcohol is chosen so that it causes no interference with the peaks to be determined. Before injecting the sample into the chromatograph, an appropriate amount of an internal standard, usually another amine derivative or an O-alkyl-N-phenylurethane, is added. UV-monitoring allows detection down to the 1 ng level. It is advisable to prepare a blank consisting of "derivatized" reagents.

REFERENCE

1. Bjorkqvist, B., *J. Chromatogr.*, 204, 109, 1981.

39. REACTION OF ALIPHATIC AMINES WITH 7-CHLORO-4-NITROBENZO-2-OXA-1,3-DIAZOLE (NBD-Cl) FOR TLC

To 20-500 $\mu\ell$ of a methanolic amine solution containing 1-20 μg of amine in a 1 to 3 mℓ measuring flask is added the four- to eight-fold equivalent of a 0.05% methanolic solution of NBD-Cl and 50-100 $\mu\ell$ of a 0.1 M aqueous $NaHCO_3$ solution. The flask is stoppered and heated at 55°C for 1.5 hr in a water bath. After cooling the solution is made up to the mark. Five $\mu\ell$ of the solution is applied to silica gel plates or polyamide sheets for thin layer chromatography.

REFERENCE

1. Klimisch, H.-J. and Stadler, L., *J. Chromatogr.*, 90, 141, 1974.

Section V
Products and Sources of Chromatographic Materials

Section V

PRODUCTS AND SOURCES OF CHROMATOGRAPHIC MATERIALS

This section contains descriptions and sources of widely-used chromatographic materials. Materials not mentioned may serve equally well in many cases as those listed. Other sources may be available for the materials in addition to those given. The sources described are not necessarily intended as the editor's endorsement.

LIQUID PHASES FOR GAS CHROMATOGRAPHY

There is considerable evidence that the number of liquid phases for gas chromatography in use exceeds a desirable level, and suggestions have been made that the use of impure (mixtures or undefined compounds (polymers and polyesters) should be discouraged.[1] McReynolds made a comprehensive tabulation of gas chromatographic liquid phase data.[2] He listed Δ I values for ten compounds each chromatographed on 226 liquid phases. Δ I is defined as the difference between the Kovats index of the compound run on squalane and the phase under consideration. The liquid phases were then arranged in increasing order of "average polarity", as determined by summing Δ I values for benzene, butanol, 2-pentanone, nitropropane, and pyridine. This method indicates the similarity and polarity order of the liquid phases. From this study the similarity of many liquid phases which are employed is evident. For example, the methyl silicones represented by SE-30, E-301, OV-1, L-46, DC 200, OV-101, DC 410, DC 401, and silicone oil (May and Baker) have been considered to be essentially identical chromatographically.

Leary et al. applied a nearest-neighbor classification to determine groups and similarities in the data of McReynolds, and proposed a system for reducing the number of gas chromatographic liquid phases.[3] A set of twelve preferred liquid phases was presented, and which of these phases could best be substitute for most of the phases currently in use were tabulated. Other lists of standard or preferred phases have been prepared.[4]

Table 1 lists McReynolds constants for some recently developed liquid phases; it supplements the list given in Volume II of the *Handbook of Chromatography*.

REFERENCES

1. Preston, S. T., *J. Chromatogr. Sci.,* 8, 18A, 1970.
2. McReynolds, W. O., *J. Chromatogr. Sci.,* 8, 685, 1970.
3. Leary, J. J., Justice, J. B., Tsuge, S., Lowry, S. R. and Isenhour, T. L., *J. Chromatogr. Sci.,* 11, 201, 1973.
4. Hawkes, S., Grossman, D., Hartkopf, A., Isenhour, T., Leary, J., Parcher, J., Wold, S., and Yancy, J., *J. Chromatogr. Sci.,* 13, 115, 1975.

Table 1
McREYNOLDS CONSTANTS FOR SOME LIQUID PHASES

Liquid phase	ΔI values[a]				
	1	2	3	4	5
SP-400	32	72	70	100	68
SP-2100	17	57	45	67	43
SP-2250	119	158	162	243	202
SP-2300	316	495	446	637	530
SP-2310	440	637	605	840	670
SP-2330	490	725	630	913	778
SP-2340	520	757	659	942	800
SP-2401	146	238	358	468	310
SF-216-PS	632	875	733	1000	680
SP-222-PS	632	875	733	1000	680
SP-1000	322	555	393	583	546
SP-1200	67	170	103	203	166
SP-1220	207	297	153	283	328
AN-600	202	368	331	483	367
OV-73	16	55	44	65	42
OV-202	146	238	358	468	310
OV-215	149	240	363	478	315
Apolane-87	21	10	3	12	25
Silar-5CP	319	495	446	637	531
Silar-7CP	440	638	605	844	673
Silar-9CP	489	725	631	913	778
Silar-10C	523	757	659	942	801

[a] 1. benzene; 2. 1-butanol; 3. 2-pentanone; 4. nitropropane; 5. pyridine.

Data from Supelco Co., Analabs and Applied Science Division, Milton Roy Company Laboratory Group catalogues.

Table 2
STATIONARY PHASES FOR GAS CHROMATOGRAPHY

OV-73

OV-73 is a 5.5% phenyl substituted methyl silicone polymer specifically treated to remove low molecular species which cause bleeding. Using a silicone gum results in a more efficient capillary column than using a silicone oil of similar chemical structure.

OV-215

OV-215 is a trifluoropropylmethyl silicone gum, which should produce more efficient capillary columns than OV-210.

OV-202

OV-202 is a low viscosity version of OV-210 with essentially the same separation characteristics and thermal stability. It should give packed columns higher efficiency, particularly at lower temperatures. It is a general liquid phase with applications in such broad areas as amines, steroids, pesticides, organo silicones, vitamins, fatty acids, and aromatics.

FFAP
Carbowax® 20M-TPA

These liquid phases are thermally stabilized versions of Carbowax® 20M. FFAP is

Table 2 (continued)
STATIONARY PHASES FOR GAS CHROMATOGRAPHY

the reaction product of Carbowax® 20M with 2-nitroterephthalic acid, Carbowax® 20M-TPA its reaction product with terephthalic acid. They are generally useful phases for the analysis of a wide variety of organic compounds such as acids, alcohols, amides, amino acids (as N-acyl, methyl esters), fatty acid methyl esters, hydrocarbons, ketones, lactones, nitriles, nitrosoamines, phenols, sulfur compounds, and terpenes. FFAP is used more frequently than Carbowax® 20M-TPA but must be used with caution since it irreversibly absorbs aldehydes and epoxides. Trace water and oxygen will degrade these phases; they are removed from the carrier gas by use of an oxygen adsorption cartridge and a molecular sieve drier for water removal.

AN-600

AN-600 is prepared from silicone XE-60 by a proprietary process and is more thermally stable than either XE-60 or OV-225, phases which have essentially the same McReynolds polarities. Both AN-600 and XE-60 are 25%-cyanoethyl, 75%-methyl silicones, while OV-225 is a 25%-cyanopropyl, 25%-phenyl, 50% methyl silicone. AN-600 should be considered as a replacement phase for XE-60 in the GC analysis of acids, alcohols, amino acids (as N-TFA, methyl esters), carbohydrate derivatives, esters, ethers, halogenated compounds, hydrocarbons, hydroxyl compounds, phenols, sulfur compounds, and steroids.

Technical information from Analabs catalogue.

SILAR PHASES

Silar-5CP
Silar-7CP
Silar-9CP
Silar-10C

Silar phases consist of a combination of phenyl and cyanoalkyl functional groups linked to a polysiloxane chain. They provide heat stability up to 275°C, selectivity at temperatures when other polar phases bleed or lose resolution and high polarity.

POLY PHASES

Poly-A 101A
Poly-A 103
Poly-A 135
Poly-I 110
Poly-S 176
Poly-S 179

Polyphases are nonsilicone polymers which have high temperature stability and good wetting characteristics. The latter property makes them particularly suitable for capillary columns.

Poly-A phases are polyamides, poly-I 110 is a polyimide, while poly-S 176 and poly-S 179 are polyphenylether sulfones.

The poly-A phases and poly-I 110 are useful for the separation of amines and alcohols. These phases are sensitive to oxidation, so precautions should be taken to eliminate contact with air.

Poly-S 176 and 179 are usable up to 400°C; their lower usable temperatures are limited by high melting points.

Table 2 (continued)
STATIONARY PHASES FOR GAS CHROMATOGRAPHY

Poly-S 176 has a lower molecular weight and lower melting point than poly-S 179. The former is more suitable for capillary columns. The poly-S phases give good separations of high molecular weight compounds.

DEXSIL® HIGH TEMPERATURE PHASES

Dexsil® 300 GC
Dexsil® 400 GC
Dexsil® 410 GC

Dexsil® polymers are linear molecules incorporating both carborane and siloxane units in the chain. The meta-carborane consists of 10 boron atoms and two carbon atoms in a three-dimensional structure. Meta-carborane acts as an "energy sink", stabilizing its surroundings against the disruptive forces of heat at very high temperatures. All three Dexsil® phases allow separations from 20 to 450°C. The differing side chains of the three phases give increased selectivity for the effective separation of many types of chemical compounds.

APOLANE-87

Apolane-87 is a tailor-made chemically pure C_{87} hydrocarbon (24,24-diethyl-19,29-dioctadecylheptatetracontane), which was prepared and characterized by Kovats and co-workers (Riedo, F., Fritz, D., Tarjan, G., and Kovats, E. Sz., *J. Chromatogr.*, 126, 63, 1976.). It is chemically pure, has no chiral centers and shows no batch-to-batch variation or structural changes during use. Its working temperature range is 30 to 280°, the upper temperature limit being determined by its pyrolytic stability. Apolane-87 is suitable for use as a nonpolar standard phase.

CHIRASIL-VAL

Chirasil-Val is a chiral stationary phase formed by coupling L-valine-*tert*-butylamide to a copolymer of dimethylsiloxane and carboxyalkylmethylsiloxane. This stationary phase has low volatility and high thermal stability, can be used at temperatures up to 260°C and can be used repeatedly for routine analysis without deterioration. Most protein amino acids are separated by temperature programming between 90 and 175°C, and separation factors are sufficient for the separation of all protein amino acid enantiomers.

Technical information from Applied Science Division catalog, Milton Roy Company Laboratory Group.

SUPELCO SP-PHASES

SP-400
SP-2100
SP-2250
SP-2300
SP-2310
SP-2320
SP-2330
SP-2340

Table 2 (continued)
STATIONARY PHASES FOR GAS CHROMATOGRAPHY

SP-2401
SP-216-PS
SP-222-PS
SP-300
SP-1000
SP-1200
SP-1220

SP-400 is a chlorophenylmethyl silicone.

SP-2100 is a methyl silicone fluid. It can be used from 0 to 350°C and can be used for a wide range of compounds, from hydrocarbons to steroids.

SP-2250 is a methyl phenyl silicone (50% phenyl). It is similar to OV-17 but has a lower viscosity and better thermal stability. When used for analysis of polar compounds it must be used with glass columns and the support must be AW-DMCS treated.

The SP-2300 series consists of four cyanosilicone phases that can be used up to 275°C. They vary from moderately polar for SP-2300 to highly polar for SP-2340. These cyanosilicones are similar to polyester phases which have been used to separate fatty acid methyl esters.

SP-2401 is a fluoropropyl silicone, similar to QF-1 and OV-210 but with lower viscosity and better thermal stability.

SP-216-PS is a high polarity polyester phase stabilized with phosphoric acid. It was developed for the separation of free fatty acids and their methyl esters in the C12 to C28 range.

SP-222-PS is a highly polar polyester phase stabilized with phosphoric acid. It was developed specifically to separate fatty acid methylesters.

SP-300, n-Lauroyl-L-valine-tert-butylamide, was developed specifically to separate D- and L-isomers of amino acids.

SP-1000 is a moderately polar phase prepared from Carbowax® 20M and a derivative of terephthalic acid. It is similar to FFAP but has greater thermal stability. It can be used to separate fatty alcohols and as a general purpose stationary phase.

SP-1200 is a low polarity ester type phase originally developed to separate C2 to C5 free acids in aqueous solution.

SP-1220 is a low polarity stationary phase developed for the separation of C2 to C5 free fatty acids. It is similar but not identical to SP-1200.

Technical information from Supelco catalog.

Table 3
SUPPORT MATERIALS FOR GAS CHROMATOGRAPHY

CHROMOSORB® DIATOMITE, AND OTHER SUPPORTS

The most commonly used supports are made from diatomite, which is also known as diatomaceous silica, diatomaceous earth and kieselguhr. Diatomite is composed of the skeletons of diatoms which consist essentially of micro-amorphous hydrous silica. The individual diatom skeletons are very small, highly porous, and have a high surface area. The four basic Chromosorb® diatomite-based grades, Chromosorb® A, G, P, and W are available in untreated and treated forms and in various mesh ranges.

Table 3 (continued)
SUPPORT MATERIALS FOR GAS CHROMATOGRAPHY

Chromosorb® A

Chromosorb® A is used in preparative-scale chromatography. It holds a maximum of 25% liquid phase, does not readily break down with handling and has a surface that is not highly adsorptive.

Chromosorb® G

Chromosorb® G is used for the separation of polar compounds. Its low surface area and good handling characteristics make it a good replacement for light, friable supports like Chromosorb® W, Chromosorb® G is used with a lower liquid phase coating; a 5% coating on Chromosorb® G corresponds to a 12% coating on Chromosorb® W.

Chromosorb® P

Chromosorb® P is prepared from Sil-O-cel C-22 insulating brick. It is a calcined diatomite, pink in color and relatively hard. Its prime use is for hydrocarbon work, but it is also employed for moderately polar compounds on an analytical and preparative scale. Chromosorb® P has a more adsorptive surface than other Chromosorb® grades.

Chromosorb® W

Chromosorb® W is a flux-calcined diatomite support processed from Celite® diatomaceous silica. It is white in color and friable compared to Chromosorb® G and P and its surface is relatively nonadsorptive. Chromosorb® W is used for the separation of polar compounds.

TREATMENTS

Nonacid Washed (NAW)

Chromosorb® grades A, G, P, and W are available in an untreated form described as nonacid washed (NAW).

Acid Washed (AW)

Chromosorb® grades G, P, and W are available with the acid washed treatment. Hydrochloric acid is used to remove soluble iron from the diatomite surface and reduce surface activity. The support is then washed to nearly neutral with deionized water.

Acid Washed Dimethyldichlorosilane Treated (AW-DMCS)

Chromosorb® grades G, P, and W are available with the combined acid washed and DMCS treatment. The DMCS treatment converts the surface silanol (Si-OH) group to silyl ethers which greatly reduces the surface activity of the support and reduces peak tailing. The AW-DMCS treatment gives the most inert surface for these three types of Chromosorb®.

Hexamethyldisilazane (HMDS)

Chromosorb® grades P and W are available with HMDS treatment on nonacid washed supports. The treatment deactivates the support by conversion of the surface silanol groups to silyl ethers, considerably reducing tailing. The HMDS-treated Chromosorb® P and W are not as inert as the AW-DMCS-treated forms but are useful where a high degree of deactivation is not necessary.

Table 3 (continued)
SUPPORT MATERIALS FOR GAS CHROMATOGRAPHY

Chromosorb® H.P.

Chromosorb® H.P. is a high quality, acid washed and silanized, flux-calcined diatomite. Its features include superior inertness and column efficiency, no catalytic surface activity and short column conditioning time. The H.P. grade is used with steroids, bile acids, alkaloids, and for analyses of pharmaceuticals and toxicological compounds. Chromosorb® W and Chromosorb® G are available in the H.P. grade.

Chromosorb® T

Chromosorb® T is made from Teflon® 6 powder. Teflon® 6 has been used as a support where a highly inert surface is needed to avoid tailing problems which may be encountered with compounds such as water, hydrazine, sulfur dioxide, etc. Teflon® 6 is a high molecular weight TFE-fluorocarbon resin. It is insoluble in common solvents and resistant to chemical attack. The material is relatively soft and fragile, and should be handled gently. If force is applied it tends to compress and become a compact solid rather than remain porous, ruining its chromatographic properties. Chromosorb® T may be chilled to improve its handling properties. During chilling condensation of atmospheric moisture will help dissipate the static charge it readily develops. Chromosorb® T is available in 30/60 and 40/60 mesh fractions; the latter is recommended for highest column efficiency.

Acid Washed Celite® 545 (Reagent Grade)

This is a high purity, flux-calcined diatomaceous silica filter aid specially prepared for chromatography and other specialized applications. It is a white powdered diatomaceous silica filter aid of high purity, inertness and uniformity, produced by a combination of acid treatment, leaching and testing. The material is recommended for use as a chromatography support.

Chromosorb® R-6470-1

Chromosorb® R-6470-1 is an ultrafine particle size diatomaceous silica developed for use in gas chromatography to coat the inside walls of capillary columns. It is produced from flux-calcined Celite® diatomaceous silica, has a surface area of approximately 5 to 6 m^2/g, has a predominant particle size of 1 to 4 μ and retains the diatom structure of Chromosorb® W.

Chromosorb® 750

Chromosorb® 750 is the most inert, nonfriable, free flowing and efficient support designed for biomedical and pesticide analysis. It is prepared from high purity diatomite crude with exhaustive acid-washing and silane treatment. The material shows minimum adsorption for polar compounds and minimum decomposition of sensitive samples. Chromosorb® 750 is 1.5 times as heavy as Chromosorb® W and is limited to a maximum liquid phase loading of 7%.

CHROMOSORB® CENTURY SERIES

These Chromosorb® porous polymers are of polyaromatic-type, cross-linked resin having a uniform rigid structure of a distinct pore size.

Chromosorb® 101

Chromosorb® 101 is a porous, styrene-divinylbenzene polymer with a surface area

Table 3 (continued)
SUPPORT MATERIALS FOR GAS CHROMATOGRAPHY

of less than 50 m²/g. The separations it gives are provided by the surface and structure characteristics of the resin. It need not be coated; however, coating can be done primarily to reduce retention time. A basic characteristic of the resin is that it has a relatively low affinity for hydroxyl-containing compounds. Columns can be operated isothermally at 275°C and can be taken to 325°C for a short time during temperature programing. Chromosorb® 101 effectively separates hydrocarbons, alcohols, fatty acids, esters aldehydes, ketones, ethers, and glycols.

Chromosorb® 102

Chromosorb® 102 is a porous styrene-divinylbenzene copolymer with a surface area in the range of 300 to 400 m²/g. Its porous structure is of an open-cell variety so that samples can readily penetrate its pores. Columns using it can be operated isothermally up to 250°C; somewhat higher temperatures can be used during temperature-programing. The chromatographic characteristics of Chromosorb® 102 are similar to those of a liquid phase. Chromosorb® 102 performs in a manner similar to that of a packed column with a high liquid phase loading. Retention times on the column are relatively high; the retention times can be reduced by coating the material with a liquid phase. The support can be used to separate light and permanent gases, and lower molecular weight compounds like acids, alcohols, glycols, ketones, esters, and hydrocarbons.

Chromosorb® 103

Chromosorb® 103 is a polyaromatic porous polymer packing material developed specifically for amines and basic compounds. It will not handle acidic materials or glycols. Although the material will deal with both alcohols and amines it is more selective for basic compounds.

Chromosorb® 104

Chromosorb® 104 is an acrylonitrile-divinylbenzene type resin with a highly polar surface. This support is used for the gas analysis of various types at subambient, ambient and higher temperatures. It will effectively separate isomeric xylenols, alcohols, ketones, nitriles, aldehydes, and hydrocarbons. Chromosorb® 104 has the highest polarity in the Chromosorb® century series porous polymers.

Chromosorb® 105

Chromosorb® 105 is a white, nonfriable, insoluble, cross-linked polyaromatic resin with an effective surface area of 600 to 700 m²/g. It is effective in separating formaldehyde from water and methanol, acetylene from lower hydrocarbons and most other classes of organic compounds of different polarity with a boiling point up to 200°C.

Chromosorb® 106

Chromosorb® 106 is a nonpolar cross-linked polystyrene resin. It is a hard, granular solid with a surface area between 700 and 800 m²/g. The surface is nonpolar and in a chromatographic column retains nonpolar organic compounds for a longer time than polar compounds. It resolves C_2-C_5 fatty acids from C_2-C_5 alcohols.

Chromosorb® 107

Chromosorb® 107 is a hard granular solid with a surface area between 400 and 500 m²/g. It is of intermediate polarity and gives efficient separations of various types of compounds and Formalin in particular.

Table 3 (continued)
SUPPORT MATERIALS FOR GAS CHROMATOGRAPHY

Chromosorb® 108

Chromosorb® 108 is a moderately polar cross-linked acrylic ester with a surface area of 100 to 200 m²/g. It is stable to 250°C. It gives good separations of gases and polar materials such as water, alcohols, aldehydes, ketones, glycols, etc.

Technical information from Mansville, Inc., publications.

ANAKROM DIATOMACEOUS EARTH SUPPORTS

The preparation of Anakrom solid supports involves the selection of a diatomaceous earth with a high degree of chemical inertness and very favorable surface characteristics. A special flux and calcination procedure is employed, and produces aggregates which, although relatively small in size, have good flow qualities. This provides a higher surface area for the uniform distribution of the liquid film. The material is sieved until a uniform particle size is approached, with the result that the pressure drop through the chromatographic column is minimized and efficiency is increased.

Anakrom ABS

Acid and alcoholic base washed and silanized. The dried support is treated with dimethyldichlorosilane, which reacts with the surface hydroxyl groups, giving an inert surface. Residual HCl, as well as unreacted -SiCl groups, are removed by repeated methanol washing. Anakrom ABS is prepared by silanizing Anakrom AB. It has a maximum phase loading capacity of 25% and an inert surface which minimizes undesirable tailing effects.

Anakrom Q

Anakrom Q is a premium acid-washed, silanized diatomaceous earth support. The washing procedure is similar to that used in preparing Anakrom A, but a special acid is used which produces a more inert surface. This support is recommended for the analysis of bile acids, steroids, drugs, and pesticides, especially those containing phosphorus. The inert surface decreases sample tailing and decomposition of sensitive compounds.

Anakrom A

Acid washed grades are prepared by HCl treatment to remove trace metals absorbed in the surface. Exhaustive washing with deionized water is followed by drying at elevated temperatures for 6 hr. The material is then sieved. Anakrom A is a general purpose support which can accept liquid phase loading of 25%. It sometimes gives better separations when coated with EGA, DEGS or OV-275 than the acid-washed silanized support. This support is also recommended for high temperature use (greater than 325°C).

Anakrom SD

Anakrom SD is prepared by treating Anakrom Q with a chemical that further deactivates the surface, ensuring minimum peak tailing during analysis. This support is recommended for the analysis of polar materials such as water, alcohols and tocopherols. The maximum recommended temperature is 280°C.

ANAKROM C22 SERIES

This series is prepared from diatomaceous earth calcined with clay binder at a tem-

Table 3 (continued)
SUPPORT MATERIALS FOR GAS CHROMATOGRAPHY

perature above 1000°C. The particles of the pink color support thus obtained are less friable than those of white supports and can hold loadings of liquid phases up to 30%.

Anakrom C22 (untreated) efficiently separates low-polarity hydrocarbons and chlorinated hydrocarbons. It is not recommended for the separation of compounds containing oxygen, amine or hydroxyl groups.

Anakrom C22A (acid-washed) gives slightly less tailing for compounds of medium polarity but is not recommended for the analysis of compounds of medium to higher polarity which contain oxygen and nitrogen.

Anakrom C22AS (acid-washed silanized). The additional silanizing treatment gives the support a low surface activity. Medium polar compounds can be analyzed. The phase loading may be as high as 30%.

Technical information from Analabs catalog.

GAS-CHROM® SUPPORTS

Diatomaceous earth forms the basis of almost all Gas-Chrom® supports. It is calcined with a small amount of flux (sodium carbonate) to produce a white porous solid.

Gas-Chrom® Q

Gas-Chrom® Q is a screened and deactivated support for the analysis of steroids, pesticides, alkaloids, pharmaceuticals, and other labile compounds. Its surface has been silane-treated to eliminate impurities and reactive sites. The use of Gas-Chrom® Q minimizes peak tailing and catalytic decomposition of sensitive compounds.

Gas-Chrom® S

This support is screened with no further treatment.

Gas-Chrom® A

Gas-Chrom® A is prepared by acid washing Gas-Chrom® S to remove the bulk of the inorganic contaminants, then water washing until neutrality is reached.

Gas-Chrom® P

This support is made by base washing Gas-Chrom® A to remove organic contaminants and then water washing until neutrality is attained.

Gas-Chrom® Z

Gas-Chrom® Z is prepared by treating Gas-Chrom® P with dimethyldichlorosilane to remove active sites on the surface.

Gas-Chrom® R

Gas-Chrom® R is prepared from Johns-Manville C-22 Firebrick material. It is made from diatomaceous earth, is calcined without a flux and is red in color. This support has a larger surface area and is generally more adsorptive than the white supports. Gas-Chrom® R can achieve the highest efficiency of all the diatomaceous supports and is recommended for the chromatography of hydrocarbons and other nonpolar compounds.

Technical information from Applied Science Division catalog, Milton Roy Company Laboratory Group.

Table 3 (continued)
SUPPORT MATERIALS FOR GAS CHROMATOGRAPHY

SUPELCOPORT

Supelcoport is an acid washed, DMCS treated diatomite support. A special procedure is used to remove fines from the support, reducing dust to a minimum. Little or no conditioning of the support is necessary once it is coated. It is available in 60-80, 80-100 and 100-120 mesh sizes.

Technical information from Supelco catalog.

PENNWALT® AMINE PACKINGS

28% Pennwalt® 223 amine packing

This packing was designed specifically for the analysis of amines. It gives excellent selectivity and yields symmetrical peaks for amines, diamines and alcohols. The packing consists of 28% Pennwalt 223 and 4% KOH on Gas-Chrom® R, 80-100 mesh.

10% Pennwalt® 223 amine packing

This 10% packing has the same separating characteristics as the 28% packing but analysis times are shorted, allowing even higher boiling amines to be analyzed.

Pennwalt® 231 GC packing

This packing was developed specifically for the analysis of ammonia, monomethylamine, methanol, dimethylamine and trimethylamine in process streams. Aqueous solutions are suggested for amine analysis for ease in dissolving monomethylamine. After each run the water may be driven off the column by keeping the temperature at 120°C for 15 min. Pennwalt® 231 GC packing should not be used at temperatures above 120°C; overnight conditioning at 120°C with a carrier flow of 40 mℓ/min is recommended.

Technical information from Applied Science Division catalog, Milton Roy Company Laboratory Group.

TABSORB

Special column packing for the analysis of amino acids as their N-trifluoroacetyl n-butyl esters.

TABSORB HAC

Special column packing for the analysis of histidine, arginine and cystine as their N-trifluoroacetyl n-butyl esters.

Obtainable from Regis Chemical Co., 8210 Austin Avenue, Morton Grove, Illinois 60053.

DURABOND COLUMNS

Durabond columns are chemically bonded fused silica capillary columns. The stationary phase is chemically bonded to the capillary wall, thus minimizing bleed and extending the temperature stability of the column. The columns tolerate large sample/solvent volumes and may be back-flushed with solvents to remove contaminants. Dur-

Table 3 (continued)
SUPPORT MATERIALS FOR GAS CHROMATOGRAPHY

abond columns are available in three film thicknesses: 0.1 μm for high boiling solutes, 0.25 μm for general purpose work and 1.0 μm for volatile solutes. The 1.0 μm film will chromatograph samples as large as 500 ng per component without appreciable loss of resolution. Two chemically bonded phases are available, DB-1 which is similar to SE-30 and DD-5 which is similar to SE-52 and SE-54. Wide-bore columns have slightly higher capacity and allow "on-column" injection with suitable needle and injection port designs.

Technical information from Applied Science "Chromatographer".

Table 4
PACKINGS FOR HIGH PRESSURE LIQUID CHROMATOGRAPHY

POLYGOSIL 60 AND POLYGOSIL 60-D SERIES

The base material is irregular silica gel particles, average pore diameter 60Å, pore volume 0.75 mℓ/g, specific surface approximately 500 m²/g, pH of a 10% aqueous suspension, 6.5 to 7.0. Polygosil 60 has a carefully controlled particle size distribution. Polygosil 60-D has broader limits of particle size distribution than Polygosil 60. The d-values give information on the particle size distribution; thus for Polygosil 60-D5, 10% are finer than 3 μm, 50% finer than 5 μm and 90% finer than 7 μm.

Groups chemically bonded to silica gel 60Å:

C_8 = octyl group
C_{18} = octadecyl group
NH_2 = amino group
CN = nitrile group
NO_2 = nitro group
$N(CH_3)_2$ = dimethylamino group

	Mean particle size	Particle size		
		d_{10}	d_{50}	d_{90}
Polygosil 60-5	5 ± 1.5 μm			
Polygosil 60-10	10 ± 1.5 μm			
Polygosil 60-15	15 ± 2 μm			
Polygosil 60-20	20 ± 2.5 μm			
Polygosil 60-D5	5 μm	3	5	7
Polygosil 60-D 10	10 μm	8	10	12
Polygosil 60-D 20	20 μm	17	20	23
Polygosil 60-D 5C_8	5 μm	3	5	7
Polygosil 60-D 10C_8	10 μm	8	10	12
Polygosil 60-D 5C_{18}	5 μm	3	5	7
Polygosil 60-D 10C_{18}	10 μm	8	10	12
Polygosil 60-D 5 NH_2	5 μm	3	5	7
Polygosil 60-D 10NH_2	10 μm	8	10	12
Polygosil 60-D 5CN	5 μm	3	5	7
Polygosil 60-D 10CN	10 μm	8	10	12
Polygosil 60-D 5NO_2	5 μm	3	5	7
Polygosil 60-D 10NO_2	10 μm	8	10	12
Polygosil 60-D 5N($CH_3)_2$	5 μm	3	5	7
Polygosil 60-D10N($CH_3)_2$	10 μm	8	10	12

Table 4 (continued)
PACKINGS FOR HIGH PRESSURE LIQUID CHROMATOGRAPHY

NUCLEOSIL PACKINGS

Nucleosil packings are spherical gel particles, totally porous.

(a) For adsorption. Average pore diameter 100 Å, pore volume 1.0 mℓ/g, specific surface 300 m^2/g

	Particle size
Nucleosil 50-5	5 ± 1.5 μm
Nucleosil 50-7	7.5 ± 1.5 μm
Nucleosil 50-10	10 ± 1.5 μm

(b) For adsorption. Average pore diameter 100 Å, pore volume 1.0 mℓ/g and/or partition specific surface 300 m^2/g.

Nucleosil 100-5	5 ± 1.5 μm
Nucleosil 100-7	7.5 ± 1.5 μm
Nucleosil 100-10	10 ± 1.5 μm

(c) For reverse phase. Base material Nucleosil 100, average pore diameter 100 Å, pore volume 1.0 mℓ/g, specific surface 300 m^2/g, C_8 = octyl group. O_{18} = octadecyl group chemically bonded to silica gel.

Nucleosil 5 C_8	5 ± 1.5 μm
Nucleosil 7 C_8	7.5 ± 1.5 μm
Nucleosil 10 C_8	10 ± 1.5 μm
Nucleosil 5 C_{18}	5 ± 1.5 μm
Nucleosil 7 C_{18}	7 ± 1.5 μm
Nucleosil 10 C_{18}	10 ± 1.5 μm

(d) For chromatography of polar compounds. Base material Nucleosil 100, average pore diameter 100 Å, pore volume 1.0 mℓ/g, specific surface 300 m^2/g, Polar phases bonded to silica gel. CN = nitrile group; NO_2 = nitro group; NH_2 = amino group; $N(CH_3)_2$ = dimethylamino group.

Nucleosil 5 CN	5 ± 1.5 μm
Nucleosil 10 CN	10 ± 1.5 μm
Nucleosil 5 NO_2	5 ± 1.5 μm
Nucleosil 10 NO_2	10 ± 1.5 μm
Nucleosil 5 NH_2	5 ± 1.5 μm
Nucleosil 10 NH_2	10 ± 1.5 μm
Nucleosil 5$N(CH_3)_2$	5 ± 1.5 μm
Nucleosil 10 $N(CH_3)_2$	10 ± 1.5 μm

(e) For ion exchange. Base material Nucleosil 100, average pore diameter 100 Å, pore volume 1.0 mℓ/g, specific surface 300 m^2/g. SA-types: cation exchangers, SO_3H groups chemically bonded, capacity approximately 1 meq/g. SB-types: anion exchangers, quaternary ammonium groups chemically bonded, capacity approximately 1 meq/g.

Nucleosil 5 SA	5 ± 1.5 μm
Nucleosil 10 SA	10 ± 1.5 μm
Nucleosil 5 SB	5 ± 1.5 μm
Nucleosil 10 SB	10 ± 1.5 μm

Technical information from Macherey-Nagel & Co. catalog.

VYDAC PACKINGS

Vydac products include spherical, totally porous (TP) materials, pellicular or solid core (SC) materials, and irregular shaped microparticles (IR) materials. The TP prod-

Table 4 (continued)
PACKINGS FOR HIGH PRESSURE LIQUID CHROMATOGRAPHY

ucts are based on spherical microparticulate silica of 10 μ particle size. The SC materials are 30 to 44 μ in particle size.

The Vydac reverse phases are made by chemically bonding octadecyl groups to a silica support through a Si-C bond. The Vydac polar bonded phase is made by chemically bonding an oxynitrile (Si-R-O-CH$_2$-CH$_2$-CN) functional group to a silica support through a Si-C type bond. Vydac ion exchangers are made by chemically bonding ion exchange groups to a spheroidal silica support through a Si-C type bond. Vydac cation exchanger has as its functional group a sulfonic acid 2(R-SO$^-_3$). Vydac anion exchanger has as its functional group a quaternary ammonium group.

Vydac adsorbents, (101 TP and 101 SC) are used to separate steroids, oil soluble vitamins, pesticides, amines, amides, phenols, esters, ethers, alcohols, ketones, aldehydes, nitro compounds, and nitriles.

Vydac reverse phase materials (201 TP and 201 SC) are used to separate water insoluble compounds such as hydrocarbons, steroids, pesticides, vitamins, halogenated hydrocarbons, esters, ethers, aldehydes, ketones, nitro compounds, and amides.

Vydac anion exchangers (301 TP and 301 SC) are used to separate nucleotides and organic acids.

Vydac cation exchangers (401 TP and 401 SC) are used to separate metals, amines, purines, pyrimidines and amino acids.

Vydac polar bonded phases (501 TP and 501 SC) are used to separate steroids, pesticides, vitamins, amines, nitriles, phenols, esters, ethers, alcohols, ketones, aldehydes, and nitro compounds.

Vydac products are from The Separations Group. Technical information from Analabs. Inc. and Packard Instrument Co. Inc. catalogs.

WATERS ASSOCIATES INC.

μBONDAPAK C$_{18}$

BONDAPAK C$_{18}$/CORASIL

BONDAPAK C$_{18}$/PORASIL B

These packings are of low polarity and may be used for partition, reverse phase separations. Typical applications include analgesics, antibiotics, antihistamines, catecholamines, nucleosides, organic acids, peptides, PTH-amino acids, and purine/pyrimidine bases. Typical solvents are methanol, water, acetonitrile, tetrahydrofuran, and PIC paired ion chromatography reagents.

μBondapak C$_{18}$

μBondapak C$_{18}$ is made by chemically bonding 10% of a C$_{18}$ group to μPorasil silica gel. It has high efficiency, moderate capacity and provides very fast analysis. It may be used both analytically and semipreparatively. This packing is available in prepacked columns.

Bondapak C$_{18}$/Corasil

Bondapak C$_{18}$/Corasil is made by chemically bonding 1.0% of a C$_{18}$ group to Corasil pellicular silica gel. It has relatively high efficiency but low capacity, and is primarily recommended for routine analytical separations where established methods exist.

Table 4 (continued)
PACKINGS FOR HIGH PRESSURE LIQUID CHROMATOGRAPHY

Bondapak C_{18}/Porasil B

Bondapak C_{18}/Porasil B is made by chemically bonding 18% of a C_{18} group to Porasil B. It has a high capacity and is primarily recommended for large-scale preparative work.

µPORASIL

CORASIL

PORASIL

These packings are of high polarity and may be used for adsorption, normal phase separations. Typical applications include anticonvulsants, essential oils, organic soluble natural products polynuclear aromatics, steroids, and fat soluble vitamins.

µPorasil

µPorasil is a silica packing material, with a mean particle size of 10 µm. It is used analytically and semipreparatively, providing rapid analyses. This packing is available in prepacked columns only.

Corasil

Corasil is made by fusing thin layers of silica to glass beads to form particles with a mean size of 10 µm. It is primarily recommended for routine analytical separations.

Porasil

Porasil is a silica with a high capacity, which is primarily recommended for large scale preparative work where high capacity is important. Porasil types A, B, C, D, E, F, and T are available.

µBondapak NH_2

µBondapak NH_2 is made by bonding 9% of an amino group to µPorasil. Its mean particle size is 10 µm. This packing material gives high speed and efficiency, moderate capacity and is used analytically and semipreparatively. It is a weak ion exchanger when an acid mobile phase is used and is normally used for adsorption with nonpolar solvents to analyze samples with different polarity groups. Aldehyde and ketone groups should be avoided. µBondapak NH_2 can also be used in the reverse phase mode. Typical applications include aromatic amines, water soluble carboxylic acids, nucleotides, phthalic acids, strongly polar compounds and water soluble vitamins. This packing is available in prepacked columns only.

µBONDAPAK CN

µBONDAPAK CN (NORMAL PHASE)

These packings are of intermediate polarity and may be used for reverse phase or normal phase separations. Typical applications include analgesics, antihistamines, drug metabolites, phenolics, polynuclear aromatics, and steroids. They are used in reverse phase chromatography in place of C_{18} columns when samples possess strong polar groups.

Table 4 (continued)
PACKINGS FOR HIGH PRESSURE LIQUID CHROMATOGRAPHY

µBondapak CN

µBondapak CN is made by bonding 9% of a cyano group to µPorasil. It has a mean particle size of 10 µm, has high speed, high efficiency and moderate capacity. This packing material is used both analytically and semipreparatively and is particularly recommended for reverse phase uses. It is available only in prepacked columns.

µBondapak CN (Normal Phase)

µBondapak CN (normal phase) is suitable for normal phase used with low to intermediate polarity solvents and is available in prepacked columns.

µBONDAPAK PHENYL

BONDAPAK PHENYL/CORASIL

BONDAPAK PHENYL/PORASIL B

These packings are of low polarity and are used for reverse phase separations. Typical applications are analgesics, antibiotics, antihistamines, narcotics, nucleosides, organic acids, peptides, polynuclear aromatics, and steroids. They are used in reverse phase separations instead of C_{18} columns when the samples have strong polar groups. Typical solvents are acetonitrile, water, methanol, and PlC reagents.

µBondapak Phenyl

µBondapak phenyl is made by bonding 10% of a phenyl group to µPorasil. It is a versatile reverse phase packing of 10 µm mean particle size with unique selectivity, high speed and moderate capacity. It may be used analytically and semipreparatively. This packing is available only in prepacked columns.

Bondapak Phenyl/Corasil

Bondapak phenyl/Corasil is made by bonding 0.8% of a phenyl group to Corasil. Its particle size range is 37 to 50 µm and it has a relatively high efficiency, although not as high as µBondapak phenyl. It is primarily recommended for routine analytical separations.

Bondapak Phenyl/Porasil B

Bondapak phenyl/Porasil B is made by binding 10% of a phenyl group to Porasil B. Its particle size is 37 to 75 µm, it has a very high capacity and is primarily recommended for large scale preparative work. It is available only as bulk packing.

µBondagel

µBondagel is made by chemically bonding an ether group to µPorasil particles of different pore sizes. It was developed for molecular weight distributions of aqueous soluble polymers, but the ether function produces a versatile packing. It is available in prepacked columns only. µBondagel gives high efficiency and moderate capacity and is used primarily for analytical separations. Typical applications include molecular weight characterizations of polymers such as dextrans, carboxymethylcellulose, polyvinyl alcohols, polyacrylic acid, and polyamides.

µPorasil GPC 60A

µPorasil GPC 60 A is part of the µPorasil silica family but when deactivated with

Table 4 (continued)
PACKINGS FOR HIGH PRESSURE LIQUID CHROMATOGRAPHY

water it is used in the gel permeation chromatography mode. It extends the range of size separation to very small molecules and is used as a complement to μBondagel. This packing provides high efficiency, moderate capacity and rapid analyses; it is used analytically and semipreparatively. μBondagel and μPorasil GPC 60 A are compatible with a wide range of polar and nonpolar solvents.

Bondapak AX/Corasil

Bondapak AX/Corasil is a strong anion exchanger made by chemically bonding 1.3% of a quaternary amine to a Corasil support. It has a particle size range of 37 to 50 μm, an exchange capacity of 10 to 15 μeq/g and may be used over the pH range 2 to 7.5. This packing provides fast separations and rapid equilibration; it has been used to separate nucleotides and organic acids. It is typically employed with buffered solutions.

Bondapak CX/Corasil

Bondapak CX/Corasil is a strong cation exchanger made by chemically bonding 1.1% of a sulfonated phenyl to a Corasil support. It has a particle size range of 37 to 50 μm, an exchange capacity of 30 to 40 μeq/g and may be used from pH 2 to pH 7. This packing provides fast separations and rapid equilibration and has been used to separate catecholamines; it is typically employed with buffered solutions.

μStyragel

μStyragel is made by cross-linking styrene and divinyl benzene to form small, rigid gel particles with carefully controlled pore sizes. It has high efficiency, high capacity, gives high speed and is used analytically and semipreparatively. Typical applications include small molecules for methods development or sample cleanup, natural products and polymers such as acrylics, amides/imides, neoprene, nonionic surfactants, polybutadiene, polycarbonates, polyethylene, polyglycols, polyisoprene, polypropylene, polystyrene, polysulfones, and polyurethane.

Styragel®

Styragel® is made in the same way as μStyragel®, but the particles are larger. It is recommended for high temperature analyses of polyolefins (polyethylene, polypropylene). In all other cases μStyragel® is recommended due to higher efficiencies and more rapid analyses.

Technical information from Waters Associates catalog.

WHATMAN

PARTISIL SILICA GELS

Partisils are high purity silica gels whose particles are irregular in shape, of narrow size range and high surface area. All are of high efficiency and have high sample capacity. Their high resolving power permits the separation of isomeric species. The high loading capacity of Partisils is important in preparative applications.

Partisil®-5

Partisil®-5 has a particle size of 5 μ; it has the best resolving power and is ideal for

Table 4 (continued)
PACKINGS FOR HIGH PRESSURE LIQUID CHROMATOGRAPHY

separation of close polarity compounds. The very high resolving power may be often required for very difficult separations.

Partisil®-10
Partisil®-10, with a particle size of 10 µ, is a general purpose HPLC silica gel.

Partisil®-20
Partisil®-20 has a particle size of 20 µ. It is free flowing and can with care be dry packed into columns, preferably using vibrator packing methods. Slurry packing usually gives considerably high efficiencies. Partisil®-20 gives a "fast" column with good resolution and wide scope.

Partisil®-5 ODS
Partisil®-5 ODS is a bonded reverse phase material; octadecyl C-18 groups are siloxane bonded to the siliceous support, Partisil®-5. Partisil®-5 ODS analytical columns are highly efficient and are very durable.

Partisil®-10 ODS
Partisil®-10 ODS is a C 18 bonded reverse phase material which differs from Partisil®-10 ODS-2 chiefly in the carbon load; this is 5% compared to the 15% in ODS-2. The lower octadecyl content means that more hydroxyl groups on the surface of the silica gel backbone are available to enter into interactions. These hydroxyl groups allow high water content mobile phases to wet the support phase, and probably are responsible for modification of the partition mechanism in this medium.

Partisil®-10 ODS-2
Partisil®-10 ODS-2 is a 10µ bonded C-18 reversed phase with a carbon load of approximately 15%. It has excellent selectivity, high efficiency and high loading capacity. Partisil 10®-ODS-2 is recommended where sample solubility in aqueous solutions is relatively high and where greater retentivity is necessary for increased resolution.

Partisil®-10 ODS-3
Partisil®-10 ODS-3 is a fully silanized C-18 reversed phase, with no residual silanols. It has the same selectivity as Partisil®-10 ODS, but with twice the efficiency. Partisil®-10 ODS-3 is highly durable and shows no bed shrinkage because of silica dissolution. It may be used for ion pair and ion suppression separations.

Partisil®-10 PAC
Partisil®-10 PAC is a microparticle bonded polar phase which is manufactured by chemically bonding cyano and amino groups to the surface hydroxyls of Partisil®-10. The chromatographic performance of Partisil®-10 PAC depends on the composition and concentration of the mobile phases with which it is used; in other words it is possible to control selectivity by simple changes in solvent. Partisil®-10 PAC is very effective in separating more highly polar compounds.

Partisil®-10 SAX
Partisil®-10 SAX is a microparticle strong anion exchanger in which quaternary nitrogen, $-NR_3-$, functional groups are Si-O-Si bonded to the surface hydroxyls of Partisil®-10 silica gel. It shows high efficiency and is very stable even at elevated

Table 4 (continued)
PACKINGS FOR HIGH PRESSURE LIQUID CHROMATOGRAPHY

temperatures although the majority of separations on this phase are run at ambient temperature. Partisil®-10 SAX is stable over the pH range 1.5 to 7.5 and is ideal for gradient elution separations. It is useful for difficult separations of anionic species and is employed in the analysis of biological samples, such as nucleotides.

Partisil®-10 SCX

Partisil®-10 SCX is a microparticle strong cation exchanger in which benzene sulfonic acid groups, $-SO_3-$, are Si-O-Si bonded to the surface hydroxyls of Partisil®-10. It is very efficient in analytical columns, highly stable structurally and chemically and stable over the pH range 1.5 to 7.5. The majority of separations on this phase can be successfully run at ambient temperature. Partisil®-10 SCX is valuable for gradient elution separations. It is used for analyzing compounds which form cations in aqueous solutions and in the analysis of nucleic acid constituents.

Pelluminas-Pellicular Aluminas

Pelluminas are manufactured by permanently bonding a layer of alumina to the prepared surface of a 37-43 µm spherical glass bead. HC Pellumina has the thicker pellicle and therefore higher sample capacity than HS Pellumina. HS Pellumina however is faster. Pellumina will not deform or flake; it is ideal where selectivity differing from that of silica gel is indicated.

Pellosils-Pellicular Silica Gels

Pellosils are manufactured by permanently bonding a silica gel layer to the prepared surface of a uniform, spherical, impermeable 37 to 43 µm glass bead. The active pellicles do not deform or flake off under high pressure, temperature or solvent action. In HC Pellosil the pellicle is thicker than in HS Pellosil, providing higher sample capacity. HS Pellosil, with a thinner pellicle, is faster.

Co: Pell PAC Pellicular Bonded Polar Phase

This packing contains cyano-amino groups that are Si-O-Si bonded to the prepared surface of a spherical glass bead. It has essentially the same chromatographic characteristics as Partisil®-10 PAC, except for differences arising from the pellicular form of Co: Pell PAC, e.g., less sample capacity and lower efficiency.

Pellamidon-Pellicular Polyamide

Pellamidon is manufactured by polymerizing nylon-6 on the prepared surface of a 37 to 43 µm glass bead. The Pellamidon active pellicle is 1 µm thick, a thickness which largely overcomes capacity limitations inherent in the pellicular structure. Pellamidon is stable in aqueous and nonaqueous solutions and does not have the compressibility disadvantages of totally porous polyamide. Adsorption and selectivity depend on the reversible formation of hydrogen bonds of the solute with the amide groups of the polymer.

Co: Pell ODS

Co: Pell ODS is a pellicular reversed-phase column packing. It is manufactured by chemically and permanently bonding a hydrocarbon, octadecylsilane, to the surface of a 30 µm glass bead. This packing gives excellent chromatographic performance in many reversed-phase applications, but is being increasingly used in ion pair chromatography.

Table 4 (continued)
PACKINGS FOR HIGH PRESSURE LIQUID CHROMATOGRAPHY

AL Pellionex WAX

This packing is a unique, pellicular, weak anion exchanger in which amino groups are attached to an aliphatic polymer backbone. The active pellicle is chemically bonded to the modified surface of a 40 μm glass bead. The packing is supplied in the Cl⁻ ionic form. The aliphatic structure reduces pi-bonding adsorption effects. The mechanism of separation makes AL Pellionex WAX suitable for the separation of higher molecular weight substrates, e.g., polyelectrolytes, nucleic acids and peptide fragments. It functions well with mild elution systems. AL Pellionex WAX is suggested for compounds which exhibit acid functionality and are too strongly retained on conventional resins. It has good chromatographic performance and high stability.

AE Pellionex SAX

This packing is a pellicular strong anion exchanger. The active shell contains a cross-linked polyaromatic resin composed in part of a polyester-type material. To this backbone quaternary ammonium groups are attached and the pellicle is chemically bonded to a surface-modified 40 μm glass bead. AE Pellionex SAX has lower inherent polarity than conventional cross-linked polystyrene matrices. It is useful where solutes have polar as well as ion exchange characteristics and/or where solutes are similar in mass or structure but differ in ionic properties. This packing exhibits high stability and good chromatographic performance.

AS Pellionex SAX

This packing is a pellicular strong anion exchanger in which quaternary ammonium groups on a 8% cross-linked polystyrene backbone are chemically bonded to 40 μm glass beads whose surface has been modified. The active pellicle is about 1 μm thick. The material is supplied in the H_2PO_4 ionic form. AS Pellionex SAX shows excellent stability and chromatographic performance. Resin exchange capacity is approximately 10 μeq/g. Elution orders are similar to those on conventional strong anion exchange resins.

HC Pellionex SCX

This packing is a pellicular strong cation exchanger. The active pellicle consists of a cross-linked polystyrene containing sulfonic acid functional groups. The active layer is chemically bonded to the modified surface of a 40 μ glass bead. The material is supplied in the H⁺ ionic form. HC Pellionex SCX is nonswelling and has an exchange capacity of approximately 60 μeq/g. Chromatographic performance and stability are good. The stability characteristics render possible excellent chromatography under gradient elution conditions. The elution order is similar to that of conventional strong cation exchange resins. Technical information from Whatman Ltd. catalog.

DU PONT

ZORBAX COLUMN PACKINGS

Zorbax column packings are based on the Zorbax microparticulate silica support.

Zorbax ODS

Zorbax ODS is a highly retentive reversed-phase packing formed by the monomolec-

Table 4 (continued)
PACKINGS FOR HIGH PRESSURE LIQUID CHROMATOGRAPHY

ular bonding of octadecylsilane groups to the surface of the particles. This packing complements the short chain Zorbax C-8. An advantage over Zorbax C-8 is its stronger retentive power for nonpolar compounds. This retention is of benefit in the separation of low polarity compounds which are essentially insoluble in water and have limited solubility in methanol or acetonitrile. Zorbax ODS is equally effective with partially aqueous mobile phases to separate compounds of moderate polarity or with some non-aqueous mobile phases for separating nonpolar compounds. Equilibration of the column packing after a change in mobile phase composition is rapid, making Zorbax ODS ideal for high speed gradient elution techniques. Zorbax ODS has been used for the separation of oligomers, phthalate esters, petroleum hydrocarbons, polynuclear aromatic hydrocarbons, chlorinated biphenyls, triglycerides, vitamins, steroids, and PTH amino acids.

Zorbax C-8

Zorbax C-8 is a reversed-phase packing formed by monomolecular bonding of octyl silane groups to the surface of the particles. It serves as an effective complement to the high performance Zorbax ODS. The packing is useful for separations over a wide range of compounds from water soluble (high polarity) to hydrocarbon soluble (low polarity). The greatest utility for Zorbax C-8 is for substances of relatively high polarity where superior resolution is often found versus Zorbax ODS. Zorbax C-8 is also very effective in gradient elution chromatography with both water methanol and water-acetonitrile mobile phases. Rapid re-equilibration with new mobile phases occurs. Zorbax C-8 has been used for the separation of substituted quinones, anticonvulsants, sulfa antibiotics, procainamide.

Zorbax NH_2

Zorbax NH_2 is a multipurpose packing formed by bonding 3-aminopropyltriethoxysilane to Zorbax support material under conditions that maximise formation of a monolayer phase. Zorbax NH_2 can be used in three entirely different modes; normal phase for low to moderately polar materials, reversed phase for polar compounds where this column is especially useful for carbohydrate separations, and as an ion exchanger for anionic solutes. In the normal phase mode Zorbax NH_2 has been used to separate positional isomers, antioxidants and pesticides, as an ion exchanger to separate organic acids, nucleotides and deoxynucleotides and in the reversed phase mode to separate sugars. Zorbax NH_2 can be used with solvents from water to hexane, but the use of strong bases and acids should be avoided. The suggested pH range is 2 to 7 in aqueous or partially aqueous mobile phases.

Zorbax CN

Zorbax CN is a polar bonded phase packing formed by monomolecular bonding of cyanopropyl groups to the surface of the particles. Maximum surface coverage is maintained to give a packing with exceptional reproducibility. Zorbax CN is used in both normal and reversed-phase liquid chromatography. In the normal mode it will separate many classes of compounds with polar functional groups such as alcohols or amines. In normal phase operation the advantages of Zorbax CN over a silica adsorbent are (1) there is no need to saturate the organic mobile phase with water for reproducible retentivity. (2) columns are much less liable to fouling from uneluted compounds and (3) rapid equilibration with mobile phases. In the normal phase Zorbax CN has been used to separate pesticides, aromatic amines and acids, essences and flavors and in the reversed-phase esters, catecholamines, nucleic acid bases and sennosides.

Table 4 (continued)
PACKINGS FOR HIGH PRESSURE LIQUID CHROMATOGRAPHY

Zorbax TMS

Zorbax TMS is a short chain reversed-phase packing formed by exhaustively reacting trimethylchlorosilane with the surface of the silica support. This produces a monolayer bonded phase with maximum surface coverage and reproducibility. Due to its short alkyl chain this packing is in general less retentive than Zorbax ODS and C-8. Consequently separations can be developed on Zorbax TMS with mobile phases containing less organic solvent. These conditions are well suited to highly polar compounds, including water soluble compounds. Zorbax TMS is also effective in reversed-phase ion pairing chromatography, in which separations of charged solutes are achieved by including ionic modifiers in the mobile phase. The packing has been used in the reversed-phase mode to separate parabens, penicillins, sulfa antibiotics, and food additives.

Technical information from Du Pont LC column reports.

J. T. Baker Chemical Co.,
222 Red School Lane,
Phillipsburg,
NJ 08865.

COLUMN PACKINGS

'Baker' high capacity large porous particle and microparticulate silica gel and 17 functional bonded phases are available. 'Baker' bonded phases are polar and reversed phase organosilanes bonded to irregular totally porous 60 Å silica gel in 5 μm, 10 μm, and 40 μm average particle diameters. Each lot is controlled for narrow particle size distribution, pore diameter, specific surface area, and polymeric functional moiety concentration.

Bulk preparative LC column packings available in 40 μm particle size are 'Baker' silica gel and 'Baker' bonded phases, octadecyl (C_{18}), octyl (C_8), cyano (CN), amino (NH_2), methyl (C_1), ethyl (C_2), butyl (C_4), hexyl (C_6), cyclohexyl (C_6H_{11}), phenyl (C_6H_5), diphenyl (C_6H_5)$_2$, 1',2'-amino (NH_2-NH), quaternary amine (N^+), carboxylic acid (COOH), aliphatic sulfonic acid (CSO_3H), aromatic sulfonic acid ($C_6H_5SO_3H$) and diol (COHCOH). HPLC analytical column packings of similar types are available in 10 μm and 5 μm particle size.

Information from J. T. Baker Chemical Co. literature.

PIC REAGENTS

Paired ion chromatography (PIC) is a method for separating ionic compounds by causing them to behave as nonionic species with some nonpolar characteristics. PIC reagents are large organic counter ions which are dissolved in the liquid chromatography mobile phase. They form an equilibrium complex with ionic samples; the resulting species may then be chromatographed using a reverse-phase system. By using PIC instead of ion-exchange techniques, problems or pH control and reproducibility can be avoided. PIC permits the simultaneous assay of acids, bases and neutral compounds.

Two types of PIC reagents are available; PIC A for use with acids and PIC B for use with bases. There are four types of PIC B: B-5, B-6, B-7 and B-8 which contain a five carbonalkyl chain, a six carbon chain, a seven carbon chain and an eight carbon

Table 4 (continued)
PACKINGS FOR HIGH PRESSURE LIQUID CHROMATOGRAPHY

chain, respectively. The length of the carbon chain influences the retention time or retention volume of the ion-pair complex formed, a longer carbon chain giving increasing retention times.

Technical information from Waters Associates catalog.

> LKB-PRODUKTER AB,
> Box 305,
> S-161 26 Bromma,
> Sweden

Ultropac resins
Ultropac resins are high-performance analytical ion-exchange resins suitable for use in amino acid analyzers.

Ultropac 13 resin
Ultropac 17 resin
Ultropac 11 resin
Ultropac 10 resin
Ultropac 8 resin, lithium form
Ultropac 8 resin, sodium form
Prewash column resin, amberlite
Ultropac 3-prewash column resin sodium
Ultropac 3-prewash column resin lithium

> BENSON COMPANY,
> P.O. Box 12812,
> Reno,
> Nevada 89510

Benson resins for amino acid analysis are controlled to assure exceptional peak resolution, low operating back pressure and reproducibility from batch to batch.

Resins for standard-bore columns, 6-9 mm I.D.
BH-4 resin, used for protein hydrolysates with single or dual-column methodologies.
BP-B5 resin, used for basic groups in protein hydrolysates or physiological fluids and peptide analysis.
BP-AN6 resin, used for acidic-neutral groups protein hydrolysate or physiological fluids, with single or dual-column methodologies.

Resins for microbore columns, 2-5 mm I.D.
B-X8 resin, 8 ± 1 μm, used for protein hydrolysates with single or dual-column methodologies.
D-X8.25 resin, 8 ± 1 μm, used for protein hydrolysates with single column methodology.
A-X8.75 resin, 5 ± 1 μm, used for protein hydrolysates with single column methodology.

Table 5
PREFORMED LAYERS FOR THIN-LAYER CHROMATOGRAPHY

WHATMAN

Plate types are layer formulations designated by the letter K plus a numerical suffix, e.g., K5, K6, etc. The type establishes product specifications, F denotes a fluorescent indicator, P denotes a preparative layer.

K6/K6F Plates

K6/K6F silica gel plates are fast running, hard surface silica gel coated glass plates. The plates are used for separating a broad range of compounds, especially pesticides and steroids.

K5/K5F Plates

The K5 silica gel formulation is designed for the separation of moderately polar to strongly polar compounds, e.g., alkaloids, polypeptides, pesticides, steroids, but it has been found to have applications beyond these, e.g., in lipid analysis, for amino acids, proteins, and sugars. K5/K5F plates are of moderate hardness. K5 plates withstand abrasive reagent sequences. The plate's temperature limitation for charring is 130°C; for reagent charring, 50% methanolic sulfuric acid or cupric acetate/phosphoric acid is recommended.

K4/K4F Plates

K4/K4F precoated glass plates use a gel produced according to Stahl and have a soft surface. A calcium sulfate binder is used which allows all destructive charring and ashing techniques to be used without temperature limitations. An activation time of 15 min at 110°C is usually sufficient to impart maximum activity.

K1/K1F Plates

K1 plates are soft surfaced media. The use of a pure, low polarity, porous silica gel with fines removed contributes to its good performance in preparative layers (Type PK1/PK1F). The plates are useful for separating polypeptides, proteins, high polarity multifunctional dyes, acids and other high molecular weight materials. K1 plates can be used with available reagent sprays but should not be heated to more than 130°C.

Linear-K Plates

Linear-K preadsorbent area TLC plates incorporate a preadsorbent sample dispensing area across the full width of the lower edge of the plates. The preadsorbent used is diatomaceous earth. After application of the sample and drying, the solvent passing through the preadsorbent extracts, concentrates the sample and presents it uniformly to the adsorbent layer of the plate as a concentrated band. The preadsorbent material can handle crude, dilute, organic, or aqueous spotting solutions. Linear-K plates give better reproducibility of R_f values and sensitivity is increased because of greater band compaction. Linear-K plates are available in analytical (250 μm layer) and preparative (1000 μm layer) forms. Ultra-fast hard surface (K6) formulation and moderately hard (K5) formulation silica gel adsorbent areas and adsorbent areas on alumina can be obtained. Each type is available with or without fluorescent indicator.

KC_{18} Plate

In the KC_{18} plate octadecylsilane is bonded to the silica gel support via extremely

Table 5 (continued)
PREFORMED LAYERS FOR THIN-LAYER CHROMATOGRAPHY

stable Si-O-Si bonds. A hydrocarbon surface is thus presented to the solutes. The reversed-phase separation mode introduced by KC_{18} plates makes TLC applicable to a wide range of nonpolar compounds, using simple solvent systems. Examples are lipid profiles, isomers, vitamins, water soluble dyes, polar amino acids, peptides. KC_{18} plates have a high loading capacity and are thus preparative as well as analytical media. They correlate very closely with HPLC reversed phase columns. KC_{18} layers are available with linear preadsorbent area.

HP-K Plates

HP-K high performance TLC plates separate nanogram and picogram quantities. The high performance of HP-K plates is achieved by a smaller particle size, an extremely narrow particle size distribution, an ultra-high purity silica gel formulation, a special, high gloss, durable, fast developing binder, and uniform thinner layers. HP-K plates are available with fluorescent indicator and linear preadsorbent area.

Avicel Cellulose TLC Plates, Type K2/K2F

K2/K2F plates have layers of micro-crystalline cellulose, which is superior to TLC cellulose in fibrous form. Although the mechanism of separation is primarily partition, a minimal amount of adsorption takes place. This dual-mechanism gives superior separations in many cases, for example, of amino acids at pH 2.2, of peptides and nucleic acids at pH 3.0 and of sugars. The K2/K2F layers are resistant to abrasion and are mechanically stable even when wet. These plates are available in analytical and preparative layer, with or without fluorescent indicator.

K3/K3F Plates

K3/K3F plates have neutral aluminium oxide layers prepared from a high quality base. They have a nonabrading surface. As packaged, K3 plates are at Brockmann activity grade IV. This imparts adsorption characteristics similar to silica gel. The plates can be activated to Brockmann 1 (0% water) by heating at 130°C for 20 min. Adsorption (azobenzene) decreases with increasing activity grade. These plates are recommended for pesticides, vitamins, polyalcohols, and strongly basic amines.

Multi-K CS5 Plates

The multi-K type CS5 plate incorporates two dissimilar, complementary media layers on a single precoated glass plate. It is designed primarily for two-dimensional development. The plate consists of a reversed-phase layer or two-dimensional development spotting strip contiguous with a silica gel analytical layer. The spotting strip consists of octadecylsilane groups bonded via Si-O-Si-C bonds to a special silica gel. The strip thus presents a hydrocarbon surface to the solutes for development in the first direction. Second dimension development takes place on the silica gel layer which uses K5F silica gel. The silica gel layer is fast and gives excellent resolution. Sample volume requirements are minimal for the multi-K CS5.

Technical information from Whatman catalog and bulletins.

MERCK

TLC plates precoated.
Available coatings are listed.

Table 5 (continued)
PREFORMED LAYERS FOR THIN-LAYER CHROMATOGRAPHY

    Silica gel 40 F_{254}
    Silica gel 60 F_{254}
    Silica gel 60 (without fluorescent indicator).
    Silica gel 60 Kieselguhr F_{254}
    Silica gel 100 F_{254}
    Aluminum oxide 60 F_{254} (Type E)
    Aluminum oxide 150 F_{254} (Type T).
    Cellulose F
    Cellulose (without fluorescent indicator)
    PEI-cellulose F
    PEI-cellulose (without fluorescent indicator)
    Kieselguhr F_{254}
    Silica gel 60 silanized (without fluorescent indicator)
    Silica gel 60 F_{254} silanized.
    Polyamide 11 F_{254}

TLC plates precoated, layer thickness 0.5 mm.
    Silica gel 60 F_{254}

TLC plates precoated with concentrating zone.
    Silica gel 60 F_{254}

Silica gel 60 (without fluorescent indicator).

TLC aluminum sheets.
    Silica gel 60 F_{254}
    Silica gel 60 (without fluorescent indicator)
    Silica gel 60/Kieselguhr F_{254}
    Aluminum oxide 60 F_{254} (Type E)
    Aluminum oxide 150 F_{254} (Type T).
    Cellulose (without fluorescent indicator).
    Cellulose F_{254}
    Polyamide 11 F_{254}
    Kieselguhr F_{254}

TLC aluminum rolls.
    Silica gel 60 F_{254}
    Cellulose (without fluorescent indicator)

TLC plastic sheets
    Silica gel 60 F_{254}
    Aluminum oxide 60 F_{254} (Type E)
    Cellulose (without fluorescent indicator)
    Silica gel 60 (without fluorescent indicator)
    Cellulose F_{254}
    PEI-cellulose F_{254}

TLC plastic rolls
    Silica gel 60 F_{254}
    Cellulose (without fluorescent indicator)

Table 5 (continued)
PREFORMED LAYERS FOR THIN-LAYER CHROMATOGRAPHY

PLC plates precoated (for preparative layer chromatography).
Silica gel F_{254}
Silica gel 60 (with fluorescent indicator)
Silica gel 60 $F_{254+366}$
Aluminum oxide 150 F_{254} (Type T)

HPTLC plates precoated for nano-TLC.
Silica gel 60 F_{254}
Silica gel 60 (without fluorescent indicator)
RP-2 F_{254s}
RP-8 F_{254s}
RP-18 F_{254s}
Cellulose (without fluorescent indicator)

HPLTC plates precoated with concentrating zone.
Silica gel 60 F_{254}
Silica gel 60 (without fluorescent indicator)

HPTLC aluminum sheets for nano-TLC.
Silica gel 60 F_{254}
Silica gel 60 (without fluorescent indicator)

Technical information from Merck catalogue and bulletins.

TLC PLATES

J. T. BAKER CHEMICAL CO.

'Baker' precoated glass-backed plates are ready to use for TLC. They are coated with abrasion-resistant, 'hard-surfaced' and inert bound layers of:

Silica gel in 250 μm analytical and 500 μm preparative thickness with high polarity for fast separations.

High performance silica gel, approximately 5 μm particle size in 200 μm analytical thickness for ultra high speed separation of nanogram and picogram sample levels.

Reversed phase octadecylsilane (C_{18}) bonded to silica gel, approximately 20 μm particle size, in 200 μm analytical thickness for well-resolved separations of nonpolar compounds.

These TLC plates are available with:

Fluorescent indicator, a noninterfering inorganic phosphor fluorescing at 254 nm.

Preadsorbent spotting area, for reduction in spotting time, elimination of most sample preparations, higher load capacity, improved sensitivity, resolution, and reproducibility.

Channeled plates for ease and speed of sample application with band development for improved resolution.

Each lot is controlled for narrow particle size distribution, pore diameter, specific surface area and activity, layer uniformity, and consistency and noninterfering inert binding.

'Baker-flex' precoated, flexible TLC sheets have an abrasion-resistant coating and

Table 5 (continued)
PREFORMED LAYERS FOR THIN-LAYER CHROMATOGRAPHY

solvent-resistant substrate for rapid, clean separations. An extra thick layer of sorbent is used on 'Baker-flex' aluminum oxide, 1B/1B-F, 200 μm, silica gel 1B/1BF, 200 μm and silica gel 1B2/1B2-F, 250 μm, sheets, which allows the loading of more than the normal amount of sample.

TLC sheets/plates coated with the following are available:

Aluminum oxide used for the separation of alcohols, alkaloids, strongly basic amines, antioxidants, heterocyclic compounds, hydrocarbons, peptides, polyalcohols, steroids, terpenes, and vitamins.

Cellulose and microcrystalline cellulose for the separation of amines, amino acids, antibiotics, carbohydrates, glycosides, hydrocarbons, inorganic ions, nucleic acids, organic acids, peptides, urea derivatives, and vitamins.

Cellulose AC-10 for the separation of antioxidants.

Cellulose CM for the separation of inorganics and metal ions.

Cellulose DEAE for the separation of nucleoside mono-, di-, and triphosphates.

Cellulose ECTEOLA for the separation of purines, pyramidines and nucleosides.

Cellulose PEI for the separation of monophosphate nucleosides.

Polyamide 6 for the separation of amines, amino acids, antioxidants, glycosides, phenols.

Silica gel 1B, silica gel 1B2, silica gel 60, and silica gel, high performance, for the separation of alcohols, aldehydes, alkaloids, amines, amino acids, amphetamines, antibiotics, antioxidants, barbiturates, carbohydrates, flavonoids, heterocyclic compounds, hydrocarbons, indoles, ketones, lipids, nitro compounds, organic acids, peroxides, phenols, plasticizers, polypeptides, steroids, terpenes, unsaturated compounds, vitamins.

Reversed phase (C_{18}) for the separation of nonpolar compounds, such as lipids, fatty acids and their esters, carotenoids, steroids, triglycerides, and cholesterol esters.

Information from J. T. Baker Chemical Co. literature.

Table 6
ADSORBENTS FOR THIN-LAYER CHROMATOGRAPHY

MERCK

The letter G is used to designate adsorbents which contain gypsum as binder. Grades designated H contain neither gypsum nor organic binders, those designated P are intended for preparative layer chromatography and those with R indicate very pure adsorbents. The letter F indicates the addition of a fluorescent indicator, the excitation wavelength of which is given as a subscript. The figure "60" after the designation silica gel indicates that this is a medium porosity silica gel with a mean pore diameter of 60 Å.

Silica gel (type 60) for thin-layer chromatography.

TLC silica gel 60 G mean particle size 15 μm.

Silica gel H (type 60) for thin-layer chromatography.

Layers prepared from silica gel H have practically the same adhesive power as a silica gel G layer. It is especially suitable for the separation of substances which cannot be separated on silica gel G on account of the added gypsum, e.g., compounds which form insoluble calcium salts.

Table 6 (continued)
ADSORBENTS FOR THIN-LAYER CHROMATOGRAPHY

TLC silica gel 60 H mean particle size 15 μm.

Silica gel macroporous type 200 for thin-layer chromatography. This silica gel has a mean pore diameter of 200 Å.

Silica gel macroporous type 500 for thin-layer chromatography.

This silica gel has a mean pore diameter of 500 Å.

Silica gel macroporous type 1000 for thin-layer chromatography.

This silica gel has a mean pore diameter of 1000 Å. The macroporous types of silica gel correspond to the Merckogel types SI 200, 500, and 1000. To determine the optimum Merckogel type SI for use in column separation, a trial on a thin-layer with macroporous silica gel is recommended.

Silica gel HF_{254} (type 60) for thin-layer chromatography.

This silica gel is mixed with an inorganic lumiphore which cannot be eluted and which when excited under short-wave UV light shows strong fluorescence. Silica gel HF_{254} is suitable for separating substances which adsorb in the medium UV range.

Separated substances can be visualized without having to be sprayed with a staining reagent.

TLC silica gel 60 HF_{254} mean particle size 15 μm.

Silica gel $HF_{254+366}$ (type 60) for thin-layer chromatography.

Apart from the inorganic fluorescent indicator, silica gel $HF_{254+366}$ contains a further (organic) additive which fluoresces at the excitation wavelength of 366 nm.

This substance may be partially eluted by some eluants. On this layer substances without intrinsic absorption can be visualized under UV light by a brightening of the layer fluorescence, e.g., lipids, saturated steroid systems.

Silica gel GF_{254} (type 60) for thin-layer chromatography.

TLC silica gel 60 F_{254} mean particle size 15 μm.

Silica gel 60 HR extra pure for thin-layer chromatography.

This silica gel contains neither gypsum nor organic binders. It is intended for separations which require particularly pure layers, e.g., where a quantitative determination or spectroscopic investigation of the separated substances is to be carried out.

Silica gel H silanized for thin-layer chromatography.

Silica gel silanized is a hydrophobic silica gel which is suitable for partition chromatographic separations with lipophilic stationary phases (reversed phase chromatography). Substances which may be separated using silica gel silanized include free fatty acids and their esters.

Silica gel HF_{254} silanized for thin-layer chromatography.

Kieselguhr G for thin-layer chromatography.

Kieselguhr G contains gypsum as binder. It is an inactive support and is especially suitable for partition chromatographic separations.

Polyamide 11 for thin-layer chromatography.

Polyamide 11 is used primarily for the thin-layer chromatographic separation of phenols and phenolic substances.

Cellulose microcrystalline for thin-layer chromatography.

This microcrystalline material produces a fine-grade layer which can be used for chromatographic separations with water as the stationary phase.

Cellulose native for thin-layer chromatography.

This native, fibrous cellulose can be used to advantage in partition chromatography with hydrophilic eluants.

CM-cellulose for thin-layer chromatography.

Carboxymethylcellulose is a cation exchanger suitable for the separation of basic and neutral proteins.

Table 6 (continued)
ADSORBENTS FOR THIN-LAYER CHROMATOGRAPHY

Aluminum oxides.

Two types of aluminum oxides, 60/E and 150/T are available which have different chromatographic properties. The figure "60" and "150" is the designation which gives the medium pore diameter of the aluminum oxide in Å.

Aluminum oxide G (type 60/E) for thin-layer chromatography.
Aluminum oxide G contains gypsum as binder.
Aluminum oxide H basic (type 60/E) for thin-layer chromatography.
Aluminum oxide H basic is free from gypsum or organic binders.
Aluminum oxide H_{254} basic (type 60/E) for thin-layer chromatography.
Aluminum oxide H_{254} contains the same insoluble inorganic luminophore as silica gel HF_{254}.
Aluminum oxide GF_{254} (type 60/E) for thin-layer chromatography.
Aluminum oxide 150 basic (type T) for thin-layer chromatography.
Aluminum oxide 150 neutral (type T) for thin-layer chromatography.
This aluminum oxide is adjusted to a neutral pH of about 7.
Aluminum oxide 150 acidic (type T) for thin-layer chromatography.
This aluminum oxide is adjusted to an acidic pH of about 4.
Technical information from E. Merck catalog.

WHATMAN

K5 silica gel
K6 silica gel

The mean pore diameter of silica gel K5 is 80 Å, that of K6 is 40 Å. K5 silica gel acts as a less polar adsorbent than K6. The adsorption characteristics of K6 silica gel are comparable to those of Stahl-type silica gel.

K3 neutral alumina.
K2 microcrystalline cellulose.

All the adsorbents are available with or without gypsum binder. They are also available with or without Whatman's inorganic UV phosphor.

Technical information from Whatman bulletin.

SCHLEICHER AND SCHUELL

Adsorbents with the designation /LS 254 contain a fluorescent indicator which has a pale green fluorescence at 254 nm.

Cellulose double acid-washed, 142 dg, with maximum length of fibers 10-50 μm.
Cellulose microcrystalline, 144 and 144/LS 254, particle size about 20 μm.
Cellulose natural fiber, 180 and 180/LS 254, particle size 2 to 25 μm.
Cellulose natural fiber acid-washed, 180a, particle size 2 to 25 μm.
pCellulose acetylated, 180/21ac and 180/45ac, particle size about 50 μm. The acetyl content is 20-25% in 180/21ac and 40-45% in 180/45ac.
Silica gel, 150 and 150/LS254, particle size less than 40 μm.
Silica gel with gypsum, 150 G and 150 G/LS 254, particle size less than 40 μm. 12% gypsum is included as binder.
Silica gel with starch, 150 S and 150 S/LS254, particle sizeless than 40 μm. 3% starch is included as binder.

Table 6 (continued)
ADSORBENTS FOR THIN-LAYER CHROMATOGRAPHY

Silica gel with gypsum, 150 G and 150 G/LS 254, particle size less than 40 μm. 12% gypsum is included as binder.

Silica gel with starch, 150 S and 150 S/LS254, particle sizeless than 40 μm. 3% starch is included as binder.

ION EXCHANGE CELLULOSE POWDERS

Anion exchangers

DEAE-cellulose, 66, capacity 0.9 ± 0.1 meq/g.

ECTEOLA-cellulose, 67, capacity 0.3 ± 0.1 meq/g

TAEA-cellulose, 83, capacity 0.9 ± 0.1 meq/g.

PEI-cellulose, 84, capacity 0.3 ± 0.1 meq/g.

QAE-cellulose, 85, capacity 0.9 ± 0.1 meq/g.

Cation exchangers

CM-cellulose, 68, capacity 0.7 ± 0.1 meq/g

P-cellulose, 69, capacity 0.9 ± 0.1 meq/g.

Technical information from Schleicher and Schuell catalogue.

Table 7
COMMERCIAL SOURCES FOR CHROMATOGRAPHY PAPERS

WHATMAN LTD.

Pure cellulose paper grades

1 Chr.   White, smooth, normally-hard surface. Recommended for general purpose chromatography uses. 0.16[a], 130[b]

2 Chr.   White, smooth, normally-hard surface. Recommended for optical or radiometric scanning. 0.18, 115.

3 Chr.   White, rougher surface than 3MM Chr. Used for separation of inorganics and for electrophoresis. Higher wet strength than all chromatography papers except 3MM, which is similar. 0.38, 130.

3MM Chr.   White surface. High wet strength. Widely used for electrophoresis. 0.33, 130.

4 Chr.   Fastest thin paper. White, smooth, normally-hard surface. For routine or repetitive chromatography where loadings are relatively low. 0.22, 180.

Table 7 (continued)
COMMERCIAL SOURCES FOR CHROMATOGRAPHY PAPERS

17 Chr. Extremely absorbent; will accept heavy loadings. Recommended for preparative paper chromatography and electrophoresis. 0.88, 190.
20 Chr. Excellent resolution at low loadings. White, smooth, normally-hard surface. Recommended for separations of samples of unknown composition. 0.16, 85.
31 ET Chr. Extremely fast paper. White, soft surface. Principal use is in electrophoresis of large molecules. 0.53, 225.

[a] Thickness in mms.
[b] Flow rate, mm/30 min (water).

Technical information from Whatman Ltd. catalog.

SCHLEICHER AND SCHUELL INC.

Acid-washed papers
Green ribbon-C. For paper chromatography. 0.342[a], 145[b], rough[c]
Orange ribbon-C. For paper chromatography and electrophoresis. 0.228, 115, smooth.
Black ribbon-C. Suitable for chromatography when R_f values are well separated. 0.193, 155, rough.
White ribbon-C. Suitable for most chromatography, electrophoresis and circular chromatography. 0.200, 100, rough.
Blue ribbon-C. Excellent for use with a 2-butanol-ammonia solvent system. 0.180, 75, rough.
Red ribbon-C. For circular and ascending chromatography. A paper with no watermark. It sometimes gives a better resolution of compounds with similar R_f values than does Blue ribbon. 0.139, 70, smooth.
507-C. For circular and ascending chromatography. An ultra-dense, hard paper which absorbs relatively little liquid. It is suitable for separating small quantities of materials with similar R_f values. 0.101, 55, smooth.
Ash-low papers.
470-C. For electrophoresis and ascending chromatography of amino acids, 0.886, 270, smooth.
470A-C. For chromatography and electrophoresis of larger quantities. 0.558, 270, very smooth.
903-C. For chromatography and electrophoresis. 0.432, 210, smooth.
593-C. For chromatographic and electrophoretic separations of amino acids and peptides. 0.340, 115, smooth.
598-C Suitable for preparative chromatography and separations in which columns composed of circles are used. 0.350, 110, rough.
2316 For the separation of larger amounts of materials and for preparative procedures. 0.340, 130, smooth.
2040b A medium-textured paper for descending chromatography. 0.250, 180, smooth.
2043bmgl For most chromatography, circular chromatography and evaluation by the elution method. 0.220, 100, smooth.
2045b For circular chromatography and descending chromatography with low absorption heights. 0.210, 65, smooth.

Table 7 (continued)
COMMERCIAL SOURCES FOR CHROMATOGRAPHY PAPERS

591-C   Smooth-textured, for one and two-dimensional chromatography. 0.175, 70, smooth.

2040a   For most chromatographic work, especially by the descending method. 0.190, 115, smooth.

2043a   For most chromatographic work. It has a special smooth, uniform surface. 0.170, 100, smooth.

2045a   For circular and ascending chromatography with low adsorption heights. 0.170, 65, smooth.

576-C   Suitable for photometric analysis and work with small samples. For circular chromatography and ascending chromatography with low absorption heights. 0.114, 55, very smooth.

[a] Thickness, mm.
[b] Absorption height of demineralized water in mm per 30 min.
[c] Surface.

Technical information from Schleicher and Schuell Inc. catalog.

MACHEREY-NAGEL AND CO., DÜREN

Chromatography papers.

MN 214 140[a], 0.28[b], 90-100[c], smooth[d]
MN 218 180, 0.36, 90-100, smooth
MN 260 90, 0.20, 130-150, smooth
MN 261 90, 0.18, 90-100, smooth
MN 827 270, 0.70 130-140[e], soft carton.
MN 866 650, 1.70, 150-160[a], soft carton
MN 214 ff 140, 0.28, 90-100, MN 214 defatted.
MN 214 AC-10, 160, 0.30, acetylated MN 214, 10% acetyl content.

[a] Weight, g/m^2.
[b] Thickness, mm.
[c] Flow rate, mm/30 min.
[d] Characteristics.
[e] Flow rate, mm/10 min.

Technical information from Macherey-Nagel and Co. catalog.

CHARACTERISTICS OF CHROMATOGRAPHY PAPER

Chromatographic resolution on paper depends on several factors, viz: (1) distance between spots and bands or R_f value, (2) the degree of band or spot spreading, (3) shape, and (4) sharpness. The choice of paper can have effects as pronounced as the control of solvent and other factors. Resolution increases as flow rate decreases, provided paper thickness and solute loadings are the same. The natural flow rate of a paper can be reduced by using a "wick" of thinner or slower paper. Increased distance of flow does not necessarily provide improved resolution, because spots increase in size with distance, particularly with thin papers. All papers have some fiber orientation

Table 7 (continued)
COMMERCIAL SOURCES FOR CHROMATOGRAPHY PAPERS

in the "machine direction" and have a higher flow rate in this direction. Papers should thus always be developed in the same direction to obtain uniform spot mobilities.

The relative importance of humidity and solvent equilibration depends largely on the requirements of each separation. Slower and thicker papers generally require longer equilibration times and tank equilibration is necessary. An increase in temperature improves resolution on fast papers provided loadings are low. The improvement is insignificant on slow papers. Increased solute loading, whether brought about by increased volume or concentration, decreases the efficiency of separation significantly on thin papers. The effect is far less marked if thicker papers are used, provided fast flow rates are avoided.

Information from Whatman Publication 814 CP/L.

Table 8
SUPPLIERS OF AUTOMATIC AMINO ACID ANALYZERS

American Instrument Company,
Division of Travenol Laboratories Inc.,
8030 Georgia Avenue,
Silver Spring,
Maryland 20910.

Beckman Instruments Inc.,
Spinco Division,
1117 California Avenue,
Palo Alto
California 94304.

Carlo Erba Strumentazione,
P.O. Box 4342
20100 Milano,
Italy.

Dionex Corporation,
1228 Titan Way,
Sunnyvale,
California 94086

Glenco Scientific Inc.,
2802 White Oak Drive,
Houston,
Texas 77007.

Joel Ltd.,
MCD Division,
1418 Nakagami,
Akishima,
Tokyo 196
Japan.

Kontron International,
Analytical Division,
Head Office,
Bernerstrasse-Sud 169,
CH-8048 Zurich,
Switzerland.

LKB Instruments Inc.,
12221 Parklawn Drive,
Rockville,
Maryland 20852.

Rank Hilger Ltd.,
Westwood,
Margate,
Kent CT9 4JL,
England.

Technicon Instruments Corporation,
511 Benedict Avenue,
Tarrytown, New York, N.Y. 10591.

Section VI
Chromatography Book Directory

Section VI

CHROMATOGRAPHY BOOK DIRECTORY

Selected books on chromatography are listed below. Readers will find reviews of new books on chromatography in, among others, *The Journal of Chromatography, The Journal of Chromatographic Science, The Journal of the American Chemical Society* and *The Journal of Chemical Education*. Lists of recent books are published at regular intervals in the *Analytical Chemistry Lab Guide* and *The Journal of Chromatography*.

Academic Press Inc., 111 Fifth Avenue, New York, N.Y. 10003.

1. Brown, P. R., *High Pressure Liquid Chromatography: (Biochemical and Biomedical Applications)*, 1973.
2. Dixon, P. F., Gray, C. H., Lim, C. K., and Stoll, M. S., *High Pressure Liquid Chromatography in Clinical Chemistry*, 1976.
3. Horvath, C., Ed., *High Performance Liquid Chromatography. Advances and Perspectives*, Vol. 1, 1980.
4. Horvath, E., Ed., *High Performance Liquid Chromatography. Advances and Perspectives*, Vol. 2, 1980.
5. Jennings, W., *Gas Chromatography with Glass Capillary Columns*, 2nd ed., 1980.
6. Lawrence, J. F., *Organic Trace Analysis by Liquid Chromatography*, 1981.
7. Ma, T. S. and Ladas, A. S., *Organic Functional Group Analysis by Gas Chromatography. (The Analysis of Organic Materials,* Belcher, R. and Anderson, D. M. W., Ed., Vol. 10) 1976.
8. Walker, J. G., Jackson, M. T., Jr., and Maynard, J. B., *Chromatographic Systems — Maintenance and Troubleshooting*, 2nd ed., 1977.
9. Zweig, G. and Sherma, J., *Paper Chromatography and Electrophoresis*, Vol. 2, 1971.

Ann Arbor Science Publishers Inc., P.O. Box 1425, Ann Arbor, Michigan 48106.

1. Budde, W. L. and Eichelberger, J. W., *Organics Analysis using Gas Chromatography/Mass Spectrometry: A Techniques and Procedures Manual*, 1979.
2. Grushka, E., Ed., *Bonded Stationary Phases in Chromatography*, 1974.
3. Niederweiser, A. and Pataki, G., Ed., *Progress in Thin-Layer Chromatography and Related Methods*, Vol. 2, 1971.
5. Niederweiser, A. and Pataki, G., Ed., *Progress in Thin-Layer Chromatography and Related Methods*, Vol. 3, 1972.
6. Niederweiser, A. and Pataki, G., Ed., *New Techniques in Amino Acid, Peptide and Protein Analysis*, 1973.
7. Scott, R. M. and Lundeen, M., *Thin-Layer Chromatography Abstracts*, 1974.

Applied Science Publishers Ltd., 22 Rippleside Commercial Estate, Barking, Essex, England.

1. Knapman, C. E. H., Ed., *Developments in Chromatography*, 1978.
2. Knapman, C. E. H., Ed., *Developments in Chromatography*, 1980.

American Society for Testing and Materials (ASTM), 1916 Race Street, Philadelphia, PA 19103.

1. Bibliography on Liquid Exclusion Chromatography (Gel Permeation Chromatography), AMD 40, 1972-1975.
2. Bibliography on Liquid Exclusion Chromatography (Gel Permeation Chromatography) AMD 40-Sl, 1977.
3. Gas Chromatographic Data Compilation. AMD 25 A Sl, 1971.
4. Liquid Chromatographic Data Compilation, AMD 41, 1975.

Avondale Division, Hewlett-Packard Co., Avondale, Pennsylvania.

1. Rowland, F. R., *The Practice of Gas Chromatography*, 1973.

Barnes and Noble Inc., 105 Fifth Avenue, New York, N.Y. 10003.

1. Simpson, C., *Gas Chromatography*, 1970.

Chapman and Hall Ltd., 11 New Fetter Lane, London, E.C.4P 4EE, England.

1. Pryde, A. and Gilbert, M. T., *Applications of High Performance Liquid Chromatography*, 1979.
2. Stock, R. and Rice, C. B. F., *Chromatographic Methods*, 3rd ed., 1974.

Dowden, Hutchinson and Ross Inc., Stroudsburg, Pennsylvania.

1. Walton, H. F., Ed., *Ion-exchange Chromatography*, 1976.

Edinburgh University Press, 22, George Square, Edinburgh EG8 9LF, Scotland.

1. Knox, J. H., Done, J. N., Fell, A. F., Gilbert, M. T., Pryde, A., and Wall, R. A., *High-Performance Liquid Chromatography*, 1978.

Elsevier Scientific Publishing Co., P.O. Box 211, Amsterdam, The Netherlands.
Elsevier North-Holland Inc., 52 Vanderbilt Avenue, New York, N.Y. 10017.

1. Deyl, Z. and Kopecky, J., Ed., *Bibliography of Liquid Column Chromatography, 1971-1973 and Survey of Applications* (Supplementary Volume No. 6, 1976, to the Journal of Chromatography), 1976.
2. Deyl, Z., Macek, K., and Janak, J., Ed., Liquid column chromatography, *Journal of Chromatography*, 3, 1975.
3. Fischer, L., *Gel Filtration Chromatography. Laboratory Techniques in Biochemistry and Molecular Biology*, Vol. 1, Part II, Work, T.S. and Burdon, R. H., Ed. 1980.
4. Frigerio, A. and McCamish, M., Ed., *Recent Developments in Chromatography and Electrophoresis*, Vol. 10, Proceedings of the 10th International Symposium on Chromatography and Electrophoresis, Venice, 1979.
5. Huber, J. F. K., Ed., Instrumentation for high-performance liquid chromatography, *Journal of Chromatography*, 13, 1978.
6. Lawrence, J. F. and Frei, R. W., Chemical derivatization in liquid chromatography, *Journal of Chromatography*, 7, 1976.
7. Macek, K., Hais, I. M., Kopecky, J., Schwarz, V., Gasparic, J., and Churacek, J., *Bibliography of Paper and Thin-layer Chromatography, 1970-1973 and Survey of Applications* (Supplementary Volume No. 5, 1976 to the Journal of Chromatography), 1976.
8. Parris, N. A., Instrumental liquid chromatography. A practical manual on high-performance liquid chromatographic methods, *Journal of Chromatography*, 5, 1976.
9. Roberts, T. R., Radiochromatography. The chromatography and electrophoresis of radio-labelled compounds, *Journal of Chromatography*, 14, 1978.
10. Scott, R. P. W., Liquid chromatography detectors, *Journal of Chromatography*, 11, 1977.
11. Sevcik, J., Detectors in gas chromatography, *Journal of Chromatography*, 4, 1976.
12. Turkova, J., Affinity Chromatography, *Journal of Chromatography*, 12, 1978.
13. Unger, K. K., Porous silica. Its properties and use as support in column liquid chromatography, *Journal of Chromatography*, 16, 1979.
14. Zlatkis, A. and Ettre, L. S., Ed., *Advances in Chromatography*, Proceedings of the Ninth International Symposium, Houston, Texas, 1974.
15. Zlatkis, A., Ed., *Advances in Chromatography*, Proceedings of the 10th International Symposium, Munich, 1975.
16. Zlatkis, A. and Kaiser, R. E., Ed., HPTLC-high performance thin-layer chromatography, *Journal of Chromatography*, 9, 1977.

Halsted Press. See Wiley & Sons, Inc.

Heinemann Educational Books Ltd., London, England.

1. Allsop, R. T. and Healey, J. A. D., *Chemical Analysis, Chromatography and Ion Exchange*, 1974.

Heyden and Sons Inc., 247 South, 41st Street, Philadelphia, Pennsylvania 19104.
Hayden and Sons Ltd., Spectrum House, Alderton Crescent, London NW 4, England.

1. Blair, K. and King, G., Ed., *Handbook of Derivatives for Chromatography*, 1977.

2. Pattison, J. B., *A Programmed Introduction to Gas-liquid Chromatography, 1,* 1973.
3. Simpson, C. F., Ed., *Practical High Performance Liquid Chromatography,* 1976.

Alfred Huthig Verlag, Heidelberg, Basel, New York.

1. Bertsch, W., Hara, S., Kaiser, R. E., and Zlatkis, A., Instrumental HPTLC. Proceedings of the First International Symposium on Instrumentalised High-performance Thin-layer Chromatography, 1980.
2. Kaiser, R. E. and Oelrich, E., *Optimierung in der HPLC (Chromatographische Methoden Series),* 1979.
3. Schwedt, G., *Chemische Reaktiondetektoren fur die Schnelle Flussigkeits-Chromatographie-Grundlagen und Anwendungen in die Spurenanalyse,* 1980.

Laboratory Data Control, Interstate Industrial Park, P.O. Box 10235, Riviera Beach, Florida.

1. Schram, S. B., *The LDC Basic Book on Liquid Chromatography,* 1980.

Marcel Dekker Inc., 270 Madison Avenue, New York, N.Y. 10016.

1. Blackburn, S., Ed., *Amino Acid Determination,* 2nd ed., Revised and Expanded, 1978.
2. Cazes, J. and Delamare, X., Chromatographic Science Series, Vol. 13, *Liquid Chromatography of Polymers and Related Materials,* 1980.
3. Domsky, I. I. and Perry, J. A., *Recent Advances in Gas Chromatography,* 1971.
4. Fried, B. and Sherma, J., *Thin Layer Chromatography — Techniques and Applications,* 1982.
5. Giddings, J. C. and Keller, R. A., Ed., *Advances in Chromatography,* Vol. 10, 1974.
6. Giddings, J. C. and Keller, R. A., Ed., *Advances in Chromatography,* Vol. 11, 1974.
7. Giddings, J. C., Grushka, E., Keller, R. A., and Cazes, J., Ed., *Advances in Chromatography,* Vol. 12, 1975.
8. Giddings, J. C., Grushka, E., Keller, R. A., and Cazes, J., Ed., *Advances in Chromatography,* Vol. 13, 1975.
9. Giddings, J. C., Grushka, E., Cazes, J., and Brown, P. R., Ed., *Advances in Chromatography,* Vol. 14, 1976.
10. Giddings, J. C., Grushka, E., Cazes, J., and Brown, P. R., Ed., *Advances in Chromatography,* Vol. 15, 1977.
11. Giddings, J. C., Grushka E., Cazes, J., and Brown, P. R., Ed., *Advances in Chromatography,* Vol. 16, 1978.
12. Giddings, J. C., Grushka, E., Cazes, J., and Brown, P. R., Ed., *Advances in Chromatography,* Vol. 17, 1979.
13. Giddings, J. C., Grushka, E., Cazes, J., and Brown, P. R., Ed., *Advances in Chromatography,* Vol. 18, 1980.
14. Gudzinowicz, B. J., Gudzinowicz, W. J., and Martin, H. F., *Fundamentals of Integrated GC-MS. Part I, Gas Chromatography,* Chromatographic Science Series, Vol. 7, 1976.
15. Gudzinowicz, B. J., Gudzinowicz, M. J., and Martin, F. M., *Fundamentals of Integrated GC-MS. Part II, Mass Spectrometry,* Chromatographic Science Series, Vol. 7, 1976.
16. Hawk, G. L., Ed., Champlin, P. B., Jordi, H. C., and Wenke, D., Assoc. Ed., *Biological/Biomedical Applications of Liquid Chromatography,* Chromatographic Science Series, Vol. 10, 1979.
17. Novak, J., *Quantitative Analysis by Gas Chromatography,* Chromatographic Science Series, Vol. 5, 1976.
18. Rajcsanyi, P. M. and Rajcsanyi, E., *High-Speed Liquid Chromatography,* 1975.

Pergamon Press Inc., Maxwell House, Fairview Park, Elmsford, N.Y. 10523.
Pergamon Press Ltd., Headington Hill Hall, Oxford OX3 OBW, England.

1. Andele, P., *Four-Language Technical Dictionary of Chromatography,* 1970.
2. Jeffery, P. G. and Kipping, P. J., *Gas Analysis by Gas Chromatography,* 1972.

The Perkin-Elmer Corporation, Norwalk, Connecticut 06856.

1. Ettre, L. S., *Introduction to Open Tubular Columns,* 1978.
2. Rhys Williams, A. T., *Fluorescence Detection in Liquid Chromatography,* 1981.

Plenum Publishing Corporation, 227 West 17th Street, New York, N.Y. 10011.

1. Frei, R. W. and Lawrence, J. F., Ed., *Chemical Derivatization in Analytical Chemistry, Vol. 1, Chromatography,* 1981.
2. Perry, S. G., Amos, R., and Brewer, P. I., *Practical Liquid Chromatography,* 1972.
3. Signeur, A. V., *Guide to Gas Chromatography Literature,* Vol. 4, 1979.

Polyscience Corporation, 6366 Gross Point Road, Niles, Illinois 60648.

1. Preston, S. T., *A Guide to the Analysis of Amines by Gas Chromatography,* 1973.

Prentice-Hall Inc., Englewood Cliffs, N.J. 07632.

1. Khm, J. X., *Analytical Ion-exchange Procedures in Chemistry and Biology,* (Theory, equipment, techniques), 1975.

Preston Publications Inc., P.O. Box 312, 6366 Gross Point Road, Niles, Illinois 60648.

1. Gas Chromatography Literature Abstracts and Index. Published Monthly.
2. Liquid Chromatography Literature, Abstracts and Index. Published bi-monthly.
3. Kovats, E., Ed., *5th International Symposium on Separation Methods; Column Chromatography,* 1970.
4. Struppe, H. G., Ed., *Aspects in Gas Chromatography,* 1971.
5. Walker, J. W., Ed., *Chromatographic Systems — Problems and Solutions,* 1980.

Regis Chemical Co., 8210 Austin Avenue, Morton Grove, Illinois 60053.

1. A User's Guide to Chromatography, Gas, Liquid, TLC, 1976.

Supelco Inc., Bellefonte, Pennsylvania.

1. Supina, W. R., *The Packed Column in Gas Chromatography,* 1974.

Van Nostrand-Reinhold Co., 450 West, 33rd Street, New York, N.Y. 10001.

1. Heftmann, E., Ed., *Chromatography: A Laboratory Handbook of Chromatographic and Electrophoretic Methods,* 3rd ed., 1975.

Varian Instrument Group, 220 Humboldt Court, Sunnyvale, California 94086.

1. GC Applications Library, 1969-1975, 1977.
2. Hadden, N., Baumann, F., MacDonald, F., Munk, M., Stevenson, R., Gere, D., Zamaroni, F., and Majors, R., *Basic Liquid Chromatography,* 1971.
3. Johnson, E. L. and Stevenson, R., *Basic Liquid Chromatography,* 1978.

John Wiley & Sons Inc., 605 Third Avenue, New York, N.Y. 10016. Wiley/Interscience - a division of John Wiley & Sons Inc.

1. David, D. J., *Gas Chromatographic Detectors,* 1974.
2. Denney, R. C., *A Dictionary of Chromatography,* 1976.
3. Done, J. N., Knox, J. H., and Loheac, J., *Applications of High-Speed Liquid Chromatography,* 1974.
4. Grob, R. L., Ed., *Modern Practice of Gas Chromatography,* 1977.
5. Hamilton, R. J. and Sewell, P. A., *Introduction to High Performance Liquid Chromatography,* 1977.
6. Kirchner, J. G. and Perry, E. S., Ed., *Thin-Layer Chromatography,* 2nd ed., 1978.
7. Knapp, D. R., *Handbook of Analytical Derivatization Reactions,* 1979.
8. Kremmer, T. and Boross, L., *Gel Chromatography: Theory, Methodology, Applications,* 1979.
9. McFadden, W. G., *Techniques in Combined Gas Chromatography/Mass Spectrometry: Applications in Organic Analysis,* 1973.
10. Mikes, O., Ed., *Laboratory Handbook of Chromatography and Allied Methods,* (Translated from the Czech), Chalmers, R. A., Translation Ed., 1979.

11. Miller, J. M., *Separation Methods in Chemical Analysis*, 1975.
12. Scott, R. P. W., *Contemporary Liquid Chromatography. Techniques of Chemistry*, Vol. 11, Weissberger, A., Series Ed., 1976.
13. Snyder, L. R. and Kirkland, J. J., *Introduction to Modern Liquid Chromatography*, 2nd ed., 1980.
14. Touchstone, J. C. and Dobbins, M. F., *Practice of Thin-Layer Chromatography*, 1978.
15. Touchstone, J. C., Ed., *Quantitative Thin-Layer Chromatography*, 1973.
16. Touchstone, J. C. and Rogers, D., Ed., *Thin-Layer Chromatography. Quantitative Environmental and Clinical Applications*, 1980.
17. Touchstone, J. C. and Sherma, J., Ed., *Densitometry in Thin-Layer Chromatography; Practice and Applications*, 1979.
18. Yau, W. W., Kirkland, J. J., and Bly, D. D., *Modern Size-Exclusion Liquid Chromatography: Practice of Gel Permeation and Gel Filtration Chromatography*, 1979.

Section VII
Reviews of Chromatographic Methods and Equipment

Section VII

REVIEWS OF CHROMATOGRAPHIC METHODS AND EQUIPMENT

This selected list of review articles supplements information that may be obtained by consulting appropriate books cited in the Book Directory.

1. Abbot, S. R., Practical aspects of normal-phase chromatography, *J. Chromatogr. Sci.*, 18, 540, 1980.
2. Bidlingmeyer, B. A., Separation of ionic compounds by reversed-phase liquid chromatography: an update of ion-pairing techniques, *J. Chromatogr. Sci.*, 18, 525, 1980.
3. Brown, P. R. and Krstulovic, A. M., Practical aspects of reversed-phase liquid chromatography applied to biochemical and biomedical research, *Anal. Biochem.*, 99, 1, 1979.
4. Chandler, C. D. and McNair, H. M., High pressure liquid chromatography equipment, *J. Chromatogr. Sci.*, 11, 468, 1973.
5. Colin, H. and Guichon, G., Introduction to reversed-phase high-performance liquid chromatography, *J. Chromatogr.*, 141, 289, 1977.
6. Cooke, N. H. C. and Olsen, K., Some modern concepts in reversed-phase liquid chromatography on chemically bonded alkyl stationary phases, *J. Chromatogr. Sci.*, 18, 512, 1980.
7. Davankov, V. A. and Semechkin, A. V., Ligand-exchange chromatography, *J. Chromatogr.*, 141, 313, 1977.
8. Deyl, Z., Advances in separation techniques in sequence analysis of proteins and peptides, *J. Chromatogr.*, 127, 91, 1976.
9. Deyl, Z., and Kopecky, J., Ed., Bibliography of liquid column chromatography 1971-1973 and survey of applications, Supplementary Volume No. 6, 1976 to the *J. Chromatogr.*
10. Drozd, J., Chemical Derivatization in gas chromatography, *J. Chromatogr.*, 113, 303, 1975.
11. Farwell, S. O., Gage, D. R., and Kagel, R. A., Current status of prominent selective gas chromatographic detectors; a critical assessment, *J. Chromatogr. Sci.*, 19, 358, 1981.
12. Haken, J. K., Polysiloxane stationary phases in gas chromatography, *J. Chromatogr.*, 141, 247, 1977.
13. Hurtubise, R. J., Lott, P. F., and Dias, J. R., Instrumentation of thin layer chromatography, *J. Chromatogr. Sci.*, 11, 476, 1973.
14. Ishii, D. and Takeuchi, T., Open tubular capillary LC, *J. Chromatogr. Sci.*, 18, 462, 1980.
15. Kirchner, J. G., Modern techniques in TLC, *J. Chromatogr. Sci.*, 13, 558, 1975.
16. Kipiniak, W., A basic problem — the measurement of height and area, *J. Chromatogr. Sci.*, 19, 332, 1981.
17. Little, C. J., Whatley, J. A., and Dale, A. D., Detection involving post-chromatographic addition of reagents, *J. Chromatogr.*, 171, 63, 1979.
18. Lochmüller, C. H. and Souter, R. A., Chromatographic resolution of enantiomers. Selective review, *J. Chromatogr.*, 113, 283, 1975.
19. Lott, P. F., Dias, J. R., and Hurtubise, R. J., Instrumentation for thin layer chromatography — an update, *J. Chromatogr. Sci.*, 14, 488, 1976.
20. Lott, P. F., Dias, J. R., and Slakck, S. C., Instrumentation for thin layer chromatography. 1978 update, *J. Chromatogr. Sci.*, 16, 571, 1978.
21. Macek, K., Hais, I. M., Kopecky, J., Schwarz, V., Gasparic, J., and Churacek, J., Bibliography of paper and thin-layer chromatography 1970-1973 and survey of applications. Supplementary Volume No. 5, 1976 to the *J. of Chromatogr.*
22. McNair, H. M. and Chandler, C. D., Gas chromatography equipment, *J. Chromatogr. Sci.*, 11, 454, 1973.
23. McNair, H. M., Gas chromatography equipment, *J. Chromatogr. Sci.*, 16, 578, 1978.
24. McNair, H. M. and Chandler, C. D., High pressure liquid chromatography equipment - II, *J. Chromatogr. Sci.*, 12, 425, 1974.
25. McNair, H. M. and Chandler, C. D., High performance liquid chromatography equipment - III, *J. Chromatogr. Sci.*, 14, 477, 1976.
26. McNair, H. M., High performance liquid chromatography equipment - IV, *J. Chromatogr. Sci.*, 16, 588, 1978.
27. Majors, R. E., High performance liquid chromatography columns and column technology. A state-of-the-art review. Part I, *J. Chromatogr. Sci.*, 18, 393, 1980.; Part II, *J. Chromatogr. Sci.*, 18, 487, 1980.
28. Majors, R. E., Recent advances in HPLC packings and columns, *J. Chromatogr. Sci.*, 18, 488, 1980.
29. Novotny, M., Capillary HPLC; columns and related instrumentation, *J. Chromatogr. Sci.*, 18, 473, 1980.

30. Pohl, C. A. and Johnson, E. L., Ion chromatography — the state-of-the-art, *J. Chromatogr. Sci.,* 18, 442, 1980.
31. Reese, C. E. and Scott, R. P. W., Microbore LC column technology, *J. Chromatogr Sci.,* 18, 479, 1980.
32. Ross, M. S. F., Pre-column derivatisation in high-performance liquid chromatography, *J. Chromatogr.,* 141, 107, 1977.
33. Seiler, N., Chromatography of biogenic amines. I. Generally applicable separation and detection methods, *J. Chromatogr.,* 143, 221, 1977.
34. Scott, C. G., Quantitation in chromatography — introduction, *J. Chromatogr. Sci.,* 19, 331, 1981.
35. Subcommittee E - 19.08 Task Group on Liquid Chromatography of the American Society for Testing and Materials (ASTM). An Evaluation of Quantitative Precision in High Performance Liquid Chromatography, *J. Chromatogr. Sci.,* 19, 338, 1981.
36. Verzele, M. and Geeraert, E., Preparative liquid chromatography, *J. Chromatogr. Sci.,* 18, 571, 1980.
37. Wolf, T., Fritz, G. T., and Palmer, L. R., An ASTM standard practice for testing fixed-wavelength photometric detectors used in liquid chromatography, *J. Chromatogr. Sci.,* 19, 387, 1981.
38. Wood, R., Cummings, L., and Jupille, T., Recent developments in ion-exchange chromatography, *J. Chromatogr. Sci.,* 18, 551, 1980.

Section VIII
Indexes

AUTHOR INDEX

A

Abbot, S. R., 267 (ref. 1)
Abe, I., 7 (ref. 1)
Abrahamsson, M., 86 (ref. 1)
Adams, R. F., 32 (ref. 1), 162, 163 (ref. 3)
Adlakha, R. C., 99 (ref. 2, 3)
Allsop, R. T., 260 (ref. 1)
Amico, V., 20 (ref. 1)
Amos, R., 262 (ref. 2)
Andele, P., 261 (ref. 1)
Anderson, R. A., 156 (ref. 24)
Ando, T., 37 (ref. 1)
Appella, C., 147 (ref. 2)
Armstrong, M. D., 110 (ref. 1)
Atkin, G. E., 156 (ref. 20, 32), 157, 158 (ref. 1)
Ayoub, L., 173 (ref. 32)

B

Bada, J. L., 167 (ref. 3)
Baltimore, B. G., 64 (ref. 1)
Bannard, R. A. B., 38 (ref. 1)
Bardocz, S., 127 (ref. 1)
Barker, H. A., 64 (ref. 1)
Barton, H., 110 (ref. 1)
Baumann, F., 262 (ref. 2)
Bayer, E., 23 (ref. 1), 24 (ref. 1), 26 (ref. 1), 167 (ref. 8, 11, 12)
Beecher, G. R., 173 (ref. 31)
Beitler, U., 11 (ref. 1), 12 (ref. 1), 13 (ref. 1)
Bellstedt, K., 128 (ref. 1)
Bengtsson, G., 161, 163 (ref. 2), 202 (ref. 1)
Benson, J. R., 173 (ref. 27)
Benson, J. V., 143 (ref. 1, 5), 146 (ref. 1), 156 (ref. 22), 157, 158 (ref. 5)
Berezin, I. V., 173 (ref. 21)
Bernard, C. W., 173 (ref. 5)
Bertsch, W., 261 (ref. 1)
Bidlingmeyer, B. A., 267 (ref. 2)
Birrell, P., 167 (ref. 6)
Bishop, C. A., 74 (ref. 1), 156 (ref. 36)
Bjorkqvist, B., 218 (ref. 1)
Blackburn, S., 261 (ref. 1)
Blair, K., 260 (ref. 1)
Blair, N. E., 166 (ref. 14), 167 (ref. 14)
Blau, K., 56 (ref. 2), 212 (ref. 1)
Bly, D. D., 263 (ref. 18)
Bogue, D. C., 156 (ref. 24)
Bohlen, P., 171 (ref. 20), 173 (ref. 11, 12, 29)
Boila, R. J., 201 (ref. 1)
Boisson, D., 29 (ref. 1), 160, 163 (ref. 1)
Bongiovanni, G., 146 (ref. 2)
Bonner, W. A., 166 (ref. 14), 167 (ref. 14, 15)
Borisov, I. L., 173 (ref. 21)
Boross, L., 262 (ref. 8)
Bowie, L., 68 (ref. 1)

Bozler, G., 64 (ref. 1)
Brewer, P. I., 262 (ref. 2)
Brooks, C. J. W., 54 (ref. 1)
Brown, P. R., 259 (ref. 1), 261 (ref. 9, 10, 11, 12, 13), 267 (ref. 3)
Brown, W. C., 93 (ref. 1)
Brown, W. E., 200 (ref. 9)
Brugger, M., 183
Bucher, D. J., 173 (ref. 14, 30)
Budde, W. L., 259 (ref. 1)
Butts, W. C., 35 (ref. 1), 53 (ref. 1), 215 (ref. 1)

C

Cancalon, P., 41 (ref. 1)
Cantera-Soler, A. M., 99 (ref. 2)
Cashaw, J. L., 52 (ref. 1)
Casselman, A. A., 38 (ref. 1)
Castensson, S., 86 (ref. 1)
Catlin, J. S., 200 (ref. 5)
Cazes, J., 261 (ref. 2, 7, 8, 9, 10, 11, 12, 13)
Champlin, P. B.. 261 (ref. 16)
Chandler, C. D., 267 (ref. 4, 22, 24, 25)
Chang, C. H., 173 (ref. 29)
Chang, D., 18 (ref. 3)
Chang, J. Y., 119 (ref. 1), 210 (ref. 1)
Chang, Y. H., 155 (ref. 6), 200 (ref. 1)
Charles, R., 6 (ref. 1), 51 (ref. 1)
Charles-Sigler, R., 167 (ref. 4, 17)
Chen, C. M., 100 (ref. 1)
Chen, E. C. M., 216 (ref. 1)
Cheng, C. N., 167 (ref. 18), 203 (ref. 2)
Chiavari, G., 40 (ref. 1)
Chimiak, A., 112 (ref. 1)
Chivari, G., 215 (ref. 1)
Churacek, J., 260 (ref. 7), 267 (ref. 21)
Clark, C. R., 216 (ref. 1)
Clarke, D. D., 163 (ref. 10)
Coffey, J. W., 144 (ref. 6)
Cohen-Solal, L., 155 (ref. 5)
Colin, H., 267 (ref. 5)
Cooke, N. H. C., 267 (ref. 6)
Corbin, J. A., 15 (ref. 1), 17 (ref. 1), 50 (ref. 1), 51 (ref. 1)
Cossu, P., 106 (ref. 1)
Crane, A. B., 155 (ref. 2)
Crawhall, J. C., 68 (ref. 1)
Creaser, E. H., 119 (ref. 1), 210 (ref. 1)
Cronin, J. R., 171 (ref. 26), 173 (ref. 25, 26)
Cronin, S. E., 167 (ref. 18), 203 (ref. 2)
Cummings, L., 268 (ref. 38)
Cunningham, T. D., 102 (ref. 1), 217 (ref. 1)
Currie, B. L., 80 (ref. 1)

D

Dairman, W., 173 (ref. 11, 12)

Dale, A. D., 267 (ref. 17)
Datko, A. H., 106 (ref. 1)
Davankov, V. A., 69 (ref. 1), 70 (ref. 1), 71 (ref. 1), 72 (ref. 1), 73 (ref. 1), 267 (ref. 7)
Daves, G. D., Jr., 18 (ref. 3)
David, D. J., 262 (ref. 1)
Davies, G., 81 (ref. 1), 82 (ref. 1)
Davis, B. A., 134 (ref. 1)
Davis, C. M., 163 (ref. 5)
Davis, T. P., 102 (ref. 1), 217 (ref. 1)
Davis, V. E., 52 (ref. 1)
Day, R., 203 (ref. 3)
De Bernardo, S. L., 173 (ref. 15)
Decuypere, J. A., 99 (ref. 1)
De Klerk, H., 155 (ref. 10)
Delamare, X., 261 (ref. 2)
De Marco, C., 106 (ref. 1)
Denney, R. C., 262 (ref. 2)
Dernini, S., 106 (ref. 1)
Desgres, J., 29 (ref. 1), 160, 163 (ref. 1)
Desideri, P. G., 125 (ref. 1), 126 (ref. 1), 130 (ref. 1), 131 (ref. 1), 132 (ref. 1)
Deyl, Z., 67 (ref. 1), 79 (ref. 1), 260 (ref. 1, 2), 267 (ref. 8, 9)
Dias, J. R., 267 (ref. 13, 19, 20)
Dierick, N. A., 99 (ref. 1)
Dixon, P. F., 259 (ref. 2)
Dobbins, M. F., 179, 263 (ref. 14)
Dochi, P. S., 211 (ref. 1)
Dohm, G. L., 173 (ref. 31)
Doi, T., 29 (ref. 1)
Domsky, I. I., 261 (ref. 3)
Done, J. N., 260 (ref. 1), 262 (ref. 3)
Donike, M., 212 (ref. 1), 213 (ref. 1)
Doshi, P. S., 56 (ref. 1, 3)
Dostovalov, I. N., 73 (ref. 1)
Drawert, F., 110 (ref. 1)
Drescher, D. G., 173 (ref. 19, 28), 201 (ref. 1)
Drozd, J., 267 (ref. 10)
Dua, V. K., 122 (ref. 1)
Dus, K., 155 (ref. 10)
Dwulet, F. E., 120
Dziewiatkowski, D. D., 68 (ref. 1)

E

Edwards, D. J., 56 (ref. 1, 2, 3), 211 (ref. 1), 212 (ref. 1)
Efron, K., 146 (ref. 4)
Eichelberger, J. W., 259 (ref. 1)
Eichhorn, M. M., 155 (ref. 15)
El Din Awad, A. M., 105 (ref. 1)
El Din Awad, O. M., 105 (ref. 1)
Ellis, J. P., Jr., 156 (ref. 29)
Endo, Y., 183
Ersser, R. S., 146 (ref. 3)
Ettre, L. S., 260 (ref. 14), 261 (ref. 1)
Eyem, J., 25 (ref. 1), 28 (ref. 2), 164 (ref. 20), 167 (ref. 2), 205 (ref. 1)

F

Farquharson, R. A., 100 (ref. 1), 216 (ref. 1)
Farwell, S. O., 267 (ref. 11)
Faupel, M., 183
Feibush, B., 11 (ref. 1), 12 (ref. 1), 13 (ref. 1), 14 (ref. 1), 167 (ref. 4, 5, 9, 17)
Felix, A. M., 173 (ref. 16, 17), 179
Felker, P., 164 (ref. 18)
Fell, A. F., 260 (ref. 1)
Felstead, G., 143 (ref. 7), 156 (ref. 26)
Felt, V., 206 (ref. 1)
Fenimore, D. C., 159, 163 (ref. 5)
Ferdinand, W., 156 (ref. 20, 32), 157, 158 (ref. 1)
Ferenczi, R., 155 (ref. 8), 200 (ref. 4)
Ferenczi, S., 111 (ref. 1)
Fischer, L., 260 (ref. 3)
Folkers, K., 18 (ref. 3)
Fong, G. W. K., 84 (ref. 1)
Forde, M. D., 173 (ref. 14)
Franc, J., 183
Francke, W., 44 (ref. 1)
Frank, H., 26 (ref. 1), 167 (ref. 11, 12)
Freeman, I. L., 144 (ref. 4)
Frei, R. W., 260 (ref. 6), 262 (ref. 1)
Frick, W., 18 (ref. 3)
Fried, B., 261 (ref. 4)
Frigerio, A., 260 (ref. 4)
Fritz, G. T., 268 (ref. 37)
Fujimara, K., 37 (ref. 1)
Furukawa, H., 91 (ref. 1, 2), 208 (ref. 1)

G

Gage, D. R., 267 (ref. 11)
Galaev, I. J., 173 (ref. 21)
Gandy, W. E., 173 (ref. 26)
Ganno, S., 156 (ref. 34), 173 (ref. 33)
Garcia, J. B., Jr., 156 (ref. 29)
Gardner, W. S., 209 (ref. 2)
Garner, M. H., 155 (ref. 12)
Garner, W. H., 155 (ref. 12)
Gasparic, J., 260 (ref. 7), 267 (ref. 21)
Gavilanes, J. G., 182
Geeraert, E., 268 (ref. 36)
Gehrke, C. W., 102 (ref. 1), 159, 163 (ref. 6, 8, 11, 12, 14), 164 (ref. 15), 217 (ref. 1)
Gehrke, C. W., Jr., 102 (ref. 1), 217 (ref. 1)
Georgiadis, A. G., 144 (ref. 6)
Gere, D., 262 (ref. 2)
Gerhardt, K. O., 102 (ref. 1), 163 (ref. 6), 217 (ref. 1)
Giddings, J. C., 261 (ref. 5, 6, 7, 8, 9, 10, 11, 12, 13)
Gil-Av, E., 6 (ref. 1), 14 (ref. 1), 51 (ref. 1), 76 (ref. 1), 167 (ref. 4, 5, 6, 17)
Gilbert, J. D., 54 (ref. 1)
Gilbert, M. T., 54 (ref. 1), 260 (ref. 1)

Giovanelli, J., 106 (ref. 1)
Gitlow, S. E., 163 (ref. 10)
Giumanini, A. G., 40 (ref. 1), 215 (ref. 1)
Glimcher, M. J., 155 (ref. 5)
Gochman, N., 68 (ref. 1)
Goldberg, W. M., 120 (ref. 1)
Goto, J., 32 (ref. 1), 33 (ref. 1), 89 (ref. 1), 205 (ref. 1), 208 (ref. 1)
Gotto, A. M., Jr., 177
Gra, C. H., 259 (ref. 2)
Gracy, R. W., 179, 182
Grob, R. L., 262 (ref. 4)
Groningsson, K., 86 (ref. 1)
Grossman, D., 223 (ref. 4)
Grushka, E., 84 (ref. 1), 259 (ref. 2), 261 (ref. 7, 8, 9, 10, 11, 12, 13)
Gubitz, G., 78 (ref. 1)
Gudzinowicz, B. J., 261 (ref. 14, 15)
Gudzinowicz, M. J., 261 (ref. 15)
Gudzinowicz, W. J., 261 (ref. 14)
Guichon, G., 267 (ref. 5)
Guire, P., 144 (ref. 3)
Gurd, F. R. N., 120, 155 (ref. 12)
Guyer, M., Jr., 183

H

Haase, L., 128 (ref. 1)
Hadden, N., 262 (ref. 2)
Hadzija, O., 63 (ref. 1)
Hais, I. M., 260 (ref. 7), 267 (ref. 21)
Hakanson, R., 188 (ref. 1)
Haken, J. K., 267 (ref. 12)
Hall, N. T., 18 (ref. 1), 164 (ref. 19)
Hamano, T., 53 (ref. 1), 214 (ref. 1, 2)
Hamilton, J. G., 144 (ref. 6)
Hamilton, K. R., 155 (ref. 4)
Hamilton, P. B., 155 (ref. 13), 156 (ref. 24, 25), 157, 158 (ref. 2), 199 (ref. 8), 200 (ref. 6, 7, 8)
Hamilton, R. J., 262 (ref. 5)
Hampai, A., 170 (ref. 18), 173 (ref. 18)
Hancock, W. S., 74 (ref. 1), 156 (ref. 36)
Haney, W. G., 83 (ref. 1)
Hanin, I., 56 (ref. 3)
Hansen, G. R., 179
Hansen, R., 144 (ref. 1)
Hapner, K. D., 155 (ref. 4)
Hara, S., 261 (ref. 1)
Haraguchi, K., 101 (ref. 1), 218 (ref. 1)
Hare, P. E., 76 (ref. 1), 143 (ref. 2), 156 (ref. 27), 171, 173 (ref. 25, 27)
Harmeyer, J., 172 (ref. 32), 173 (ref. 32)
Harris, J. U., 88 (ref. 1), 156 (ref. 37)
Hartkopf, A., 223 (ref. 4)
Hascall, V., 68 (ref. 1)
Hasegawa, A., 214 (ref. 1)
Hasegawa, M., 89 (ref. 1), 208 (ref. 1)
Hawk, G. L., 261 (ref. 16)
Hawkes, S., 223 (ref. 4)

Hayashi, T., 216 (ref. 1)
Healey, J. A. D., 260 (ref. 1)
Hearn, M. T., 74 (ref. 1), 156 (ref. 36)
Heftmann, E., 262 (ref. 1)
Heimer, E. P., 104 (ref. 1)
Heimler, D., 125 (ref. 1), 126 (ref. 1), 130 (ref. 1), 131 (ref. 1), 132 (ref. 1)
Heitefuss, R., 156 (ref. 30)
Helm, R., 67 (ref. 1)
Hempel, K., 156 (ref. 35)
Henderickx, H. K., 99 (ref. 1)
Herbst, M. M., 64 (ref. 1)
Hill, D. W., 147 (ref. 1), 209 (ref. 1)
Hinshaw, J. V., Jr., 47 (ref. 1), 48 (ref. 1)
Hixson, C. V., 99 (ref. 4)
Hood, L. E., 147 (ref. 3), 156 (ref. 38)
Hoopes, E. A., 167 (ref. 3)
Horhold, H. H., 128 (ref. 1)
Horiba, M., 5 (ref. 1)
Horvath, C., 152, 156 (ref. 39), 259 (ref. 3, 4)
Hoshika, J., 39 (ref. 1, 2)
Hoshika, Y., 214 (ref. 1, 2, 3, 4)
Houpert, Y., 146 (ref. 6)
Howard, G. C., 200 (ref. 9)
Howard, P. Y., 9 (ref. 1)
Hsu, K. T., 80 (ref. 1)
Huber, J. F. K., 260 (ref. 5)
Hudson, B. G., 144 (ref. 3)
Hughes, G. J., 119 (ref. 1)
Hugli, T. E., 155 (ref. 9)
Hunkapiller, M. W., 147 (ref. 3), 156 (ref. 38)
Hurtubise, R. J., 267 (ref. 13, 19)
Husek, P., 206 (ref. 1)

I

Ikegawa, S., 208 (ref. 1)
Imai, K., 173 (ref. 29)
Inglis, A. S., 201 (ref. 2)
Irreverre, F., 98 (ref. 1)
Isenhour, T. L., 223 (ref. 3, 4)
Ishii, D., 267 (ref. 14)
Ishizuka, T., 92 (ref. 1)
Ito, K., 91 (ref. 1, 2), 208 (ref. 1)
Iwata, T., 32 (ref. 1), 33 (ref. 1), 205 (ref. 1)

J

Jackson, M. T., Jr., 259 (ref. 8)
James, L. B., 173 (ref. 8, 9)
Janak, J., 260 (ref. 2)
Jazai, I., 111 (ref. 1)
Jeffery, P. G., 261 (ref. 2)
Jellenz, W., 78 (ref. 1)
Jennings, W., 259 (ref. 5)
Jiminez, M. H., 179
Johnson, A. J., 8 (ref. 1), 156 (ref. 37)
Johnson, D. F., 173 (ref. 22, 23, 24)
Johnson, E. L., 262 (ref. 3), 268 (ref. 30)

Johnson, H. D., 102 (ref. 1), 217 (ref. 1)
Johnson, K., 68 (ref. 1)
Johnson, N. D., 147 (ref. 3), 156 (ref. 38)
Johnson, R. D., 167 (ref. 1)
Jones, D., 144 (ref. 1)
Jones, G. R. N., 193
Jonsson, J., 28 (ref. 2), 164 (ref. 20)
Jordi, H. C., 261 (ref. 16)
Jungclaus, G. A., 42 (ref. 1), 215 (ref. 1)
Jupille, T., 268 (ref. 38)
Justice, J. B., 223 (ref. 3)

K

Kagel, R. A., 267 (ref. 11)
Kai, M., 101 (ref. 1), 218 (ref. 1)
Kaiser, F. E., 164 (ref. 15)
Kaiser, R. E., 260 (ref. 16), 261 (ref. 1,2)
Kamei, A., 91 (ref. 2)
Kamen, M. D., 155 (ref. 10)
Karger, B. L., 81 (ref. 1), 82 (ref. 1, 2), 94 (ref. 1), 152, 156 (ref. 39), 207 (ref. 1)
Karsai, T., 127 (ref. 1)
Kasahara, Y., 77 (ref. 1)
Kawai, S., 216 (ref. 1)
Kedenburg, C. P., 146 (ref. 5)
Keglevic, D., 63 (ref. 1)
Keller, R. A., 261 (ref. 5, 6, 7, 8)
Khm, J. X., 262 (ref. 1)
Khulbe, K. C., 134 (ref. 1)
King, G., 260 (ref. 1)
Kinoshita, T., 77 (ref. 1), 90 (ref. 1), 178, 190 (ref. 1), 209 (ref. 1)
Kinsey, S., 203 (ref. 3)
Kipiniak, W., 267 (ref. 16)
Kipping, P. J., 261 (ref. 2)
Kirchner, J. G., 262 (ref. 6), 267 (ref. 15)
Kirkland, J. J., 263 (ref. 13, 18)
Kirkman, M. A., 164 (ref. 16)
Kitahara, H., 5 (ref. 1), 29 (ref. 1)
Klemm, D., 128 (ref. 1)
Klimisch, H. J., 219 (ref. 1)
Klingman, J. D., 41 (ref. 1)
Knapman, C. E. H., 259 (ref. 1, 2)
Knapp, D. R., 262 (ref. 7)
Knox, J. H., 260 (ref. 1), 262 (ref. 3)
Koni, M., 204 (ref. 1)
Konig, W. A., 25 (ref. 1), 30 (ref. 1), 44 (ref. 1, 2), 167 (ref. 2, 7, 16), 205 (ref. 1)
Kopecky, J., 260 (ref. 1, 7), 267 (ref. 9, 21)
Kossiva, D., 155 (ref. 5)
Koudelkova, V., 183
Kovacs, K., 155 (ref. 8), 200 (ref. 4)
Kovats, E., 262 (ref. 3)
Kremmer, T., 262 (ref. 8)
Krstulovic, A. M., 267 (ref. 3)
Kruse, K., 30 (ref. 1), 205 (ref. 1)
Kruse, W., 44 (ref. 1)
Kuo, K. C., 102 (ref. 1), 163 (ref. 6, 11), 164 (ref. 14, 15), 217 (ref. 1)

Kuwata, K., 36 (ref. 1)

L

Ladas, A. S., 259 (ref. 7)
Lai, C. Y., 180
Lamberts, B. L., 144 (ref. 2)
Larsson, L. I., 188 (ref. 1)
Lawrence, J. F., 259 (ref. 6), 260 (ref. 6), 262 (ref. 1)
Leary, J., 223 (ref. 3, 4)
Lee, H. M., 172 (ref. 30), 173 (ref. 14, 30)
Lee, K. S., 173 (ref. 19, 28), 201 (ref. 1)
Lee, M. C., 173 (ref. 14)
Lee, P. L. Y., 59 (ref. 1)
Leimer, K., 163 (ref. 12)
Leimgruber, W., 173 (ref. 11, 15)
LePage, J. N., 81 (ref. 1), 82 (ref. 1)
Lepri, L., 125 (ref. 1), 126 (ref. 1), 130 (ref. 1), 131 (ref. 1), 132 (ref. 1)
Levenson, S. M., 173 (ref. 5)
Lian, J. B., 155 (ref. 5)
Liao, T. H., 143 (ref. 3), 155 (ref. 11), 156 (ref. 18)
Lim, C. K., 259 (ref. 2)
Lin, J. K., 210 (ref. 1)
Lindeberg, E. G., 182, 187
Lindner, W., 82 (ref. 1), 207 (ref. 1)
Little, C. J., 267 (ref. 17)
Liu, T. Y., 200 (ref. 1, 2)
Lnge, H. W., 156 (ref. 35)
Lochmuller, C. H., 47 (ref. 1), 48 (ref. 1), 49 (ref. 1), 267 (ref. 18)
Loheac, J., 262 (ref. 3)
Longton, R. W., 144 (ref. 2)
Lott, P. F., 267 (ref. 13, 19, 20)
Lowry, S. R., 223 (ref. 3)
Lui, T. Y., 155 (ref. 6)
Lundeen, M., 259 (ref. 6)
Lustenberger, N., 156 (ref. 35)

M

Ma, T. S., 259 (ref. 7)
MacDonald, F., 262 (ref. 2)
Macek, K., 67 (ref. 1), 260 (ref. 2, 7), 267 (ref. 21)
MacKenzie, S. L., 159, 160, 163 (ref. 4), 164 (ref. 17), 204 (ref. 1)
Maeda, M., 156 (ref. 34), 172 (ref. 33), 173 (ref. 33)
Mague, T. H., 21 (ref. 1)
Majors, R. E., 262 (ref. 2), 267 (ref. 27, 28)
Makita, M., 204 (ref. 1)
Malhotra, S. S., 18 (ref. 2), 34 (ref. 2), 161, 163 (ref. 9)
Mancheva, I. N., 118 (ref. 1)
Mann, R. S., 134 (ref. 1)
Martin, 150

Martin, F. M., 261 (ref. 15)
Martin, H. F., 261 (ref. 14)
Martin, R. J., 41 (ref. 1), 214 (ref. 1)
Matsubara, H., 200 (ref. 3)
Matsuki, Y., 53 (ref. 1), 214 (ref. 1, 2)
Maynard, J. B., 259 (ref. 8)
McCamish, M., 260 (ref. 4)
McFadden, W. G., 262 (ref. 9)
McHugh, W., 83 (ref. 1)
McMahon, D. T. W., 201 (ref. 2)
McNair, H. M., 267 (ref. 4, 22, 23, 24, 25, 26)
McReynolds, W. O., 223 (ref. 2)
Melancon, S. B., (ref. 1)
Mellet, M., 171 (ref. 20), 173 (ref. 20)
Mendez, E., 180, 182
Metcalfe, L. D., 41 (ref. 1), 214 (ref. 1)
Metrione, R. M., 122 (ref. 1)
Michael, G., 206 (ref. 1)
Mikes, O., 262 (ref. 10)
Miles, B. J., 156 (ref. 31)
Miller, J. M., 263 (ref. 11)
Miller, N., 82 (ref. 2)
Miller, O. N., 144 (ref. 6)
Miller, S. L., 203 (ref. 3)
Miller, W. H., III, 209 (ref. 2)
Milligan, L. P., 201 (ref. 2)
Mitchell, S. C., 191 (ref. 1)
Mitsuhashi, Y., 53 (ref. 1), 214 (ref. 2)
Miyamoto, A. K., 202 (ref. 1), 203 (ref. 1)
Mizobuchi, Y., 75 (ref. 1)
Mondino, A., 146
Moodie, I. M., 164 (ref. 21)
Moore, S., 150, 152, 155 (ref. 9, 11, 16), 156 (ref. 23, 28), 173 (ref. 1, 2, 10)
Mori, Y., 91 (ref. 1), 208 (ref. 1)
Mudd, S. H., 106 (ref. 1)
Munk, M., 262 (ref. 2)
Murayama, K., 62 (ref. 1), 178, 190 (ref. 1)
Murren, C., 143 (ref. 7), 156 (ref. 26)
Musha, S., 7 (ref. 1)
Myasoedov, N. F., 73 (ref. 1)

N

Nagy, B., 200 (ref. 7)
Nagy, S., 18 (ref. 1), 164 (ref. 19)
Nair, B. M., 201 (ref. 1)
Nakamura, H., 113 (ref. 1), 114 (ref. 1), 115 (ref. 1), 117 (ref. 1), 133 (ref. 1), 135 (ref. 1), 136 (ref. 1), 179, 180, 181, 182, 189 (ref. 1), 211 (ref. 1)
Nakamura, K., 105 (ref. 1)
Nakamura, S., 208 (ref. 1)
Nakaparksin, S., 167 (ref. 6)
Nambara, T., 32 (ref. 1), 33 (ref. 1), 89 (ref. 1), 205 (ref. 1), 208 (ref. 1)
Navratil, J. D., 95 (ref. 1), 96 (ref. 1)
Neuberger, M. R., 200 (ref. 2)
Nicholson, G. J., 26 (ref. 1), 167 (ref. 7, 11, 12)
Niece, R. L., 58 (ref. 1), 155 (ref. 7), 173 (ref. 7)

Niederweiser, A., 259 (ref. 3, 4, 5)
Nimura, M., 209 (ref. 1)
Nimura, N., 77 (ref. 1), 90 (ref. 1)
Noe, V., 146 (ref. 2)
Novak, J., 261 (ref. 17)
Novotny, M., 267 (ref. 29)

O

Odham, G., 161, 163 (ref. 2), 202 (ref. 1)
Oelrich, E., 261 (ref. 2)
Ogata, T., 101 (ref. 1), 218 (ref. 1)
Ogura, H., 90 (ref. 1), 209 (ref. 1)
Ohkura, Y., 101 (ref. 1), 218 (ref. 1)
Ohno, T., 216 (ref. 1)
Oi, N., 5 (ref. 1), 29 (ref. 1)
Olsen, K., 267 (ref. 6)
Onishi, Y., 156 (ref. 34), 173 (ref. 33)
Oriente, G., 20 (ref. 1)
Oro, J., 155 (ref. 14), 167 (ref. 6)
Osborne, R. M., 144 (ref. 2)
Oulevey, J., 156 (ref. 30)
Oyama, V. I., 167 (ref. 1)

P

Padieu, P., 29 (ref. 1), 160, 163 (ref. 1)
Page, R. C., 144 (ref. 1)
Palmer, L. R., 268 (ref. 37)
Parcher, J., 223 (ref. 4)
Parr, W., 9 (ref. 1), 23 (ref. 1), 24 (ref. 1), 167 (ref. 8)
Parris, N. A., 260 (ref. 8)
Pataki, G., 259 (ref. 3, 4, 5)
Patthy, A., 111 (ref. 1)
Pattison, J. B., 261 (ref. 2)
Peacock, I., 173 (ref. 31)
Pearce, R. J., 28 (ref. 3), 164 (ref. 22)
Peltzer, E. T., 167 (ref. 3)
Penke, B., 155 (ref. 8), 200 (ref. 4)
Penkina, V. I., 73 (ref. 1)
Perini, F., 99 (ref. 4)
Perry, E. S., 262 (ref. 6)
Perry, J. A., 261 (ref. 3)
Perry, S. G., 262 (ref. 2)
Persson, B. A., 94 (ref. 1)
Petrenik, O. V., 73 (ref. 1)
Pisano, J. J., 113 (ref. 1), 114 (ref. 1), 115 (ref. 1), 117 (ref. 1), 133 (ref. 1), 136 (ref. 1), 147 (ref. 2, 4), 181, 182, 189 (ref. 1)
Pizzarello, S., 173 (ref. 26)
Pleterski, J., 23 (ref. 1), 24 (ref. 1), 167 (ref. 8)
Pocklington, R., 34 (ref. 1), 161, 164 (ref. 13)
Pohl, C. A., 268 (ref. 30)
Pollock, G. E., 167 (ref. 1, 18), 202 (ref. 1), 203 (ref. 1, 2, 3)
Polonski, T., 112 (ref. 1)
Prandi, C., 124 (ref. 1), 178
Preston, S. T., 223 (ref. 1), 262 (ref. 1)

Prochazka, Z., 158 (ref. 3)
Pryde, A., 260 (ref. 1)

R

Raffaele, I., 146
Rahm, J., 157, 158 (ref. 3, 4)
Rahn, W., 25 (ref. 1), 167 (ref. 2), 205 (ref. 1)
Rajcsanyi, E., 261 (ref. 18)
Rajcsanyi, P. M., 261 (ref. 18)
Reese, C. E., 268 (ref. 31)
Reese, M. W., 155 (ref. 1)
Rhoad, J. E., 17 (ref. 1)
Rhys Williams, A. T., 261 (ref. 2)
Rice, C. B. F., 260 (ref. 2)
Riguetti, P., 144 (ref. 3)
Rinaldi, A., 106 (ref. 1)
Riolo, R. L., 68 (ref. 1)
Robel, E. J., 155 (ref. 2, 3)
Roberts, T. R., 260 (ref. 9)
Robinson, D., 88 (ref. 1), 156 (ref. 37)
Robinson, G. W., 143 (ref. 3), 155 (ref. 17), 156 (ref. 18)
Rogers, D., 263 (ref. 16)
Rogers, L. B., 15 (ref. 1), 17 (ref. 1), 50 (ref. 1), 51 (ref. 1)
Rosen H., 173 (ref. 5)
Rosmus, J., 79 (ref. 1)
Ross, M. S. F., 268 (ref. 32)
Roth, M., 170 (ref. 18), 173 (ref. 18)
Rowland, F. R., 259 (ref. 1)
Roxburgh, C. M., 201 (ref. 2)

S

Sadow, J. B., 99 (ref. 4)
Saeed, T., 22 (ref. 1), 166 (ref. 13), 167 (ref. 13)
Sakakibara, E., 91 (ref. 2)
Sallmann, H. P., 173 (ref. 32)
Salnikow, J., 143 (ref. 3), 155 (ref. 11), 156 (ref. 18)
Samejima, K., 217 (ref. 1)
Sandmann, R. A., 83 (ref. 1)
Sandra, P., 22 (ref. 1), 167 (ref. 13)
Santi, W., 78 (ref. 1)
Sarkar, S. K., 18 (ref. 2), 34 (ref. 2), 161, 163 (ref. 9)
Sawicki, E., 183
Scarponi, F. L., 40 (ref. 1), 215 (ref. 1)
Schiltz, E., 179, 182
Schmidt, D. E., 207 (ref. 1)
Schmidt, G. J., 32 (ref. 1), 162, 163 (ref. 3)
Schnackerz, K. D., 179, 182
Schneider, J. A., 68 (ref. 1)
Schram, S. B., 261 (ref. 1)
Schubert, K., 128 (ref. 1)
Schulze, U., 44 (ref. 1)
Schwabe, C., 200 (ref. 5)
Schwarz, V., 260 (ref. 7), 267 (ref. 21)
Schwedt, G., 261 (ref. 3)
Schwerdtfeger, E., 173 (ref. 4)
Scott, C. G., 268 (ref. 34)
Scott, R. M., 259 (ref. 6)
Scott, R. P. W., 260 (ref. 10), 263 (ref. 12), 268 (ref. 31)
Segura, R., 177
Seid, R. C., Jr., 173 (ref. 30)
Seile, N., 268 (ref. 33)
Seitz, D. E., 81 (ref. 1), 82 (ref. 1)
Semechkin, A. V., 267 (ref. 7)
Sesaki, R. M., 200 (ref. 3)
Sevcik, J., 260 (ref. 11)
Sewell, P. A., 262 (ref. 5)
Shepherd, S. L., 156 (ref. 33)
Sherma, J., 179, 259 (ref. 9), 261 (ref. 4), 263 (ref. 17)
Shimada, K., 89 (ref. 1), 208 (ref. 1)
Shimizu, I., 194
Shindo, N., 62 (ref. 1)
Shono, T., 75 (ref. 1)
Siest, G., 146 (ref. 6)
Sievers, S., 44 (ref. 1, 2)
Siezen, R. L., 21 (ref. 1)
Signeur, A. V., 262 (ref. 3)
Simons, S. S., Jr., 173 (ref. 22, 23, 24)
Simpson, C. F., 260 (ref. 1), 261 (ref. 3)
Simpson, R. J., 200 (ref. 2)
Singh, V. P., 138 (ref. 1)
Sjoquist, J., 28 (ref. 2), 164 (ref. 20)
Skewes, H. B., 155 (ref. 14)
Slakck, S. C., 267 (ref. 20)
Sletten, K., 155 (ref. 10)
Small, H., 93 (ref. 1)
Smith, G. G., 159, 163 (ref. 7), 167 (ref. 10)
Snyder, L. R., 152, 156 (ref. 39), 263 (ref. 13)
Somack, R., 87 (ref. 1)
Sood, S. P., 83 (ref. 1)
Sorimachi, K., 67 (ref. 1)
Souter, R. A., 267 (ref. 18)
Souter, R. W., 45 (ref. 1), 47 (ref. 1), 49 (ref. 1)
Spackman, D. H., 170 (ref. 10), 173 (ref. 10)
Srivastava, S. P., 122 (ref. 1)
Stadler, L., 219 (ref. 1)
Stein, S., 173 (ref. 11, 12, 29)
Stein, W. H., 150, 152, 155 (ref. 11, 16), 156 (ref. 23, 28), 173 (ref. 1, 2, 10)
Stelling, D., 143 (ref. 7), 156 (ref. 26)
Stevens, T. S., 93 (ref. 1)
Stevenson, R., 262 (ref. 2), 262 (ref. 3)
Stock, R., 260 (ref. 2)
Stoll, M. S., 259 (ref. 2)
Stolting, K., 30 (ref. 1), 205 (ref. 1)
Stone, J., 173 (ref. 12)
Struppe, H. G., 262 (ref. 4)
Stuart, J. D., 147 (ref. 1), 209 (ref. 1)
Sugiura, T., 216 (ref. 1)
Sunahara, H., 92 (ref. 1)
Sundler, F., 188 (ref. 1)
Supina, W. R., 262 (ref. 1)
Suzuki, T., 77 (ref. 1)

Svedas, V. J. K., 171 (ref. 21), 173 (ref. 21)
Synge, 150

T

Tabor, C. W., 98 (ref. 1)
Tabor, H., 98 (ref. 1)
Taguchi, K., 32 (ref. 1), 33 (ref. 1), 205 (ref. 1)
Takahashi, S., 173 (ref. 6)
Takayanagi, H., 201 (ref. 2)
Takeda, H., 163 (ref. 8)
Takeuchi, T., 208 (ref. 1), 267 (ref. 14)
Takeuchi, Y., 91 (ref. 1)
Tamura, Z., 117 (ref. 1), 179, 180, 189 (ref. 1)
Tanaka, K., 92 (ref. 1)
Tanaka, M., 75 (ref. 1)
Tapuhi, Y., 82 (ref. 2), 207 (ref. 1)
Tarallo, P., 146 (ref. 6)
Tayco, J., 60 (ref. 1)
Tenaschuk, D., 159, 160, 163 (ref. 4), 164 (ref. 17), 204 (ref. 1)
Terkelsen, G., 173 (ref. 16, 17)
Thomson, A. R., 156 (ref. 31)
Tomita, T., 105 (ref. 1)
Touchstone, J. C., 179, 263 (ref. 14, 15, 16, 17)
Tringali, C., 20 (ref. 1)
Tsuge, S., 223 (ref. 3)
Tsuji, A., 156 (ref. 34), 173 (ref. 33)
Tsuzuki, S., 179
Turkova, J., 260 (ref. 12)
Tyihak, E., 111 (ref. 1)

U

Uchytil, B., 128 (ref. 1), 129 (ref. 1)
Uchytil, I., 181
Udenfriend, S., 170 (ref. 11), 173 (ref. 11, 12, 29)
Uebori, M., 36 (ref. 1)
Ui, N., 67 (ref. 1)
Unger, K. K., 260 (ref. 13)
Upadhyay, R. K., 138 (ref. 1)

V

Vajpai, U., 138 (ref. 1)
Vancikova, O., 67 (ref. 1)
Vandemark, F. L., 32 (ref. 1), 162, 163 (ref. 3)
Van Dort, M. A., 166 (ref. 15), 167 (ref. 15)
Vergnes, J. P., 144 (ref. 4)
Vervaeke, I. F., 99 (ref. 1)
Verzele, M., 22 (ref. 1), 167 (ref. 13), 268 (ref. 36)
Villanueva, V. R., 99 (ref. 2, 3)
Vincendon, G., 28 (ref. 1)
Vladovska Yulchovska, Y. G., 118 (ref. 1)
Voelter, W., 143 (ref. 6), 156 (ref. 21), 173 (ref. 13)
Von Arx, E., 183

W

Wagner, F. W., 156 (ref. 33)
Walker, J. G., 259 (ref. 8)
Walker, J. W., 262 (ref. 5)
Wall, R. A., 260 (ref. 1)
Walsh, M. J., 52 (ref. 1)
Walters, F. H., 147 (ref. 1), 209 (ref. 1)
Walton, H. F., 95 (ref. 1), 96 (ref. 1), 260 (ref. 1)
Waring, R. H., 191 (ref. 1)
Weigele, M., 173 (ref. 11, 15)
Weinova, H., 158 (ref. 3)
Weinstein, S., 14 (ref. 1)
Wells, M. M., 216 (ref. 1)
Wenke, D., 261 (ref. 16)
Whatley, J. A., 267 (ref. 17)
Whitford, J. H., 163 (ref. 5)
Wiesmer, I., 126 (ref. 1)
Wiesnerova, L., 26 (ref. 1)
Wilk, S., 163 (ref. 10)
Williams, C. H., 102 (ref. 1), 217 (ref. 1)
Wilson, T. D., 147 (ref. 1), 209 (ref. 1)
Wittmer, D. P., 83 (ref. 1)
Wold, S., 223 (ref. 4)
Wolf, P. L., 146 (ref. 4)
Wolf, T., 268 (ref. 37)
Wonnacott, D. M., 159, 163 (ref. 7), 167 (ref. 10)
Wood, R., 268 (ref. 38)

Y

Yamamoto, Y., 179
Yamanaka, K., 52 (ref. 1)
Yamazaki, Y., 36 (ref. 1)
Yamomoto, S., 204 (ref. 1)
Yancy, J., 223 (ref. 4)
Yang, C., 23 (ref. 1), 24 (ref. 1), 167 (ref. 8)
Yau, W. W., 263 (ref. 18)
Yoda, R., 179
Young, J. C., 121 (ref. 1)
Younger, D. R., 163 (ref. 6)
Zamaroni, F., 262 (ref. 2)
Zanetta, J. P., 28 (ref. 1)
Zarkadas, C. G., 65 (ref. 1)
Zech, K., 143 (ref. 6), 156 (ref. 21), 173 (ref. 13)
Zeman, S., 107 (ref. 1)
Zemanova, E., 107 (ref. 1)
Zimmerman, C. L., 147 (ref. 2, 4)
Zlatkis, A., 163 (ref. 5), 260 (ref. 14, 15, 16), 261 (ref. 1)
Zolotarev, Yu. A., 69 (ref. 1), 70 (ref. 1), 71 (ref. 1), 72 (ref. 1), 73 (ref. 1)
Zoltan, S., 111 (ref. 1)
Zumwalt, R. W., 163 (ref. 11), 164 (ref. 14), 164 (ref. 15)
Zweig, G., 259 (ref. 9)

COMPOUND INDEX

A

N-Acetyl amino acid isopropyl esters, 31—32
N-Acetyl-5-hydroxy-tryptamine, 113
N-Acetyl-L-(−)-leucine, 117
N^2-Acetyllysine, 58
N-Acetyl-5-methoxy-tryptamine, 113
N-Acetylphenylalanine, 53
N-Acetyl-L-(+)-phenylalanine, 117
N-Acetyl-L-tryptophan, 113
N-Acetyl-L-(+)-tryptophan, 117
$O^{4'}$-Acetyltyrosine, 58
N-Acyl DL-alanine esters, 32
Adrenaline, 94
Agmatine, 97, 98, 101, 127
Alanine, 8—11, 14, 18—20, 23, 25—28, 30, 31, 33—35, 58, 60, 61, 63, 69—74, 74—80, 82, 83, 88, 89, 91, 101, 104, 111, 118, 119, 121, 162, 186, 189, 190
α-Alanine, 81
β-Alanine, 19, 20, 28, 31, 35, 58—61, 69
D-Alanine, 4, 5, 7, 29, 90
DL-Alanine, 105
L-Alanine, 4, 5, 7, 29, 59, 90
R,S-Alanine, 43
Alanine n-butyl ester, 13
L-Alanine-tert-butyl ester, 83
Alanine cyclopentyl ester, 13
Alanine ethyl ester, 13
Alanine isopropyl ester, 13
Alanine methyl ester, 13
Alanine 3-pentyl ester, 13
Alanine n-propyl ester, 13
Alaninol, 63
L-Alanyl-L-alanine, 83
Aliphatic amines, 36, 37
N,N'-alkylated amines, 127—128
Allocystathionine, 31
Allohydroxylysine, 61
D-Allohydroxylysine, 31
D-Allo-δ-hydroxylysine, 59
L-Allo-δ-hydroxylysine, 59
δ-Allohydroxy-DL-lysine, 65
Allohydroxyproline, 19, 69, 70—73
D-Allohydroxyproline, 110
L-Allohydroxyproline, 72
Alloisoleucine, see also Isoleucine, 25—28, 30, 58, 60, 76, 82
D-Alloisoleucine, 5, 7, 31
L-Alloisoleucine, 5, 7
Allothreonine, 69, 70, 76
N-Allyl-2,4-dinitroaniline, 107
N-Allyl-2,6-dinitroaniline, 107
N-Allyl-2,4,6-trinitroaniline, 107
Aminoadipic acid, 31
α-Aminoadipic acid, 19, 59—61
DL-2-Aminoadipic acid, 21
4-Aminoazobenzene, 192

2-Aminobenzoic acid, 192
4-Aminobenzoic acid, 192
m-Aminobenzoic acid, 130, 131, 191
o-Aminobenzoic acid, 130, 131, 191
p-Aminobenzoic acid, 130, 131, 191
4-Aminobiphenyl, 192
Aminobutyric acid, 30, 69—72
Amino-n-butyric acid, 8, 9
2-Aminobutyric acid, 20, 58, 83
3-Aminobutyric acid, 83
4-Aminobutyric acid, 20, 83
α-Aminobutyric acid, 10, 11, 19, 25, 28, 35, 60, 61, 81, 119
α-Amino-n-butyric acid, 31, 76
β-Amino-n-butyric acid, 31
ε-Amino-n-butyric acid, 31
γ-Aminobutyric acid, 18, 19, 28, 35, 59—61, 134
D-α-Aminobutyric acid, 6
DL-3-Aminobutyric acid, 20
L-α-Aminobutyric acid, 6
L-α-Amino-n-butyric acid, 59
ε-Aminocaproic acid, 19, 62
ε-Amino-n-caproic acid, 31
6-Amino-4-chloro-3-toluene-sulfonic acid, 130, 131
3-Aminocyclohexane 51
1-Amino-1-cyclopropanecarboxylic acid, 19
2-Aminodecane, 51
4-Amino-4-dimethyl-aminoazobenzene, 192
2-Amino-1,3-dimethyl-benzene, 191
2-Amino-1,4-dimethyl-benzene, 191
3-Amino-1,2-dimethyl-benzene, 191
4-Amino-1,2-dimethyl-benzene, 191
4-Amino-1,3-dimethyl-benzene, 191
4-Amino-3,5-dimethylbenzoic acid, 130, 131
3-Amino-2,2'-dimethylbutane, 123
2-Aminodiphenylamine, 132
4-Aminodiphenylamine, 132, 192
Aminoethanol, 58
S-Aminoethylcysteine, 58
1-(2-Aminoethyl)piperazine, 38
α-Amino-γ-guanidinobutyric acid, 97
α-Amino-β-guanidinopropionic acid, 97
α-Amino-β-guanidino-propionic acid, 62
L-α-Aminoguanidino-propionic acid, 59
2-Aminoheptane, 51
(±)-2-Aminoheptane, 44
R, S-2-Aminoheptane, 43
D-α-Aminoheptanoic acid, 6
L-α-Aminoheptanoic acid, 6
(±)-2-Aminohexane, 44
R,S-2-Aminohexane, 43
α-Aminohexanoic acid, 10, 11
D-α-Aminohexanoic acid, 6
L-α-Aminohexanoic acid, 6
2-Amino-5-hydroxybenzoic acid, 191
3-Amino-2-hydroxybenzoic acid, 191
3-Amino-4-hydroxybenzoic acid, 191
DL-4-Amino-3-hydroxybutyric acid, 20

1-Amino-2-hydroxy-naphthalene-4-sulfonic acid, 192
2-Aminoisobutyric acid, 20
α-Aminoisobutyric acid, 19, 31, 35, 83
β-Aminoisobutyric acid, 19, 28, 31, 35, 60, 61
DL-3-Aminoisobutyric acid, 20
DL-β-Aminoisobutyric acid, 59
δ-Aminolevulinic acid, 31, 61
(±)-2-Amino-6-methylheptane, 44
R,S-2-Amino-6-methylheptane, 43
D-γ-Amino-ε-methylheptanoic acid, 6
L-γ-Amino-ε-methylheptanoic acid, 6
(±)-2-Amino-5-methylhexane, 44
R,S-2-Amino-5-methylhexane, 43
D-γ-Amino-δ-methylhexanoic acid, 6
L-γ-Amino-δ-methylhexanoic acid, 6
2-Amino-3-methylpentane, 123
(±)-2-Amino-3-methylpentane, 44
2-Amino-4-methylpentane, 123
2-Amino-5-nitrotoluene, 191
4-Amino-2-nitrotoluene, 191
4-Amino-3-nitrotoluene, 191
2-Aminononane, 51
2-Aminooctane, 51
(±)-2-Aminooctane, 44
R,S-2-Aminooctane, 43
α-Aminooctanoic acid, 10, 11
D-α-Aminooctanoic acid, 6
L-α-Aminooctanoic acid, 6
(+)-2-Aminopentane, 44
(−)-2-Aminopentane, 44
R,S-2-Aminopentane, 43
D-α-Aminopentanoic acid, 6
L-α-Aminopentanoic acid, 6
2-Aminophenol, 192
m-Aminophenol, 133, 190
o-Aminophenol, 190
p-Aminophenol, 133, 190
o-Aminophenylarsonic acid, 130, 131
p-Aminophenylarsonic acid, 130, 131
m-Aminophenylsulfonic acid, 130, 131
o-Aminophenylsulfonic acid, 130, 131
p-Aminophenylsulfonic acid, 130, 131
α-Aminopimelic acid, 19
1-Amino-2-propanol, 41
3-Amino-1-propanol, 41
Aminopropylcadaverine, 98
N-(3-Aminopropyl) cyclohexylamine, 124
N-(3-Aminopropyl)-1,3-diaminopropane, 97
N,N-bis(3-Aminopropyl-1,3-diaminopropane, 97
2-Aminopyridine, 192
2-Aminotoluene-5-sulfonic acid, 130, 131
4-Aminotoluene-3-sulfonic acid, 130, 131
α-Aminovaleric acid, 10, 12
D-γ-Aminovaleric acid, 6
L-γ-Aminovaleric acid, 6
Ammonia, 42, 58—61, 63, 65, 93, 95, 97, 98, 100
Ammonium hydrogen carbonate, 177
Amphetamine, 55, 125
(±)-Amphetamine, 44
(R)-(−)-Amphetamine, 54

(S)-(+)-Amphetamine, 54
n-Amylamine, 39, 42, 92, 125
tert-Amylamine, 42
Angiotensin II, 101
Aniline, 100, 128—132, 134, 191
4-Anisidine, 192
m-Anisidine, 132, 190
o-Anisidine, 132, 190
p-Anisidine, 132, 191
Anserine, 62
Apolane-87, 226
Arginine, 18, 21, 24, 25, 27, 28, 31, 34, 35, 58, 60, 62, 63, 65, 66, 78—80, 82, 88, 95, 97, 98, 111, 118, 119, 121, 127, 162, 186, 189
L-Arginine, 59, 105
Arginine-vasopressin, 186
Arginylglutamic acid, 58
L-Arginyl-L-tryptophyl-L-glycine, 188
Ascorbic acid, 153, 169
Asparagine, 20, 31, 35, 58, 60, 61, 69—73, 76—79, 82, 88, 111, 118, 119, 121, 162, 186, 189
DL-Asparagine, 105
L-Asparagine, 59
Aspartic acid, 10, 11, 18—20, 24—28, 30, 31, 33—35, 58, 60, 61, 69—74, 76—78, 80—82, 88, 89, 101, 104, 118, 119, 162, 186, 189
D-Aspartic acid, 5, 7, 22, 90
L-Aspartic acid, 5, 7, 22, 59, 90
Azetidine carboxylic acid, 69
L-Azetidine carboxylic acid, 71
L-Azetidine-2-carboxylic acid, 19

B

Baikiain, 19
Benzidine, 132, 134
α-N-Benzoyl-DL-alanine, 116
Benzoyl-DL-arginine-p-nitroanilidine·HCl, 116
N-Benzoyl-DL-phenylalanine, 116
Benzylamine, 55, 100, 125
N-Benzyloxycarbonyl- L-alanine-p-nitrophenyl ester, 116
N-Benzyloxycarbonyl-L-arginine·HCl, 116
N-Benzyloxycarbonyl-L-glutamic acid dibenzyl ester·HCl, 116
N-Benzyloxycarbonyl-glycine, 116
N-Benzyloxycarbonyl-glycine-glycine-L-leucine, 116
α-N-Benzyloxycarbonyl-L-lysine benzyl ester-p-tosylate, 116
N-Benzyloxycarbonyl-nitro-L-arginine, 116
N-Benzyloxycarbonyl-L-phenylalanine, 116
N-Benzyloxycarbonyl-L-proline-p-nitrophenyl ester, 116
N-Benzyloxycarbonyl-L-valine, 116
5-Benzyloxytryptamine·HCl, 114, 115
DL-5-Benzyloxytryptophan, 114, 115
DL-6-Benzyloxytryptophan, 114, 115
DL-7-Benzyloxytryptophan, 115

N,N'-Bis(3-aminopropyl)-1,3-diaminopropane, 97
N,N"-Bis-(carboxyethyl)-1,4-diaminobutane, 97
Bis(diphenyl)indenonylthiourea (BITU), 118
2,5-Bis(methylamino)toluene, 128, 129
BITU, see Bis(diphenyl)indenonylthiourea
Bradykinin, 186
m-Bromoaniline, 130—132
o-Bromoaniline, 130—132
p-Bromoaniline, 130—132
Brucine, 191
1,4-Butanediamine, 41, 100
Butylamine, 123
n-Butylamine, 36, 37, 39, 42, 92, 93, 95, 122, 125, 134
sec-Butylamine, 36, 37, 42
t-Butylamine, 122
tert-Butylamine, 36, 42
n-Butyl 1-dihydroteresantalinyl, 32
tert-Butyldimethylsilyl ester, 208
N-n-Butyl-2,4-dinitroaniline, 107
N-n-Butyl-2,6-dinitroaniline, 107
n-Butyl d-isoketopinyl, 32
tert-Butyloxycarbonyl-L-glutamine, 116
tert-Butyloxycarbonyl-L-glutamine-γ-benzyl ester DCS salt, 116
tert-Butyloxycarbonyl-L-glutamine-p-nitrophenyl ester, 116
tert-Butyloxycarbonyl-glycine TCP ester, 116
tert-Butyloxycarbonyl-L-valine, 116
tert-Butyloxycarbonyl-L-valine TCP ester, 116
n-Butyl 1-teresantalinyl ester, 32
N-n-Butyl-2,4,6-trinitroaniline, 107
α-n-Butyric acid, 82

C

Cadaverine, see 1,5-Diaminopentane
N-d-10-Camphorsulfonyl amino acid p-nitrobenzyl esters, 91, 208
Canavanine, 62
Carbamylputrescine, 99
N-Carbobenzoxy-L-tryptophan, 113
S-Carboxyamidomethylglutathionylspermidine, 97
N-Carboxyethyl-1,4-diaminobutane, 97
N,N"-bis-(Carboxyethyl)-1,4-diaminobutane, 97
Carboxymethylcysteine, 119
S-Carboxymethylcysteine, 21, 27, 58, 61, 68, 110, 189
S-Carboxymethyl-cysteine-glycine, 110
Carnosine, 60, 62, 101
L-Carnosine, 59
Catecholamines, 133, 136
CH₃-cysteine, 120
Chirasil-Val, 226
m-Chloroaniline, 130, 131
o-Chloroaniline, 130, 131
p-Chloroaniline, 130, 131, 132
N-Chloro-5-dimethylaminoaphthalene-1-sulfonamide (NCDA), 177—178, 190
Chloroform-formic acid, 190

L-α-Chloroisovaleryl methyl esters, 30
7-Chloro-4-nitrobenzo-2-oxa-1,3-diazole (NBD-Cl), 219
DL-p-Chlorophenylalanine, 133, 136
p-Chlorophenylethylamine, 55
N-Chlorosuccinimide, 170—171
Cinchonine, 191
Citrulline, 21, 35, 58, 60, 61, 78, 82, 101
L-Citrulline, 59
CM-cysteine, 119
Creatine, 62
Creatinine, 59, 62
β-Cyclodextrin polyurethane, 75
Cyclohexylamine, 93, 123
N-Cyclohexyl-2,4-dinitroaniline, 107
N-Cyclohexyl-2,6-dinitroaniline, 107
N-Cyclohexyl hexamethylenediamine-1,6, 127
N-Cyclohexyl propylenediamine-1,3, 127
N-Cyclohexyl-2,4,6-trinitroaniline, 107
Cycloleucine, 28
Cystamine, 97
Cystathionine, 21, 28, 35, 60, 61, 68
allo-Cystathionine, 31
D-Cystathionine, 59
L-Cystathionine, 59
L-Cystathionine sulfone, 106
L-Cystathionine sulfoxide, 106
Cysteic acid, 20, 35, 58, 60, 61, 68, 76, 82, 118, 189
DL-Cysteic acid, 68
Cysteine, 18—20, 23, 25, 27, 31, 34, 35, 58, 59, 79, 89, 111, 121, 162, 172, 190, 206
  modification of, 200—201
CM-Cysteine, 119
L-Cysteine, 105, 106
Cysteine-S-S-homo-cysteine, 28
Cysteine-NEM, 68
L-Cysteine-D-penicillamine mixed disulfide, 68
Cystine, 18, 21, 26—28, 35, 58, 60, 61, 78, 110, 172, 186, 206
  modification of, 200—201
D-Cystine, 5
L-Cystine, 5, 68, 106, 117
Cystinyl-bis-diglycine, 110
Cystinylglycine, 110

D

Dabsyl chloride, see 4-Dimethylaminoazobenzene-4'-sulfonyl chloride
DABTH-serine, 119
DAH, see Diaminohexanoic acid
DANABITC, see 4-N,N-Dimethylaminonaphthyl-azobenzene-4'-isothiocyanate
DANABTH, see 4-N,N-Dimethylaminophthylazo-benzene-4'-thiohydantoin
Dansyl chloride, see 1-Dimethylaminoaphthalene-5-sulfonyl chloride
2,4-DAP, see 2,4-Diaminopentanoic acid
DCTFA, see 1,3-Dichlorotetrafluoroacetone

Decylamine, 124
n-Decylamine, 125
Dehydroserine, 189
Dehydrothreonine, 189
Diaminoadipic acid, 112
Diaminoazelaic acid, 112
Diaminobenzidine, 134
3,4-Diaminobenzoic acid, 130, 131, 134
3,5-Diaminobenzoic acid, 130, 131, 134, 191
1,3-Diaminobutane, 38
1,4-Diaminobutane (putrescine), 38, 95—97, 99, 101, 126, 127
2,3-Diaminobutane, 38
2,4-Diaminobutyric acid, 19, 20
Diaminocarboxylic acids, 112
1,2-Diaminocyclohexane, 38
4,4'-Diaminodiphenyl, 134
4-4'-Diaminodiphenyl-amine, 132
1,2-Diaminoethane, 125
1,7-Diaminoheptane, 99, 126
1,6-Diaminohexane, 95, 96, 126
3,5-Diaminohexanoic acid, 64
erythro-2,5-Diaminohexanoic acid, 63
erythro-3,5-Diaminohexanoic acid, 63
threo-2,5-Diaminohexanoic acid, 63
threo-3,5-Diaminohexanoic acid, 63
1,4-Diamino-2-hydroxybutane, 97
1,2-Diamino-2-methylpropane, 38
1,8-Diaminooctane, 126
1,5-Diaminopentane (cadaverine), 38, 95—98, 101, 126
2,4-Diaminopentanoic acid, 63
Diaminopimelic acid, 112
2,6-Diaminopimelic acid, 63
α,α'-Diaminopimelic acid, 19
α,ε-Diaminopimelic acid, 31
DL-2,6-Diaminopimelic acid, 21
1,2-Diaminopropane, 37, 96, 126
1,3-Diaminopropane, 38, 95—97, 99, 101, 126
Diaminopropionic acid, 25
Diaminosebacic acid, 112
Diaminosuberic acid, 112
Diaminosuccinic acid, 112
2,4-Diaminotoluene, 132
2,6-Diaminotoluene, 132
3,4-Diaminotoluene, 132
o-Dianisidine, 132
Dibenzylamine, 134
Dibutylamine, 122
Di-n-butylamine, 53, 123
N,N'-Di-sec-butyl-1,4-phenylenediamine, 128, 129
N,N-Di-n-butyl-2,4,6-trinitroaniline, 107
2,4-Dichloroaniline, 130—132, 191
2,5-Dichloroaniline, 191
3,4-Dichloroaniline, 191
1,3-Dichlorotetrafluoroacetone (DCTFA), 205
N,N'-Dicyclohexyl hexamethylenediamine-1,6, 127
N,N'-Dicyclohexyl trimethylhexamethylene diamine-1,6, 127
N,N-Di-dinitrophenyl-L-lysine, 116
Diethanolamine, 93, 123, 126

Diethylamine, 36, 37, 39, 42, 53, 92, 93, 100, 123
N,N-Diethyl-2,4-dinitroaniline, 107
N,N-Diethyl-2,6-dinitroaniline, 107
Diethylenetriamine, 38, 100, 122, 123, 126
Di-2-ethylhexylamine, 124
N,N'-Diethyl-1,4-phenylenediamine, 128, 129
Diethyltriamine, 122
N,N-Diethyl-2,4,6-trinitroaniline, 107
3,3'-Diglycolic acid, 134
N,N'-Diglycylcystine, 110
Dihydroxyphenylalanine, 61, 89
3,4-Dihydroxyphenylalanine (DOPA), 21, 53, 76, 78, 101
L-3,4-Dihydroxyphenylalanine (L-DOPA), 133, 136
β-(3,4-Dihydroxyphenyl)alanine, 75
3,4-Dihydroxyproline, 20
Diiodohistidine, 67, 120
3,3'-Diiodothyronine, 67
3,5-Diiodothyronine, 67
3',5'-Diiodothyronine, 67
Diiodotyrosine, 58, 62, 67
3,5-Diiodotyrosine, 21, 120
Diisobutylamine, 53, 123
Diisopropanolamine, 93
Diisopropylamine, 37, 53, 123
N,N-Di-isopropyl-2,4-dinitroaniline, 107
N,N'-Diisopropyl isophorondiamine, 127
N,N'-Diisopropyl-1,4-phenylenediamine, 128, 129
N,N'-Diisopropylpropylenediamine-1,3, 127
N,N'-Diisopropyl trimethylhexamethylene diamine-1,6, 127
DIITC, see 2-p-Isothiocyanophenyl-3-phenylindenone
2,4-Dimethoxyaniline, 191
2,5-Dimethoxyaniline, 191
3,5-Dimethoxyaniline, 191
Dimethoxybenzidine, 134
3,4-Dimethoxyphenylethylamine, 52, 56
Dimethoxytetrahydrofuran-p-dimethylaminobenzaldehyde, 178
Dimethylamine, 36, 37, 39, 42, 53, 93, 123
4-Dimethylaminoazobenzene-4'-sulfonyl chloride (dabsyl chloride), 209—210
4-N,N-Dimethylaminoazobenene-4'-thiohydantoins, 119
5-Dimethylamino-1-naphthalene-5-sulfonyl-alanine, 80
5-Dimethylamino-1-naphthalene-5-sulfonyl-arginine, 80
5-Dimethylamino-1-naphthalene-5-sulfonyl-aspartic acid, 80
1-Dimethylamino-1-naphthalene-5-sulfonyl-chloride (dansyl chloride), 207
5-Dimethylamino-1-naphthalene-5-sulfonyl-glutamic acid, 80
1-Dimethylamino-1-naphthalene-5-sulfonyl-glycine, 116
5-Dimethylamino-1-naphthalene-5-sulfonyl-glycine, 80
5-Dimethylamino-1-naphthalene-5-sulfonyl-hydroxyproline, 80

5-Dimethylamino-1-naphthalene-5-sulfonyl-isoleucine, 80
5-Dimethylamino-1-naphthalene-5-sulfonyl-leucine, 80
5-Dimethylamino-1-naphthalene-5-sulfonyl-methionine, 80
1-Dimethylamino-1-naphthalene-5-sulfonyl-NH$_2$, 121
5-Dimethylamino-5-naphthalene-5-sulfonyl-NH$_2$, 80
1-Dimethylamino-1-naphthalene-5-sulfonyl-OH, 121
5-Dimethylamino-1-naphthalene-5-sulfonyl-OH, 80
1-Dimethylamino-1-naphthalene-5-sulfonyl-L-β-phenylalanine, 116
5-Dimethylamino-1-naphthalene-5-sulfonyl-phenylalanine, 80
5-Dimethylamino-1-naphthalene-5-sulfonyl-proline, 80
5-Dimethylamino-1-naphthalene-5-sulfonyl-serine, 80
5-Dimethylamino-1-naphthalene-5-sulfonyl-threonine, 80
5-Dimethylamino-1-naphthalene-5-sulfonyl-tryptophan, 80
5-Dimethylamino-1-naphthalene-5-sulfonyl-tyrosine, 80
1-Dimethylamino-1-naphthalene-5-sulfonyl-valine, 116
4-N,N-Dimethylaminoaphthylazobenzene-4'-isothiocyanate, 210
4,N,N-Dimethylaminonaphthylazobenzene-4'-thiohydantoin, 210
2,4-Dimethylaniline, 132
2,6-Dimethylaniline, 132
asym-Dimethylarginine, 66
sym.-Dimethylarginine, 67
N^G-Dimethylarginine, 58, 105
N^G,N^G-Dimethylarginine, 62, 65, 97, 105
N^G,N$^{G'}$-Dimethylarginine, 65
N^G,N^G-Dimethyl-L-arginine (DMA), 111
(R)-(+)-N,α-Dimethylbenzylamine, 54
(S)-(−)-N-α-Dimethylbenzylamine, 54
N,N-Dimethyl-1,3-diaminopropane, 38
N,N-Dimethyl-2,4-dinitroaniline, 107
N,N-Dimethyl-2,6-dinitroaniline, 107
sym.-Dimethylethylenediamine, 38
N,N-Dimethylglycine·HCl, 117
N,N'-Dimethyl isophorondiamine, 127
N,N-Dimethyllaurylamine, 40
Dimethyllysine, 66
N^6-Dimethyllysine, 58
N-ε-Dimethyllysine, 97
ε-N,ε-N-Dimethyllysine, 65
N,N-Dimethyl-1,4-phenyleledíamine, 128
N,N'-Dimethyl-1,4-phenyleledíamine, 128, 129
N,N'-Dimethyl trimethylhexamethylene diamine-1,6, 127
N,N-Dimethyl-2,4,6-trinitroaniline, 107
N,N-Dimethyltryptamine, 113
Di-1-naphthylamine, 194
Di-1,2-naphthylamine, 194
Di-2-naphthylamine, 194

N,N'-Di-2-naphthyl-1,4-phenylenediamine, 129
2,4-Dinitro-4'-amino-diphenylamine, 132
2,4-Dinitroaniline, 107, 132
2,6-Dinitroaniline, 107
Dinitrophenyl, 147
N-2,4-Dinitrophenyl-L-alanine, 116
N-2,4-Dinitrophenyl-L-phenylalanine, 116
N-(2,4-Dinitrophenyl)piperidine, 107
N-(2,6-Dinitrophenyl)piperidine, 107
N-Dinitrophenyl, O-trimethylsilyl derivatives of amines, 211—212
N-2,6-Dinitro-4-trifluoromethyl, o-trimethylsilyl derivatives of amines, 211
p,p'-Dioctyldiphenylamine, 194
3,7-Dioctylphenothiazine, 194
Diphenylamine, 128, 129, 134, 194
N,N'-Dihenylbenzidine, 134
N,N'-Diphenyl-1,4-phenylenediamine, 128, 129
Di-n-propylamine, 39, 53, 123
N,N-Di-n-propyl-2,4-dinitroaniline, 107
N,N-Di-n-propyl-2,6-dinitroaniline, 107
Dipropylenetriamine, 126
N,N-Di-n-propyl-2,4,6-trinitroaniline, 107
N,N'-Disubstituted ureas, 218
Divinylbenzene (DVB)-styrene copolymer, 157
Djenkolic acid, 35
DMA, see N^G,N^G-Dimethyl-L-arginine
DML, see N^ε,N^ε-D,L-lysine hydrochloride
Dodecylamine, 124
n-Dodecylamine, 125
DOPA, see 3,4-Dihydroxyphenylalanine
Dopamine, 52, 55, 94, 101, 102
Dopamine·HCl, 133, 136
Dopamine(tris-dansyl), 134
DVB, see Divinylbenzene

E

EAE, 188
EGA, see Ethlene glycol adipate
Epinephrine, 52
Ethanethiol, 171
Ethanolamine, 41, 59—61, 99, 123, 125
Ethionine, 33, 76, 78
Ethylamine, 36, 37, 39, 42, 92, 123, 125
Ethyldiethanolamine, 123
N-Ethyl-2,4-dinitroaniline, 107
N-Ethyl-2,6-dinitroaniline, 107
Ethylenediamie, 37, 38, 100, 122, 123, 126
Ethylene glycol adipate (EGA), 160
Ethylenepropylenetriamine, 126
Ethylethanolamine, 123
2-Ethylhexylamine, 123
2-Ethylhexylethanolamine, 124
N-Ethylmorpholine, 123
S-(N)-(Ethylsuccinimido)-glutathionylspermidine, 97
N-Ethyl trimethylhexamethylenediamine-1,6, 127
N-Ethyl-2,4,6-trinitroaniline, 107
Eugenol, 191

F

Fluorescamine, 83, 170, 171, 178—180, 210—211
Fluorescamine-L-alanine, 116
Fluorescamine-L-alanine-L-histidine, 116
Fluorescamine-glycine-L-histidine-glycine, 116
Fluorescamine-perchloric acid, 180—181
p-Fluorophenylalanine, 76
5-Fluorotryptamine·HCl, 114, 115
6-Fluorotryptamine·HCl, 114, 115
DL-4-Fluorotryptophan, 114, 115
DL-5-Fluorotryptophan, 114, 115
DL-6-Fluorotryptophan, 114, 115
Formaldehyde, 188
Formaldehyde HCl, 188
Formaldehyde-ozone, 188

G

Galactosamine, 58, 62
Glucagon, 135, 188
Glucosamine, 58, 61, 63
Glucosaminitol, 63
Glutamic acid, 10, 11, 18, 19, 21, 24—28, 30, 31, 33—35, 58, 60, 61, 63, 69—74, 76—78, 80, 82, 88, 89, 91, 101, 104, 118—121, 162, 186, 189, 190
D-Glutamic acid, 5, 7, 22, 90
L-Glutamic acid, 5, 7, 22, 59, 90
p-Glutamic acid, 120
L-Glutamic acid dibenzyl ester·HCl, 116
L-Glutamic-γ-hydrazide, 117
Glutamine, 21, 31, 35, 58, 60, 61, 69—73, 76—79, 82, 88, 111, 118—120, 186, 189
L-Glutamine, 59, 110
Glutathione, 58, 61, 101
Glutathione-NEM, 68
Glutathionylspermidine, 97
Glycerophosphoethanolamine, 60
Glycine, 5, 7, 18—20, 22, 23, 25, 27, 28, 31, 34, 35, 58—61, 63, 69—72, 74, 79, 80, 88, 101, 104, 105, 110, 111, 116, 118—121, 123, 162, 186, 189, 190
N-Glycine-S-carboxymethylcysteine, 110
Glycine-glycine-L-tryptophan, 113
Glycine-p-nitroanilide, 116
Glycine-L-tryptophan, 113
Glycine-L-tryptophan-glycine, 113
Glycocyanomine, 31
Glycylaspartic acid, 58
Glycyglycine, 58, 186
Glycyltryptophan, 58
L-Glycyl-L-tryptophan, 188
Guanidine, 100

H

Heptadecylamine, 124

N-Heptafluorobutyric-2-aminoethylbenzene, 48
N-Heptafluorobutyric(dl)2-aminoethylbenzene, 49
N-Heptafluorobutyric-2-aminohexane, 50
Heptafluorobutyric anhydride (HFBA), 163
N-Heptafluorobutyric L-leucyl dl-α-methylphenethylamine, 45
N-Heptafluorobutyric L-prolyl dl-p-chloro-α-methylphenethylamine, 45
N-Heptafluorobutyric L-prolyl dl-α-ethylphenethylamine, 45
N-Heptafluorobutyric L-prolyl dl-α-methylbenzylamine, 45, 46
N-Heptafluorobutyric L-prolyl dl-α-methylphenethylamine, 45, 46
N-Heptafluorobutyric L-prolyl dl-1-methyl-3-phenylpropylamine, 45
N-Heptafluorobutyric L-valyl dl-α-methylphenethylamine, 45
1,7-Heptanediamine, 100
Heptylamine, 123
n-Heptylamine, 92, 125
Hexadecylamine, 124
Hexamethyldisilazane (HMDS), 206, 228
Hexamethylenediamine, 99, 126
1,6-Hexanediamine, 101
Hexylamine, 123
n-Hexylamine, 42, 92, 125
N-Hexyl trimethylhexamethylenediamine-1,6, 127
HFBA, see Heptafluorobutyric anhydride
Hippuric acid, 61
Histamine, 41, 52, 55, 95—97, 99, 101, 102, 125, 135
  detection of, 210—211
Histamine(bis-dansyl), 134
Histamine dihydrochloride, 135
Histidine, 18, 21, 25, 27, 28, 31, 35, 58, 60, 61, 63, 65, 66, 69—72, 74, 78, 79, 82, 88, 89, 95, 97, 99, 101, 104, 111, 118—121, 135, 12, 186, 189
  detection of, 210—211
L-Histidine, 59, 135
L-Histidine-L-alanine, 135
L-Histidine ethyl ester hydrochloride, 135
L-Histidine-glycine, 135
DL-Histidine-DL-histidine, 135
L-Histidine hydroxamate, 135
L-Histidine-L-leucine, 135
L-Histidine-L-lysine hydrobromide, 135
L-Histidine methyl ester dihydrochloride, 135
L-Histidine-L-phenylalanine, 135
L-Histidine-L-serine, 135
Histidine thiohydantoin, 116
L-Histidine-L-tyrosine, 135
L-Histidinol dihydrochloride, 135
L-Histidinol phosphate, 135
HMDS, see Hexamethyldisilazane
Homoarginine, 28, 97
Homocarnosine, 60, 62
Homocitrulline, 61
Homocysteic acid, 61

L-Homocysteine, 106
L-Homocysteine-L-cysteine mixed disulfide, 68
Homocysteine thiolactone, 62
L-Homocysteine thiolactone, 106
Homocystine, 28, 35, 61
DL-Homocystine, 21
L-Homocystine, 59, 68, 106
Homohypotaurine, 106
L-Homolanthionine, 106
Homoserine, 19, 20, 34, 58, 61, 69
L-Homoserine, 106
Homoserine lactone, 58
sym-Homospermidine, 99
Homotaurine, 106
Homothiotaurine, 106
Hydrazine, 169, 191
Hydrochloric acid, 150
5-Hydroxy-4-aminopentanoic acid, 63
3-Hydroxyanthranilic acid, 113
α-Hydroxycarboxylic acid esters, 29
6-Hydroxydopamine·HBr, 133, 136
5-Hydroxydopamine·HCl, 133, 136
5-Hydroxyindole-3-acetic acid DCA salt, 112
Hydroxylysine, 97
4-Hydroxylysine, 35
5-Hydroxylysine, 21, 31, 61
δ-Hydroxy-DL-lysine, 65
allo-Hydroxylysine, 161
δ-allo-Hydroxy-DL-lysine, 65
D-allo-Hydroxylysine, 31
D-allo-δ-Hydroxylysine, 59
DL-5-Hydroxylysine, 21
DL-allo-Hydroxylysine, 21
L-allo-δ-Hydroxylysine, 59
6-Hydroxymelatonin, 113
4-Hydroxy-3-methoxybenzylamine, 56
5-Hydroxypipecolic acid, 19
Hydroxyproline, 18, 19, 27, 31, 34, 35, 60, 61, 68—72, 79, 80, 104, 110, 111, 118, 119, 162
Hydroxy-L-proline, 59
4-Hydroxyproline, 20, 28, 58
4-Hydroxy-L-proline, 121
allo-Hydroxyproline, 19, 69—73
D-allo-Hydroxyproline, 110
L-Hydroxyproline, 69, 110
L-allo-Hydroxyproline, 72
4-Hydroxy-1-pyrrolin-2-carboxylic acid, 110
6-Hydroxytryptamine creatinine sulfate complex, 114, 115
5-Hydroxytryptamine oxalate salt, 113
5-Hydroxytryptophan, 78
5-Hydroxy-DL-tryptophan, 113
5-Hydroxytryptophol, 113
3-Hydroxytyramine, 125
Hypotaurine, 106

I

Iminodiacetic acid, 69—72

Indole, 41
Indoleacetic acid, 188
Indole-3-acetic acid, 112
DL-3-Indoleacetic acid, 112
3-Indoleacetone, 112
3-Indoleacetonitrile, 112
3-Indolepyruvic acid, 112
Indomethacin, 113
Indoxyl-β-D-glucoside, 113
Indoxyl sulfate, 113
Iodide, 67, 120, 181
Iodohistidine, 120
Iodotyrosine, 120
3-Iodotyrosine, 21
Isoamylamine, 39, 42, 123, 125
Isobutylamine, 36, 37, 39, 41, 42, 123, 125
N-Isobutyl-2,4-dinitroaniline, 107
N-Isobutyloxycarbonyl amino acid methyl esters, 204
Isoglutamine, 63
Isoleucine, see also Alloisoleucine, 8, 9, 18—20, 23, 25—28, 30, 31, 33—35, 58, 60, 61, 69—74, 76—78, 80, 82, 83, 88, 89, 91, 101, 111, 118—121, 162, 186, 189, 190
allo-Isoleucine, 25—28, 30, 58, 60, 76, 82
D-Isoleucine, 5, 7, 90
D-allo-Isoleucine, 5, 7, 31
DL-Isoleucine, 105
L-Isoleucine, 5, 7, 59, 90
L-allo-Isoleucine, 5, 7
Isopentylamine, 41
Isophorondiamine, 126
Isopropylamine, 36, 37, 39, 42, 92, 123
N-Isopropyl-2,4-dinitroaniline, 107
2-p-Isothiocyanophenyl-3-phenylindenone (DIITC), 118
Isovaline, 69

K

Kainic acid, 19
Kynurenic acid, 113
Kynurenine, 27, 35, 61, 75
DL-Kynurenine sulfate, 113

L

Laevulinic acid, 61
Lanthionine, 28, 35, 61
Leucine, 8—11, 14, 18—20, 23, 25—28, 30, 31, 33—35, 58, 60, 61, 69—74, 76—83, 88, 89, 91, 101, 104, 111, 118—121, 162, 186, 189, 190
D-Leucine, 4, 5, 7, 22, 29, 90
DL-Leucine, 105
D-tert-Leucine, 6
L-Leucine, 4, 5, 7, 22, 29, 59, 90
L-tert-Leucine, 6
n-Leucine, 120

tert-Leucine, 8, 9, 10, 11, 14
Leucine n-butyl ester, 13
Leucine cyclopentyl ester, 13
Leucine ethyl ester, 13
Leucine isopropyl ester, 13
Leucine methyl ester, 13
L-Leucine-β-nephthylamide·HCl, 116
L-Leucine-p-nitroanilide·HCl, 116
Leucine 3-pentyl ester, 13
Leucine n-propyl ester, 13
L-Leucine-L-tryptophan, 113
L-Leucine-L-tryptophan-L-leucine, 113
Leucyltyrosine, 58
Lysine, 18, 21, 24—28, 30, 31, 34, 35, 58, 60, 61, 63, 65, 66, 69—74, 78, 79, 82, 88, 89, 95, 97, 99, 101, 104, 111, 118, 120, 121, 162, 186, 189, 190
β-Lysine, 63, 64
D-Lysine, 6, 7
D-β-Lysine, 64
L-Lysine, 6, 7, 59, 105
L-β-Lysine, 64
N$^\epsilon$,N$^\epsilon$-D,L-Lysine hydrochloride (DML), 111
L-Lysine-L-trypophan-L-lysine, 113
Lysyllysine, 58

M

Mannosamine, 62
2-Mercaptoethanol, 170—172
Metanephrine, 52, 94
Metaraminol, 55
(R)-(−)-Methamphetamine, 54
(S)-(+)-Methamphetamine, 54
Methionine, 10, 11, 18—20, 24—28, 30, 31, 33—35, 58, 60, 61, 68—74, 76, 78—80, 82, 88, 89, 91, 101, 111, 118—121, 162, 186, 189, 190
D-Methionine, 4, 5, 7
DL-Methionine, 105
L-Methionine, 4, 5, 7, 59, 106
Methionine sulfone, 19, 28, 31, 58, 61, 82, 118, 189
DL-Methionine sulfone, 68
Methionine sulfoxide, 20, 31, 58, 61
DL-Methionine sulfoxide, 68
3-Methoxy-4-amino-diphenylamine, 132
4-Methoxy-4′-amino-diphenylamine, 132
5-Methoxy-3-indole-acetic acid, 112
N-(−)-α-Methoxy-α-methyl-1-naphthalene acetylamino acid methyl esters, 208
p-Methoxyphenylethylamine, 55
3-Methoxypropylamine, 123
5-Methoxytryptamine-HCl, 113
5-Methoxy-DL-tryptophan, 113
5-Methoxytryptophol, 113
3-Methoxytyramine, 52, 55, 94
3-Methoxytyramine(bis-dansyl), 134
3-Methoxytyramine·HCl, 133, 136
N-Methyl-DL-alanine, 117
Methylamine, 36, 39, 42, 92, 122, 123, 125, 127
2-Methylamino-5-aminotoluene, 128, 129
N^G-Methylarginine, 58
N-Methylbenzylamine, 134
(R)-(+)-α-Methylbenzylamine, 54
(S)-(−)-α-Methylbenzylamine, 54
2-Methylbutylamine, 123
N-Methyl-n-butylamine, 42
S-methylcysteine, 20
N-Methyl-1,3-diaminopropane, 38
Methyl 1-dihydroteresantalinyl ester, 32
N-Methyl-2,4-dinitroaniline, 107
N-Methyl-2,6-dinitroaniline, 107
DL-α-Methyl-DOPA, 133, 136
L-3-O-Methyl-DOPA, 133, 136
4-O-Methyldopamine, 136
N-Methylethylenediamine, 37
4-Methyl-α-ethyl-m-tyramine HCl, 133, 136
1-Methylhistamine, 55, 99
1,4-Methylhistamine dihydrochloride, 135
N^n-Methylhistidine, 58
1-Methylhistidine, 60, 61, 65, 66, 97, 111
3-Methylhistidine, 60, 62, 65, 66, 11
DL-4-Methylhistidine, 135
L-1-Methylhistidine, 59
L-3-Methylhistidine, 59, 135
DL-2-Methylhistidine dihydrochloride hydrate, 135
3-Methylindole, 112
Methyl d-isoketopinyl ester, 32
N-Methylisoleucine, 30
α-Methylleucine, 6, 14
α-Methyl tert-leucine, 14
N-Methyl-DL-leucine, 117
N^6-Methyllysine, 58
N-ε-Methyllysine, 61
N-Methylmethionine, 19
DL-α-Methylnorepinephrine HCl, 136
α-Methylnorleucine, 6, 14
α-Methylnorvaline, 6, 14
p-Methylphenylalanine, 34
2-Methyl-1,4-phenylenediamine, 128, 129
Methyl 1-teresantalinyl ester, 32
Methylthiohydantoin-DL-asparagine, 117
Methylthiohydantoin-glycine, 116
Methylthiohydantoin-DL-histidine, 117
3-Methylthiopropylamine, 41
N-Methyl-2,4,6-trinitroaniline, 107
7-Methyltryptamine, 114, 115
N-Methyltryptamine, 113
5-Methyltryptamine HCl, 114, 115
4-Methyl-DL-tryptophan, 114, 115
5-Methyl-DL-tryptophan, 114, 115
6-Methyl-DL-tryptophan, 114, 115
7-Methyl-DL-tryptophan, 114, 115
DL-α-Methyltryptophan, 114, 115
O-Methyl-L-tyrosine HCl, 133, 136
α-Methylvaline, 6, 14
N-Methylvaline, 30
N-Methyl-DL-valine, 117
3-Methoxytyramine, 52
Metol, 191

MITU, see Mono(diphenyl)indenonylthiourea
MMA, see N^G-Monomethyl-L-arginine
MML, see N^ε-Monomethyl-D,L-lysine hydrochloride
Monoacetyldiaminobutane, 97
Monoacetyl-1,5-diaminopentane, 97
Monoacetyl-1,3-diaminopropane, 97
N^1-Monoacetylspermidine, 97
N^8-Monoacetylspermidine, 97
Monocarbamyl-1,4-diaminobutane, 97
Mono(diphenyl)indenonylthiourea (MITU), 118
Monoethanolamine, 93, 100, 126
Monoethylamine, 93
Monoglycylcystine, 110
Monoiodohistidine, 67, 120
3-Monoiodothyronine, 67
3'-Monoiodothyronine, 67
Monoiodotyrosine, 62, 67
3-Monoiodotyrosine, 120
Monoisopropanolamine, 93
Monomethylamine, 93
N^G-Monomethylarginine, 62, 105
N^G-Monomethyl-L-arginine (MMA), 111
Monomethyllysine, 66
N-ε-Monomethyllysine, 28, 97
ε-N-Monomethyllysine, 65
N^ε-Monomethyl-D,L-lysine hydrochloride (MML), 111
Morpholine, 53, 123
Muramic acid, 63
Muramicitol, 63

N

α-Naphthol, 191
β-Naphthol, 191
1-Naphthylamine, 192
2-Naphthylamine, 192
a-Naphthylamine, 130, 131
α-Naphthylamine, 132
1-Naphthylamino-4-sulfonic acid, 130, 131
1-Naphthylamino-7-sulfonic acid, 130, 131
α-(1-Naphthyl)ethylamine, 51
N,N'-2-Naphthyl-1,4-phenylenediamine, 128
NBD-Cl, see 7-Chloro-4-nitrobenzo-2-oxa-1,3-diazole
NCDA, see N-Chloro-5-dimethylaminoaphthalene-1-sulfonamide
Ninhydrin, 169, 171
2-Nitroaniline, 192
3-Nitroaniline, 192
4-Nitroaniline, 193
m-Nitroaniline, 130—132
o-Nitroaniline, 130—132
p-Nitroaniline, 130—132
O-Nitrophenylsulfenyl-L-alanine DCA salt, 116
O-Nitrophenylsulfenyl-glycine DCA salt, 116
O-Nitrophenylsulfenyl-L-proline, DCA salt, 116
N-Nitroso-4-hydroxy-L-proline, 121
N-Nitroso-L-proline, 121
N-Nitrososarcosine, 121

Nonylamine, 124
Noradrenaline, 94, 101, 125
Norepinephrine, 52, 55, 97, 102
L-Norepinephrine HCl, 133, 136
Norleucine, 8, 9, 19, 20, 27, 31, 33, 58, 60, 61, 69—73, 76—78, 81, 82, 89
L-Norleucine, 59
Normetanephrine, 52, 55, 94, 102
Normetanephrine(bis-dansyl), 134
DL-Normetanephrine HCl, 133, 136
sym-Norspermidine, 99
sym-Norspermine, 99
Norvaline, 8, 9, 19, 20, 33, 61, 69—72, 76—78, 82, 89

O

1,8-Octanediamine, 100
Octopamine, 102, 125
m-Octopamine, 55
p-Octopamine, 55
p-Octopamine(bis-dansyl), 134
DL-Octopamine HCl, 133, 136
Octylamine, 123
n-Octylamine, 92, 125
tert-Octylamine, 123
p-Octyldiphenylamine, 194
N-Octylphenyl-2-naphthylamine, 194
Orcinol, 191
Ornithine, 21, 25—28, 30, 31, 35, 58, 60, 61, 69—73, 78, 82, 89, 97, 101, 127, 162
D-Ornithine, 5, 7
DL-Ornithine, 65
L-Ornithine, 5, 7, 59
Oxoproline, 110
2-Oxyglutaric acid, 110

P

PE-cysteine, 119
D-Penicillamine disulfide, 68
Pentadecylamine, 124
N-Pentafluoropropionyl-DL-alanine ethyl ester, 16, 17
N-Pentafluoropropionyl amino acid isopropyl esters, 23, 24
N(O,S)-Pentafluoropropionyl amino acid isopropyl esters, 26
1,5-Pentanediamine, 41, 100
Pentylamine, 123
Peptavlon, 188
Perchloric acid, 181
N-Permethylated C_{10}-C_{18} polyamines, 40
N-Permethylation of polyamines, 215
Phenol, 191
Phenothiazine, 194
Phenylalanine, 10, 11, 14, 18, 19, 21, 24—28, 30, 31, 33—35, 52, 58, 60, 61, 65, 69—80, 82, 88, 89, 91, 101, 111, 118—121, 134, 162, 186, 189, 190

α-Phenyl-α-alanine, 69
β-Phenyl-β-alanine, 69
D-Phenylalanine, 4, 5, 7, 22, 90
DL-Phenylalanine, 105
L-Phenylalanine, 4, 5, 7, 22, 59, 90, 117, 133, 136
(R)-(+)-Phenylalanine methyl ester, 54
(S)-(−)-Phenylalanine methyl ester, 54
L-Phenylalanine modified polyacrylamide, 73
L-Phenylalanyl-L-tryptophan, 188
L-Phenylalanine-L-tryptophan-acetate, 113
N-Phenyl-N'-sec-butyl-1,4-phenylenediamine, 128, 129
N-Phenyl-2,4-dinitroaniline, 107
o-Phenyleneamine, 133
1,4-Phenylenediamine, 128, 129
m-Phenylenediamine, 130—133
o-Phenylenediamine, 130—132
p-Phenylenediamine, 100, 130—133
Phenylethanolamine, 55, 134
Phenylethylamine, 41, 52, 55—56, 94, 99
1-Phenylethylamine, 125
2-Phenylethylamine, 125
α-Phenylethylamine, 51
β-Phenylethylamine, 134
(−)-α-Phenylethylamine, 29, 44
(+)-α-Phenylethylamine, 29, 44
R,S-α-Phenylethylamine, 43
Phenylglycine, 10, 14, 69—73, 75, 77, 82, 89
D-Phenylglycine, 6, 90
L-Phenylglycine, 6, 90
(R)-(−)-Phenylglycine methyl ester, 54
(S)-(+)-Phenylglycine methyl ester, 54
Phenylhydantoin, 84—88
Phenylhydrazine, 191
N-Phenyl-N'-isopropyl-1,4-phenylenediamine, 128, 129
Phenyl-2-naphthylamine, 128, 129
N-Phenyl-1-naphthylamine, 194
N-Phenyl-2-naphthylamine, 194
Phenylpropranolamine, 55
(+)-α-Phenylpropylamine, 29
(−)-α-Phenylpropylamine, 29
Phenylserine, 69, 78
Phenylthiocarbamyl-L-alanine, 116
Phenylthiocarbamyl-L-alanine- L-histidine, 116
Phenylthiocarbamyl-glycine-L-histidine-glycine, 116
Phenylthiohydantoin, 147
Phenylthiohydantoin-alanine, 86, 87
Phenylthiohydantoin-DL-alanine, 84, 116
Phenylthiohydantoin-arginine, 86, 87
Phenylthiohydantoin-L-arginine, 84, 116
Phenylthiohydantoin-asparagine, 86, 87
Phenylthiohydantoin-L-asparagine, 84
Phenylthiohydantoin-aspartic acid, 86, 87
Phenylthiohydantoin-DL-aspartic acid, 84
Phenylthiohydantoin-(S-carboxymethyl)-cysteine, 86, 87
Phenylthiohydantoin-(S-carboxymethyl)-L-cysteine, 84
Phenylthiohydantoin-cysteic acid, 86, 87
Phenylthiohydantoin-L-cysteic acid, 84

Phenylthiohydantoin-glutamic acid, 86, 87
Phenylthiohydantoin-L-glutamic acid, 84
Phenylthiohydantoin-glutamine, 86, 87
Phenylthiohydantoin-L-glutamine, 84
Phenylthiohydantoin-glycine, 84, 86, 87
Phenylthiohydantoin-histidine, 87
Phenylthiohydantoin-histidine monohydrochloride, 86
Phenylthiohydantoin-L-histidine monohydrochloride, 84
Phenylthiohydantoin-L-hydroxyproline, 84
Phenylthiohydantoin-isoleucine, 84, 86, 87
Phenylthiohydantoin-leucine, 86, 87
Phenylthiohydantoin-L-leucine, 84
Phenylthiohydantoin-methionine, 86, 87
Phenylthiohydantoin-DL-methionine, 84
Phenylthiohydantoin-methionine sulfone, 86
Phenylthiohydantoin-L-methionine sulfone, 84
Phenylthiohydantoin-(S-methyl)-cysteine, 87
Phenylthiohydantoin-S-methyl-L-cysteine, 84
Phenylthiohydantoin-norleucine, 84, 87
Phenylthiohydantoin-phenylalanine, 86, 87
Phenylthiohydantoin-DL-phenylalanine, 84, 116
Phenylthiohydantoin-N^ϵ-phenylthiocarbamoyllysine, 86
Phenylthiohydantoin-(δ-phenylthiocarbamyl)-L-lysine, 84
Phenylthiohydantoin-proline, 86, 87
Phenylthiohydantoin-L-proline, 84
Phenylthiohydantoin-serine, 86, 87
Phenylthiohydantoin-DL-serine, 84
Phenylthiohydantoin-(ϵ-ϕ-thiocarbamyl)-lysine, 87
Phenylthiohydantoin-threonine, 86, 87
Phenylthiohydantoin-DL-threonine, 84
Phenylthiohydantoin-tryptophan, 86, 87
Phenylthiohydantoin-DL-tryptophan, 84
Phenylthiohydantoin-tyrosine, 86, 87
Phenylthiohydantoin-L-tyrosine, 84
Phenylthiohydantoin-valine, 86, 87
Phenylthiohydantoin-DL-valine, 84
N-Phenyl-N'-tosyl-1,4-phenylenediamine, 128, 129
Phosphoethanolamine, 59—61
Phosphoserine, 58—61
o-Phthalaldehyde, 170—172, 182
o-Phthalaldehyde derivatives, 102
  of amines, 217
o-Phthalaldehyde detection sensitivity, 186—187
o-Phthalaldehyde/ethanethiol derivatives of amino acids, 209
o-Phthaldialdehyde-ethanethiol, 147
N-Phthalyl-glycine, 116
N-Phthalyl-L-leucine, 116
N-Phthalyl-L-phenylalanine, 116
Pipecolic acid, 27, 31, 35, 61
DL-Pipecolic acid, 20
Piperazine, 38, 100, 123
Polyacrylamide, 73
Polyamines, 40, 126—127
Polystyrene, 157

Proline, 10, 11, 14, 18—20, 23, 25—28, 30, 31, 33—35, 58, 60, 61, 69—72, 74, 78—80, 82, 88, 89, 104, 111, 118, 119, 121, 162, 186, 189, 190
D-Proline, 4, 5, 7, 22, 90
L-Proline, 4, 5, 7, 22, 59, 70, 90, 110, 121
(R)-(+)-Proline methyl ester, 54
(S)-(−)-Proline methyl ester, 54
L-Proline-L-tryptophan, 113
Prolylglycine, 58
L-Prolyl-L-tryptophan, 188
1,3-Propane sultone, 200
Propanolamine, 123
Propylamine, 123
n-Propylamine, 36, 37, 39, 42, 92, 125
Propyldiethanolamine, 123
N-n-Propyl-2,4-dinitroaniline, 107
N-n-Propyl-2,6-dinitroaniline, 107
Propylenediamine, 123, 126
N-n-Propyl-2,4,6-trinitroaniline, 107
Putrescine, see 1,4-Diaminobutane
S-β-(4-Pyridylethyl)cysteine, 62
Pyrocatechol, 191
Pyrogallol, 191
L-Pyroglutamic acid, 116
Pyrolidine-2,5-dicarboxylic acid, 19
Pyrrol-2-carboxylic acid, 110
Pyrrolidine, 53

R

Resorcinol, 191

S

Sarcosine, 19, 20, 31, 35, 59—61, 121
Selenocysteine, 31
Selenohomohypotaurine, 106
Selenohomotaurine, 106
Selenohypotaurine, 106
Selenomethionine, 31
Selenotaurine, 106
Serine, 10, 11, 18—20, 23, 25—28, 30, 31, 34, 35, 58, 60, 61, 68—74, 76, 78—82, 88, 89, 111, 118, 119, 121, 162, 186, 189
D-Serine, 5, 7, 90
DL-Serine, 105
L-Serine, 5, 7, 59, 90, 105
Serotonin, 55, 101, 102
Serylglycine, 58
Skatole, 41
Sodium borohydride, 169
Spermidine, 95—97, 99, 101, 126, 127
Spermine, 95—97, 99, 101, 126, 127
Stannous chloride, 169
Stannous chloride-dimethylaminobenzaldehyde, 182—183
Stearylamine, 124
Styrene-divinylbenzene (DVB) copolymer, 157

Sulfanilic acid, 134, 193
Sulfapyridine, 134
Sulfonamides, 53
Sulfonated polystyrene, 151, 157
S-3-Sulfopropylcysteine, 200
O^4-Sulfotyrosine, 58

T

Taurine, 59, 60, 61, 106
TCNQ, see 7,7,8,8-Tetracyanoquinodimethane
TDM, see 4,4′-Tetramethyldiamino-diphenylmethane
N-1-Teresantalinyl DL-amino acid esters, 33
Tetra-n-butylammonium Br, 93
7,7,8,8-Tetracyanoquinodimethane (TCNQ), 183
Tetradecylamine, 40—41, 124
n-Tetradecylamine, 125
Tetraethylammonium Br, 93
Tetraethylenepentamine, 38, 123
Tetragastrin, 188
Tetramethylammonium Br, 93
4,4′-Tetramethyldiamino-diphenylmethane (TDM), 183
β-(2-Thienyl)-DL-alanine, 61
β-(2-Thienyl)-DL-serine, 61
Thiodyglycol, 153
2-Thiohydantoin, 119
Thiotaurine, 106
Thiourea, 90, 209
Threonine, 10, 11, 18—20, 23, 25—28, 30, 31, 34, 35, 58, 60, 61, 68—74, 76, 78—82, 88, 89, 101, 111, 118—121, 162, 186, 189, 190
allo-Threonine, 69, 70, 76
D-Threonine, 5, 7
L-Threonine, 5, 7, 59
Threonylmethionine, 58
Thyroglobulin, 188
Thyroxine, 67, 120
Titanous chloride, 169
TML, see $N^ε,N^ε,N^ε$-Trimethyl-D,L-lysine dihydrochloride
TNBS, see 2,4,6-Trinitrobenzene sulfonate
o-Tolidine, 132, 133
f-Toluenesolfonyl-L-arginine, 116
f-Toluenesolfonyl-DL-arginine methyl ester·HCl, 116
4-Toluidine, 193
m-Toluidine, 130—132
o-Toluidine, 130—132
p-Toluidine, 130—132
Tri-n-butylamine, 93
Tri-n-butylphosphine, 200
Tridecylamine, 124
Triethanolamine, 93, 123, 126
Triethylamine, 37, 39, 92, 93, 123
Triethylenediamine, 38
Triethylenetetramine, 38, 100, 122, 126
N-Trifluoroacetyl-alanine esters, 13
N-Trifluoroacetyl-DL-alanine ethyl ester, 16, 17

N-Trifluoroacetyl-DL-alanine isopropyl ester, 15—17, 50
N-Trifluoroacetyl-L-alanylamides of chiral amines, 44
N-Trifluoroacetyl L-alanyl dl-p-chloro-α-methylphenethylamine, 46
N-Trifluoroacetyl L-alanyl dl-0,α-dimethylphenethylamine, 47
f-Trifluoroacetyl L-alanyl dl-α-ethylphenethylamine, 46
N-Trifluoroacetyl L-alanyl dl-α-ethylphenethylamine, 45
N-Trifluoroacetyl L-alanyl dl-α-methylbenzylamine, 45, 46
N-Trifluoroacetyl L-alanyl dl-1-methylhexylamine, 46
N-Trifluoroacetyl L-alanyl dl-α-methyl,3,4-(methylenedioxy)-phenethylamine, 46
N-Trifluoroacetyl L-analyl dl-α-methylphenethylamine, 46
N-Trifluoroacetyl L-alanyl dl-1-methyl-3-phenylpropylamine, 45, 46
N-Trifluoroacetyl amines, 51
N-Trifluoroacetyl-amino acid esters, 14
N-Trifluoroacetyl-amino acid isopropyl amides, 4—5
N-Trifluoroacetyl-amino acid isopropyl esters, 4—10, 24, 43, 203—204
 enantiomeric, 23
N-Trifluoroacetyl-amino acid trimethylsilyl esters, 206
N-Trifluoroacetyl-2-amino-3,3-dimethylbutane, 50
N-Trifluoroacetyl-2-aminoethylbenzene, 48
N-Trifluoroacetyl(dl)2-aminoethylbenzene, 49
(R,S)-N-Trifluoroacetyl-2-aminoethylbenzene, 47
N-Trifluoroacetyl-2-aminoethylcyclohexane, 48
(R,S)-N-Trifluoroacetyl-2-aminoethylcyclohexane, 47
N-Trifluoroacetyl-2-aminoheptane, 48, 50
(R,S)-N-Trifluoroacetyl-2-aminoheptane, 47
N-Trifluoroacetyl-2-aminohexane, 50
N-Trifluoroacetyl-3-aminohexane, 50
N-Trifluoroacetyl-2-aminooctane, 48
N-Trifluoroacetyl-DL-2-aminooctane, 50
N-Trifluoroacetyl(dl)2-aminooctane, 49
(R,S)-N-Trifluoroacetyl-2-aminooctane, 47
N-Trifluoroacetyl(dl)2-amino-4-phenylbutane, 49
N-Trifluoroacetyl(dl)2-amino-3-phenylpropane, 49
N-Trifluoroacetyl n-butyl esters, 18—20
 of amino acids, 161
N-Trifluoroacetyl derivatives, 42
 of amines, 215
N-Trifluoroacetyl isopropylamide, 4
N-Trifluoroacetyl isopropyl esters, 4, 7, 23, 24, 43
N-Trifluoroacetyl-leucine esters, 13
N-Trifluoroacetyl L-leucyl dl-p-chloro-α-methylphenethylamine, 46
N-Trifluoroacetyl L-leucyl dl-0,α-dimethylphenethylamine, 47
N-Trifluoroacetyl L-leucyl dl-α-ethylphenethylamine, 45, 46
N-Trifluoroacetyl L-leucyl dl-α-methylbenzylamine, 45, 46
N-Trifluoroacetyl L-leucyl dl-1-methylhexylamine, 46
N-Trifluoroacetyl L-leucyl dl-α-methyl-3,4- (methylenedioxy)-phenethylamine, 46
N-Trifluoroacetyl L-leucyl dl-α-methylphenethylamine, 46
N-Trifluoroacetyl L-leucyl dl-1-methyl-3-phenylpropylamine, 45, 46
N-Trifluoroacetyl-α-methylbenzylamine, 50
N-Trifluoroacetyl-3-methylcyclohexamine, 50
N-Trifluoroacetyl L-prolyl dl-p-chloro-α-methylphenethylamine, 45, 46
N-Trifluoroacetyl L-prolyl dl-0,α-dimethylphenethylamine, 46
N-Trifluoroacetyl L-prolyl dl-α-ethylphenethylamine, 45, 46
N-Trifluoroacetyl L-prolyl dl-α-methylbenzylamine, 45, 46
N-Trifluoroacetyl l-prolyl dl-1-methylhexylamine, 46
N-Trifluoroacetyl L-prolyl dl-α-methyl-3,4-(methylenedioxy)-phenethylamine, 46
N-Trifluoroacetyl L-prolyl dl-α-methylphenethylamine, 45, 46
N-Trifluoroacetyl L-prolyl dl-1-methyl-3-phenylpropylamine, 45, 46
N-Trifluoroacetyl secondary amines, 43
N-Trifluoroacetyl-DL-valine ethyl ester, 16, 17
N-Trifluoroacetyl L-valyl dl-p-chloro-α-methylphenethylamine, 46
N-Trifluoroacetyl 1-valyl dl-α-methylbenzylamine, 46
N-Trifluoroacetyl L-valyl dl-α-methylbenzylamine, 45, 46
N-Trifluoroacetyl L-valyl dl-1-methylhexylamine, 46
F-Trifluoroacetyl L-valyl dl-α-methyl-3,4-(methylenedioxy)-phenethylamine, 46
N-Trifluoroacetyl L-valyl dl-α-methylphenethylamine, 46
N-Trifluoroacylation-O-trimethylsilylation of phenol alkylamines, 212
N-Trifluoroaceylation-O-trimethylsilylation of phenol hydroxyamines, 212
3,3′,5-Triiodothyronine, 67, 120
3,3′,5′-Triiodothyronine, 67
Triisopropanolamine, 93, 124
Trimethylamine, 36, 37, 39, 93, 122, 123
Trimethylhexamethylenediamine, 126
Trimethyllysine, 66, 97
ε-N-ε-N,ε-N-Trimethyllysine, 65
N^6-Trimethyllysine, 58
N^E-Trimethyllysine, 105
$N^ε,N^ε,N^ε$-Trimethyl-D,L-lysine dihydrochloride (TML), 111
N,N,N'-Trimethyl-1,4-phenylenediamine, 128, 129
2,4,6-Trinitroaniline, 107
2,4,6-Trinitrobenzene sulfonate (TNBS), 183
2,4,6-Trinitrobenzene sulfonic acid, 172
N-(2,4,6-Trinitrophenyl)piperidine, 107
Tripropylenetetramine, 126

Tryptamine, 41, 52, 55, 99, 125, 134
Tryptamine HCl, 113, 114, 115
Tryptophan, 18, 21, 24—27, 30, 31, 34, 35, 58, 60, 62, 69—75, 77—80, 82, 88, 91, 101, 104, 111—115, 118—120, 162, 186, 188—190
    derivatives of, 112—115
D-Tryptophan, 6, 90
L-Tryptophan, 6, 59, 90, 113—115, 117, 118
L-Tryptophan-L-alanine, 114, 115
L-Tryptophan-β-alanine, 114, 115
DL-Tryptophanamide HCl, 114, 115
DL-Tryptophan benzyl ester HCl, 114, 115
DL-Tryptophan butyl ester HCl, 114, 115
DL-Tryptophan ethyl ester HCl, 114, 115
L-Tryptophan-L-glutamine acid, 114, 115
L-Tryptophan-glycine, 114, 115
L-Tryptophan-glycine-glycine, 114, 115
L-Tryptophan hydroxamate, 114, 115
L-Tryptophan-L-isoleucine, 114, 115
1-Tryptophan-L-leucine, 114, 115
L-Tryptophan-α-L-lysine, 114, 115
L-Tryptophan-L-methionine-L-asparagine-L-phenylalanine NH_2 HCl, 115
1-Tryptophan-L-methionine-L-phenylalanine NH_2 HCl, 114
DL-Tryptophan methyl ester HCl, 114, 115
DL-Tryptophan octyl ester HCl, 114, 115
L-Tryptophan-L-phenylalanine, 114, 115
Tryptophan thiohydantoin, 116
L-Tryptophan-L-tryptophan, 114
L-Tryptophan-L-tyrosine, 114, 115
L-Tryptophan-L-valine, 114, 115
Tryptophol, 113
L-Tryptophyl-L-alanine, 188
L-Tryptophyl-L-glutamic acid, 188
L-Tryptophyl-L-glycine, 188
L-Tryptophyl-L-phenylalanine, 117, 188
L-Tryptophyl-L-tyrosine, 188
Tyramine, 41, 52, 94, 99, 101, 102, 125
m-Tyramine, 55
o-Tyramine, 55
p-Tyramine, 55
p-Tyramine(bis-dansyl), 134
Tyramine HCl, 133, 136
Tyrosine, 18, 19, 21, 24—28, 31, 34, 35, 53, 58, 60, 61, 65, 68—77, 80, 82, 88, 89, 91, 101, 118—121, 162, 186, 189
D-Tyrosine, 7, 90
L-Tyrosine, 7, 59, 90, 105, 117, 133, 136
m-Tyrosine, 76, 78
o-Tyrosine, 76, 78
p-Tyrosine, 78

U

Undecylamine, 124
Urea, 59, 60, 61, 101

V

Valine, 8—10, 12, 14, 18—20, 23, 25—28, 30, 31, 33—35, 58, 60, 61, 69—74, 76—79, 82, 83, 88, 89, 101, 104, 111, 118—121, 162, 186, 189, 190
D-Valine, 4, 5, 7, 22, 29, 90
L-Valine, 4, 5, 7, 22, 29, 59, 90
R,S-Valine, 43
(R)-(−)-Valine methyl ester, 54
(S)-(+)-Valine methyl ester, 54
L-Valine-β-naphthylamide, 116
L-Valine-L-tryptophan, 113
L-Valyl-L-tryptophan, 117

X

1,3,6-Xylidene-4-sulfonic acid, 130, 131
m-Xylylenediamine, 126
p-Xylylenediamine, 126
(+)-α-(2,5-Xylyl)ethylamine, 29
(−)-α-(2,5-Xylyl)ethylamine, 29

SUBJECT INDEX

A

Acidic amino acid analysis, 61
Acid washed dimethyldichlorosilane treatment (AW-DMCS), 228
Acylation, 162—163
    first, 206
    second, 206
Adsorbents for TLC, 250—253
Alcohols, see also specific alcohols, 63
Aliphatic amines, 36, 92, 122—124, 219
    N,N-disubstituted ureas of, 218
    lower, 37, 39
Aliphatic fluorescamine derivatives, 217
L-Allohydroxyproline, 72
Amine derivatives
    N-dinitrophenyl, O-trimethylsilyl, 211—212
    N-2,6-dinitro-4-trifluoromethyl, O-trimethylsilyl, 211
    enantiomeric, 47—51
    fluorescamine, 100—101
    4-methoxybenzamide, 216
    o-phthalaldehyde, 217
    m-toluoyl, 216
    N-trifluoroacetyl, 215
    trimethylsilyl, 215
Amine methyl ester derivatives, 54
Amine packings, 233
Amino acid analysis, 150, 256
Amino acid derivatives, 118, 120, 121, 147
    enantiomers of, 50—51, 165—167, 204—205
    ligand exchange chromatography of, 69—73, 78
    resolution of, 77
    separation of, 76
Amino acid ethyl esters, 90
Amino acid methyl ester derivatives, 54
DL-Amino acid methyl esters
    L-α-chloroisovaleryl derivatives of, 30
    29L-α-chloroisovaleryl derivatives of, 205
Amino acid-phenylthiohydantoins, 150
DL-Amino acids
    derivatives of, 22
    racemic, 105
D-Amino acid separation from L-amino acids, 89
Amino alcohols, 63
Aminophenols, 190—191
Amino sugars, 63
Ammonium hydrogen carbonate, 177
Ammonium tungstophosphate, 130
Anakrom C22 series, 231—232
Anakrom diatomaceous earth supports, 231
Anils, 137—138
Antioxidants, 153
Aromatic amines, 107, 128, 129, 133, 190—191
    ammonium tungstophosphate and, 130
    biologically significant, 52—53
    N,N'-disubstituted ureas of, 218
    primary, 132
    secondary, 194
    tungstophosphate layers and, 131
Aromatic amino acids on β-cyclodextrin polyurethane resins, 75
Arylmonoamines, 192—193
Automated ion-exchange chromatography, 97—98
Automatic analyzers, 169—173
    suppliers of, 256
Automatic ion-exchange chromatography, 58—61
    sulfur-containing amino acids and, 68
AW-DMCS, see Acid washed dimethyldichlorosilane treatment

B

Bases, see also specific bases, 39, 213—214
Basic amino acids
    analysis of, 61
    methylated, 65
Biogenic amines, 52
    as o-phthalaldehyde derivatives, 102
Biologically significant aromatic amines, 52—53
μ-Bondapak, 237—239
Bondapak phenyl/Corasil, 238—239
Book directory, 259
Borate buffer, 209
Buffer-methyl cellosolve, 169
Buffers, see also specific buffers, 153
    borate, 209
    change in, 58
    citrate, 153, 172
    lithium, 59, 60
    sodium, 58

C

Capillary columns, 159
Catecholamines
    3-O-methyl derivatives of, 133
    primary, 136
Cation-exchange chromatography of iodoamino acids, acids, 67
Cellosolve, 169
Cellulose powders, 253
Chelate additives, 82
Chelate complexation, 77
Chelate eluants, 81
Chiral amines, 44
Chiral eluants, 76
Chiral metal chelate additives, 82
Chiral N-trifluoroacetyl amines, 51
L-α-Chloroisovaleryl derivatives of DL-amino acid methyl esters, 30
29-L-α-Chloroisovaleryl derivatives of DL-amino acid methyl esters, 205
Chromatography, see specific types

Chromphoric labeling of amino acids, 209—210
Chromosorb, 227—231
Citrate buffers, 153, 172
Collagen hydrolysate amino acids, 144
Column packings, 244
    Zorbax, 242—244
Columns
    capillary, 159
    dimensions of, 151
    Durabond, 233—234
    glass, 159
    polyamide, 79
    single, 59
Commercial sources for chromatography papers, 253—256
Complexation, 77
Condensation, 206
Copper, 73
Corasil, 237
Cyanide, 169
β-Cyclodextrin polyurethane resins, 75

D

Dansyl amino acids, 79, 80, 207
    optical isomers of, 81
    reversed phase separation of enantiomers of, 82
    Dansyl derivatives, 112, 121—122, 134
Derivatization, 197—219
Detection reagents, 177—183, 185—194
Detectors, 161
Diamines, 37—38, 95—98
    aliphtic, 217
    primary, 125—126
Diamino acids, 63—64
Diaminodicarboxylic acids in free form, 112
3,5-Diaminohexanoic acid lactams, 64
Diastereoisomeric pentafluoropropionyl-DL-amino acid (+)-3-methyl-2-butyl esters, 25
Diastereomeric N-acyl DL-alanine esters, 32
Diastereomeric N-1-teresantalinyl DL-amino acid esters, 33
Diastereomers
    N-d-10-camphorsulfonyl amino acid p-nitrobenzyl esters, 91
    enantiomeric amines as, 45—47
    separation of, 165
Diasteriomeric methoxy-α-methyl-1-naphthaleneacetylamino acid methyl ester derivatives, 25
Diatomaceous earth supports, 231
Diatomite, 227—228
Dicarboxylic amino acid derivatives, 205—206
Dimethylalkylamines, 214
4-N,N-Dimethylaminoazobenzene-4'-thiohydantoins, 119
4-N,N-Dimethylaminonaphthylazobenzene-4'-thiohydantoin (DANABTH)
    amino acids, 210
Du Pont, 242—244
Durabond columns, 233—234

E

EGA polymer, 161
Eluants
    chiral, 76
    metal chelate, 81
Enantiomeric amine methyl ester derivatives, 54
Enantiomeric amines
    as diastereomers, 45—47
    derivatives of, 47—50
Enantiomeric amino acid derivatives, 16
Enantiomeric amino acid ethyl esters, 90
Enantiomeric amino acid methyl ester derivatives, 54
Enantiomeric derivatives of amino acids and amines on α-hydroxycarboxylic acid esters, 29
Enantiomeric N-pentafluoropropionyl amino acid isopropyl esters, 23, 24
Enantiomeric N-trifluoroacetyl amino acid isopropyl esters, 23, 24
Enantiomers
    amine derivatives, 50—51
    amino acid, 69—73, 76—78, 165—167, 204—205
    amino acid derivatives, 50—51
    chiral N-trifluoroacetyl amines, 51
    dansyl amino acids, 82
    ligand exchange chromatography of, 69—73, 78
    N(O,S)-pentafluoropropionyl amino acid isopropyl esters, 26
    resolution of, 77
    reversed phase separation of, 82
    separation of, 76, 82, 205, 209
    N-trifluoroacetyl-alanine esters, 13
    N-trifluoroacetyl-DL-alanine isopropyl esters, 15
    N-trifluoroacetyl-amino acid esters, 14
    N-trifluoroacetyl-amino acid isopropyl amides, 4—5
    N-trifluoroacetyl-amino acid isopropyl esters, 4—10
    N-trifluoroacetyl-leucine esters, 13
Enzymes, see also specific enzymes
    proteolytic, 150
Esterification, 161—162
Extraction, 206

F

First acylation, 206
Flame ionization detector, 161
Flow rate, 152—153
Fluorescamine-amino acid reaction products, 83
Fluorescamine derivatives, 115, 135, 217—218
    aliphatic diamines, 217
    aliphatic polyamines, 217
    amines, 100—101
    primary catecholamines, 136
Fluorescence, 188

intensities of, 190
Fluorogenic reaction, 210—211
Fluorogenic reagents, 170
Fluorogens, 171
Fluorophore preparation, 210—211
Food sample amino acids, 201
Fragmentography, 161
Free amino acids
 in natural waters, 201—202
 reverse-phase high performance liquid chromatography of, 74

G

Gas chromatography (GC), 3—56, 159—167, 201, 203—206, 211—215
 liquid phases for, 223
 stationary phases for, 224—227
 support materials for, 227—234
Gas-Chrom supports, 232
GC, see Gas chromatography
Gels, 239—242
Glass columns, 159
Glutamyl derivatives of β-lysine, 64
Glycine derivatives, 110
Gradient mobile phases, 88
Gradient reverse-phase HPLC, 87
Ground rock samples, 202—203

Y

N-Heptafluorobutyryl amino acid isobutyl esters, 28—29, 162
N-Heptafluorobutyryl derivatives, 161
N-Heptafluorobutyryl esters, 20—28
HFB, see N-Heptafluorobutyryl
High performance ion pair partition chromatography, 94
High performance liquid chromatography (HPLC), 76 89, 147, 154—155, 207—209, 216—218
 packings for, 234—245
 reverse-phase, see Reverse-phase high performance liquid chromatography
High temperature phases, 226
Histidyl peptides, 135
HPLC, see High performance liquid chromatography
Hydrolysate amino acids, 142—143
Hydrolysis, 201
 peptides, 198—200
 proteins, 198—200
Hydrophilics, 74
Hydrophobics, 74
Hydroxy amino acids, 68
α-Hydroxycarboxylic acid esters, 29

I

Intensities, 190
Iodamino acids, 67
Iodine, 181
Iodohistidines, 120
Iodotyrosines, 120
Ion-exchange cellulose powders, 253
Ion-exchange chromatography, 63—65, 142—146, 151—154
 automatic, see Automatic ion-exchange chromatography
 O-sulfate esters of hydroxy amino acids, 68
Ion-exchange cleanup, 201
Ion-exchange paper chromatography, 105
Ion-exchange resins, 157—158
Ion-exchange thin layer chromatography, 127
Ionization detector, 161
Isocratic reverse-phase HPLC of phenylthiohydantoin amino acids, 86
Isomers
 optical, 81
 positional, 40—41

J

J. T. Baker Chemical Co., 249—250

L

Lactams of 3,5-diaminohexanoic acid, 64
LC, see Liquid chromatography
Ligand exchange chromatography, 95, 96, 154, 172
 amino acid enantiomers, 69—73, 78
 lower aliphatic amines, 37
Liquid chromatography (LC), 57—102, 141—156, 209
 high performance, see High performance liquid chromatography
 reverse-phase high performance, see Reverse phase high performance liquid chromatography
Liquid phases, 160—161
 for gas chromatography, 223
Lithium buffers, 59, 60
Lower aliphatic amines, 39
 ligand-exchange chromatography of, 37

M

Macherey-Nagel and Co., Duran, 255
Mass fragmentography, 161
Merck, 247—252
Metal chelate additives, 82
Metal chelate eluants, 81
4-Methoxybenzamide derivatives of amines, 216
Methoxy-α-methyl-1-naphthalene-acetylamino acid methyl ester derivatives, 89
Methylated amino acids, 65—67, 105
 derivatives of, 111

3-O-Methyl derivatives of catecholamines, 133, 136
Methyl ester method, 207
Micro-procedure, 211
Mixed chelate complexation, 77
Mobile phases, 88
Monoamines, 125—126

N

Natural waters, 201—202
NAW, see Nonacid washed
Neutral amino acid analysis, 61
Ninhydrin reagents, 169—170
Nitro derivatives, 107
Nitrogen-selective flame ionization detector, 161
Nitrosamino acids, 121
Nonacid washed (NAW), 228
Nonprotein amino acids, 19—20
Nucleosil packings, 23

O

Optical isomers of dansyl amino acids, 81
Optically active terpene derivatization reagents, 207—208

P

Packings
 column, 242—244
 HPLC, 234—245
 Nucleosil, 235
 Pennwalt amine, 233
 Vydac, 235—236
Paper chromatography (PC), 103—107, 177—183
 ion-exchange, 105
Papers for chromatography, 253—256
Partisil silica gels, 239—242
PC, see Paper chromatography
Pennwalt amine packings, 233
Pentafluoropropionyl-amino acid-3-methyl-2-butyl esters, 205
Pentafluoropropionyl-DL-amino acid-3-methyl-2-butyl esters, 25
Peptide hydrolysis, 198—200
Peptidoglycans, 63
Persulfates, 133—134
Phenol alkylamines, 212
Phenol hydroxyamines, 212
Phenylethylamines, 133
 derivatives of, 55—56
Phenylthiohydantoin (PTH) amino acids, 84—85, 87, 88, 189, 190
 isocratic reverse-phase HPLC of, 86
Phenylthiohydantoins, 150
Physiological fluid amino acids, 145—146
PIC reagents, 244—245
Polyamide columns, 79

Polyamines, 38, 95—99, 126, 217—218
 aliphatic, 217
 by ion-exchange TLC, 127
 N-permethylation of, 215
 tosyl derivatives of, 216
Polygosil, 234
Poly phases, 225—226
Polystyrene resins, 69—72
 sulfonated, 151
Porasil, 237
Positional isomer derivatives, 40—41
Preformed layers for TLC, 246—250
Primary amines, 125
 trifluoroacetyl derivatives of, 212
 Primary aromatic amines, 132
Primary arylmonoamines, 192—193
Primary catecholamines, 136
Primary diamines, 125—126
Primary monoamines, 125—126
Protein hydrolysis, 198—200
Proteolytic enzymes, 150
PTH, see Phenylthiohydantoin
Purity of reagents, 199—200
Pyridoxal method, 154
n-Pyridoxylidene amino acids, 172

R

Racemic DL-amino acids, 105
Reagents, see also specific reagents
 detection, 177—183, 185—194
 fluorogenic, 170
 ninhydrin, 169—170
 optically active terpene derivatization, 207—208
 PIC, 244—245
 Purity of, 199—200
Relative fluorescence intensities, 190
Relative molar response (RMR), 161
Resins, see also specific resins, 151—152
 β-cyclodextrin polyurethane, 75
 ion-exchange, 157—158
 polystyrene, see Polystyrene resins
 sulfonated polystyrene, 151
Resolution of amino acid enantiomers, 77
Reverse phase high performance liquid chromatography, 80, 81, 83, 84—85, 88
 free amino acids, 74
 gradient, 87
 isocratic, 86
Reverse phase separation of enantiomers of dansyl amino acids, 82
RMR, see Relative molar response
Rock samples, 202—203

S

Samples
 application of, 152
 food, 201

ground rock, 202—203
preparation of, 150—151, 197—219
soil, 202—203
Schiff bases, 39
of amines, 213—214
Schleicher and Schuell Inc., 252—255
Second acylation, 206
Secondary amines
aromatic, 194
sulfonamide, 53, 214
trifluoroacetyl derivatives of, 212
Selenotaurines, 106
Sensitivity of o-phthalaldehyde detection, 186—187
Separation
amino acid enantiomers, 76
L- and D-amino acids, 89
diastereomers, 165
enantiomers, 205, 209
reversed phase, 82
Silar phases, 225
Silica gels, 239—242
Single column systems, 59
Soap chromatography, 132
Sodium buffers, 58
Soil samples, 202—203
Stationary phases
for gas chromatography, 224—227
tripeptide bonded, 84—85
Sugars, 63
O-Sulfate esters, 68
Sulfonamides, 53, 214
Sulfur-containing amino acids, 106
automatic ion-exchange chromatography of, 68
Supelcoport, 233
Supelco SP-phases, 226—227
Support materials for gas chromatography, see also specific supports, 227—234
Synthesis of derivatives, 161—163

T

Tabsorb, 233
Temperature, 153—154
Terpene derivatization reagents, 207—208

Thin-layer chromatography (TLC), 109—138, 177—183, 209—210, 219
adsorbents for, 250—253
ion-exchange, 127
plates for, 249—250
preformed layers for, 246—250
2-Thiohydantoins of amino acids, 119—120
Thiourea derivatives
of amino acids, 209
of enantiomeric amino acid ethyl esters, 90
TLC, see Thin-layer chromatography
TMS amino acids, see Trimethylsilyl amino acids
m-Toluoyl derivatives, 100, 216
Trifluoroacetyl amines, 41
Trifluoroacetyl derivatives of amines, 212
Trimethylsilyl (TMS) amino acids, 161
Trimethylsilyl (TMS) derivatives, 34—35, 52—53, 215
of amines, 215
Trimethylsilylation reactions, 212—213
Tripeptide bonded stationary phase, 84—85
Tungstophosphate layers, 131
Two-column systems, 60

U

Ureas, 218

V

Vydac packings, 235—236

W

Waters Associates, Inc., 236—239
Whatman Ltd., 239—242, 246—247, 252—254

Z

Zorbax column packings, 242—244